T0092412

POLISH PHILOSOPHERS OF SCIENCE AND NATURE
IN THE 20TH CENTURY

POZNAŃ STUDIES
IN THE PHILOSOPHY OF THE SCIENCES AND THE HUMANITIES

VOLUME 74

EDITORS

This book has been partly sponsored by the Committee for Scientific Research
(Komitet Badań Naukowych)

The address: prof. L. Nowak, Cybulskiego 13, 60-247 Poznań, Poland.
Fax: (061) 8477-079 or (061) 8471-555
E-mail: epistemo@main.amu.edu.pl

POLISH ANALYTICAL PHILOSOPHY
Volume III

Editors:
Jacek Juliusz Jadacki (editor-in-chief)
Leszek Nowak   *Jan Woleński*
Jerzy Perzanowski   *Ryszard Wójcicki*

POLISH PHILOSOPHERS OF SCIENCE AND NATURE IN THE 20TH CENTURY

Edited by

Władysław Krajewski

Amsterdam – New York, NY 2001

The paper on which this book is printed meets the requirements of "ISO 9706:1994, Information and documentation - Paper for documents - Requirements for permanence".

ISSN 0303-8157
ISBN: 90-420-1497-0
©Editions Rodopi B.V., Amsterdam – New York, NY 2001
Printed in The Netherlands

CONTENTS

6

II. SCIENTISTS

APPENDIX
POLISH PHILOSOPHY OF SCIENCE. THREE COMMENTARIES

INTRODUCTION

The aim of the present volume is to introduce prominent Polish philosophers of the 20th century as well as their significant accomplishments in the field of the philosophy of science and philosophy of nature. What follows is a short survey of the Polish traditions in this area.

1. The Origins of the Polish Philosophy of Science and Philosophy of Nature, and Its Later Development Until the 19th Century

The first known Polish philosopher was Witelon (ca. 1230 - ca. 1314), a physicist and the author of *Perspectiva* and *De causa primaria poenitentiae et de natura daemonum*. The issues which he addressed in those treatises included the problems of various modes of knowledge. However, the emergence of a true philosophical milieu in Poland is connected with the founding of the Cracow Academy in 1364.

In the 15th century, aspects of methodology and the philosophy of science were essential parts of treatises that contained comments on Aristotle's *Analytics, Topics, Physics*, etc. The issues raised by Cracow philosophers included questions concerning the operations of defining, classifying and proving. The problems of induction were also analyzed. Five personages should be mentioned in this context: Andrzej Wężyk (ca. 1377 - 1430), the author of *Exercitium librorum Physicorum*; Benedykt Hesse from Cracow (ca. 1389 - 1456), who discussed, among others, the problem of movement in *Quaestiones super libros Physicorum* (where he adopted the conception of impetus); Jan from Słupcza (1408-1488), the author of a commentary on Aristotle's *De caelo et mundo*; Jan Schilling from Głogów (ca. 1445 - 1507), who was not only a philosopher but also an astronomer and geographer and who considered, among others, the problem of *principium individuationis* in his treatise *Exercitium veteris artis*; and Jakub from Gostynin (ca. 1450 - 1506), who analyzed the notion of cause in his commentary on Proclos' *Liber de causis*.

In the 16th century, the following philosophers dealt with issues in philosophy of science: Michał Twaróg from Bystrzyków (ca. 1450 - ca. 1520) in *Quaestiones in tractatus Parvorum logicalium Petri Hispani* (where he discussed the problem of *suppositio*, among others); Michał Falkener from Wrocław (ca. 1460-1534), a philosopher and astronomer in *Epitoma figurarum in libros Physicorum et De anima Aristotelis*; Wojciech Nowopolczyk (1508-1559) in "Oratio de laude physices"; Jakub Górski (ca. 1525 - 1585) in *Commentariorum artis dialecticae libri decem. De revolutionibus*, the work of the greatest scientist of the century, Nicolaus Copernicus (1473-1543), can also be considered a contribution to natural philosophy: it was of great methodological significance, as it had overcome both dogmatism and crude empiricism. Among 17th-century philosophers, Adam Burski (ca. 1560 - 1611) occupies a particularly prominent place. In his *Dialectica Ciceronis*, Burski advocated stoic empiricism and recommended the inductive method (prior to Francis Bacon). Bacon's methodology was disseminated in Poland by Jan Jonston (1603-1675). General methodology was one of the principal subjects of the famous manuals of those times: *Logica* by Marcin Śmiglecki (ca. 1562 - ca. 1618), and *Systema Logicae tribus libris adornatum* by Bartłomiej Keckermann (1572-1609). With regard to the issues of natural philosophy, the question of vacuum was debated particularly intensely by Polish thinkers: this is revealed by a comparison of positions adopted in this context by Jan Brożek (1585-1652) in his *Peripateticus cracoviensis*, Wojciech Wijuk-Kojałowicz (1605-1677) in *Oculus ratione correctus*, and Ferdynand Ohm-Januszowski (1639-1712) in *Summa philosophica*. In *Prelectiones philosophicae in octo libors Physicorum* by Tomasz Młodzianowski (1622-1686), modes of existence of matter were described. On the other hand, Adam Kwiryn Krasnodębski (1628-1702) was occupied by some problems of the philosophy of physics in *Philosophia Aristotelis explicata*, and Adam Kochański (1631-1700) raised questions in the field of the philosophy of mathematics in *Analecta mathematica* (co-authored by K. Schott).

In the 18th century, Marcin Świątkowski (ca. 1620-1790), the author of *Prodromus Polonus eruditae veritatis*, emerged as a precursor of the philosophy of science; Stanisław Konarski (1700-1777) in *De arte bene cogitandi ad artem bene dicendi* analyzed modes of reasoning; the philosophy of physics was one of subjects of *Commentariorum philosophiae, logicae scilicet metaphysicae, physicae generalis et particularis* by Antoni Skorulski (1715-1789), *Propositiones philosophicae ex physica recentiorum* by Antoni Wiśniewski (1718-1774), and *Carmina* by Ignacy Wilczek (1728- after 1788). The notion of causality was analyzed in detail by Benedykt Dobszewicz (1722-1794) in *Placita recentiorum philosophorum explanata*, whereas Samuel Chróścikowski (1730-1799) wrote on features of the matter in *Physics Confirmed by Experiments*. On the other hand, Ignacy Włodek (1723-1780) proposed an original classification of the sciences in his *Two Books on Liberal Arts in General and in Particular*.

In the Polish philosophy of science, the first half of the 19th century is associated with the name of Jan Śniadecki (1756-1830), a mathematician and philosopher who advanced empiricism in his papers collected in *Miscellaneous Writings*. His brother, Jędrzej Śniadecki (1768-1838), a chemist and biologist, the author of *A Theory of Organic Beings* and "A Speech on the Uncertainty of Sentences and Sciences Founded on Experience" appealed to the unity of experience and reason. The classification of the sciences preoccupied Józef Łęski (1760-1825) in *A Treatise on the Science of Nature*. In Poland of that time, epistemology founded on common sense had a prominent advocate in the person of Anioł Dowgird (1776-1835), the author of *An Exposition of the Natural Rules of Thinking*. The renaissance of methodological empiricism was the mark of the epoch. Michał Wiszniewski (1794-1865) published *Bacon's Method of Explaining Nature*, and Dominik Szulc (1797-1860), a student of Jan Śniadecki, emerged as a precursor of Polish positivism in his *On the Source of Modern Knowledge*. In contrast to these trends, an eminent mathematician, Józef Hoene-Wroński (1778-1853), developed an extremely speculative system of all branches of philosophy, including the philosophy of nature (despite his professional mastery of mathematics, his philosophical views were rather obscure).

In the second half of the 19th century, positivism and scientism became prominent in Poland as in other European countries. Wojciech Urbański (1820-1889) a physician, was one of the early positivists. His papers were published as *Minor Writings*. Władysław Kozłowski (1832-1899), a psychologist and sociologist, whose works were published posthumously in *Philosophical and Psychological Writings* was a mature advocate of scientism. The program of the positivist movement was formulated by Julian Ochorowicz (1850-1917) in *An Introduction to and an Overview of Positive Philosophy*. The program was criticized by Stefan Pawlicki (1839-1916) in *Studies on Positivism*. In *A Few Remarks on the Foundations and Limits of Philosophy*, he discussed, among others, the relations between philosophy and the sciences. Another adversary of positivism was Marian Morawski (1845-1901), a Catholic philosopher and the author of the treatises *Philosophy and Its Task* and *Teleology in Nature*. In that period, an important role in the development of the anti-speculative philosophy of science was played by physicians. The first impulse was given by Ferdynand Dworzaczek (1804-1876) in "On the Philosophy of Medicine". The classic volumes in the Polish philosophy of medicine are the treatise *On the Method of Searching for Medical Indications* written by Tytus Chałubiński (1820-1889), a physician and naturalist, and *The Logic of Medicine, or the Principles of the General Methodology of the Medical Sciences*, written by Władysław Biegański (1857-1917). In another book, *Neo-teleology*, Biegański defended the teleological method of explaining biological phenomena referring to a purpose or a function. His adversary was Adam Mahrburg (1860-1913), an opponent of teleological explanations, and a defender of empirical instrumentalism; his works were collected in *Philosophical Writings*. More general questions were raised in the methodological text *On the Method of Scientific Study* written by another physician, Henryk Hoyer (1834-1907). Other topics in philosophy of science discussed in Poland during that period that should be mentioned include the problem of prediction, investigated by Stanisław Kramsztyk (1841-1906) in *Studies of Nature: Essays in Physics, Geophysics and Astronomy*, and the problem of causality, discussed by Aleksander Raciborski (1845-1920) in *The Concept of Causality in J.S. Mill's "System of Logic Ratiocinative and Inductive"*. The philosophy of mathematics occupied Samuel Dickstein (1851-1939) in *Mathematics and Reality*, and the methodology of history was studied by Zofia Daszyńska-Golińska (1866-1934) in *Methodological Essays*.

Many of the philosophers mentioned above were still involved in research in the early 20th century, but their views are not discussed in our volume. Only the great Polish physicist of the beginning of our century, **Marian**

Smoluchowski[1] (1872-1917), who authored important ideas in the methodology of statistical physics, and **Czesław Białobrzeski** (1878-1953), who developed an Aristotelian approach to quantum mechanics, are presented.

2. The Polish Philosophy of Science and Philosophy of Nature from the Beginning of the 20th Century to 1945

In the 20th century, many professional philosophers in Poland dealt seriously with philosophy of science and philosophy of nature.

An original approach to general methodology was introduced by Leon Petrażycki (1877-1931) in *New Foundations of Logic and Classification of Abilities*. Leon Chwistek (1884-1944) published the inspiring book *The Limits of Science* (Chwistek 1949),[2] where he presented the conception of scientific theories as results of a schematization of reality, and outlined the so-called rational meta-mathematics. Napoleon Cybulski (1854-1919) discussed the problem of reductionism in the work *On Contemporary Vitalism and Mechanism*. The problem of determinism was discussed by Władysław Horodyski (1885-1920), the author of *The Notion of a Causal Relation*, Edward Stamm (1886-1940), a mathematician, the author of "Causality and the Functional Relation," **Adam Wiegner** (1889-1967), the author of *Remarks on Indeterminism in Physics*, and **Joachim Metallmann** (1889-1942), the author of *Determinism of the Natural Sciences*. Helena Konczewska (1877-1959) supplied a multi-faceted analysis of the notion of substance in *Le problème de la substance*. The mind-body problem occupied Władysław Mieczysław Kozłowski (1858-1935) in *Natural Science and Philosophy*. The evolutionary paradigm was defended by Józef Nusbaum-Hilarowicz (1859-1917) in *The Idea of Evolution in Biology*, and by Tadeusz Garbowski (1869-1940) in *Organism and Society*. Stanisław Kobyłecki (1864-1939) supported the idea of separating scientific facts and their interpretations in *Postulates of Experimental Psychology*. Józefa Kodisowa (1865-1940), in papers collected in *Philosophical Studies* and Władysław Heinrich (1869-1957) in the work "On the Methodology of the Sciences" developed the position of empirio-criticism. Investigation in the field of the methodology of medicine was continued by Edmund Biernacki (1866-1911) in *Principles of Medical Knowledge*, and by Władysław Szumowski (1875-1954) in *Logic for Physicians*. A distinctive approach to the sociology of science was presented by Florian Znaniecki (1882-1958) in *The Object and Aims of the Science of Knowledge*. Bronisław Malinowski (1884-1942), a sociologist and ethnographer, employed the functional method of explanation in sociology and ethnography (cf. his papers collected in *"A Scientific Theory of Culture" and Other Essays*).

---

[1] The names of philosophers and scientists discussed in this volume are printed in boldface.
[2] References are provided only for English-language editions of the works of Polish philosophers.

Each of scholars mentioned above worked alone, so to speak. The situation changed in the case of Kazimierz Twardowski (1866-1938). Twardowski was primarily interested in the philosophy of psychology not in the philosophy of natural sciences. As Franz Brentano's student, Twardowski was awarded his doctoral degree in Vienna and was a professor of philosophy at the University of Lvov[3] from 1895 onward. He had many students and was a founder of the "Lvov School," which became known as the "Lvov-Warsaw School" (Szkoła Lwowsko-Warszawska) after World War I when many of its members became professors of the University of Warsaw. It was a philosophical movement that instilled analytical and scientific philosophy in Poland. An enemy of "philosophical systems", Twardowski fostered analysis of specific problems and concepts in a clear, precise and consistent way. The majority of philosophers presented in this volume are his disciples or disciples of his disciples.

The Lvov-Warsaw School attached great importance to formal logic and had an excellent "logical branch". Among the representatives of Polish logic were: Stanisław Leśniewski (1886-1939), who developed some original formal systems including that of "ontology" and "mereology", **Jan Łukasiewicz** (1878-1956), the author of many-valued logic, Alfred Tarski (1902-1983), famous for his definition of truth, and Andrzej Mostowski (1913-1975), who investigated the foundations of mathematics. All of them also engaged in an in-depth analysis of the philosophical problems arising from their formal works. Especially sensitive to those issues was Łukasiewicz (see Łukasiewicz 1970).

The "philosophical branch" of the Lvov-Warsaw School (which also used logic to a great extent) continued Twardowski's tradition. Below is a list of its most distinguished members. Bronisław Bandrowski (1879-1914) dealt with the problem of induction. **Zygmunt Zawirski** (1882-1948) was one of the most outstanding Polish philosophers of physics, a forerunner of historical methodology, a pioneer of axiomatizing empirical sciences (especially physics), who initiated discussions on the logical foundations of quantum mechanics in Poland. Outside Poland, he is known primarily as the author of the monograph *L'évolution de la notion du temps*. **Tadeusz Kotarbiński** (1886-1981) is the author of a materialistic and nominalistic ontological conception known as *reism* (or *concretism*). His book *Elementy* was an academic textbook and, at the same time, an exposition of reism and his other original views. The English translation of a new version of *Elementy* appeared under the title *Gnosiology* (Kotarbiński 1966). **Kazimierz Ajdukiewicz** (1890-1963) published in *Erkenntnis*, presenting his profound epistemology known as *radical conventionalism* (cf. Ajdukiewicz 1978). **Tadeusz Czeżowski** (1889-1981) analyzed various methodological and epistemological concepts, adopting rather

---

[3] This town has a complicated history. For many centuries it belonged to Poland (the Polish name is 'Lwów'). After the partition of Poland in the 18th century it belonged to Austria (then to its province of Galicia, which was granted autonomy) under the German name *Lemberg*. In the years 1918-1939, Lwów was again a part of the Republic of Poland. Throughout that period, it was an important center of Polish culture. Since 1939, it belongs to Ukraine (formerly a Soviet Republic, now an independent state) and bears the Ukrainian name 'Lviv'. Its population before 1939 was predominantly Polish and Jewish, and now is Ukrainian and Russian. In English, the name 'Lvov' is used.

unorthodox positions in many fields. Edward Poznański (1901-1976), jointly with Aleksander Wundheiler (1902-1957), analyzed the notion of truth in physics. Jan Rutski (1903-1939) contributed a logical analysis of statistical dependencies. **Izydora Dąmbska** (1904-1983), the youngest of Twardowski's disciples, wrote about the status of conventions and scientific laws. Adolf Lindenbaum (1904-1941) was an eminent logician; his wife **Janina Hosiasson-Lindenbaumowa** (1899-1942) examined the reliability of the method of induction. **Henryk Mehlberg** (1904-1979) wrote about time and other general issues in philosophy of science. **Janina Kotarbińska** (T. Kotarbiński's wife, 1901-1997) analyzed the concepts of chance, of laws of nature, of determinism, and others. **Seweryna Łuszczewska-Romahnowa** (1904-1978) worked in the fields of semantics and methodology and constructed a systematic theory of classification. **Maria Kokoszyńska-Lutmanowa** (1905-1981) was an insightful critic of relativism; she analyzed semantic and methodological concepts, making valuable distinctions. The couple Stanisław Ossowski (1897-1963) and Maria Ossowska (1896-1974) worked in semiotics. They were among the pioneers of the science of science. Later on, Ossowski became a prominent sociologist and Ossowska an eminent ethicist.

The Lvov-Warsaw School co-operated closely with the Vienna Circle. Both schools were akin in many respects. Nevertheless, there were essential differences between them. While members of the Lvov-Warsaw School made many efforts to highlight those dissimilarities, commentators often overlooked them. In particular, the Vienna Circle dismissed any metaphysics, whereas the Lvov-Warsaw School was opposed to speculative metaphysics not to metaphysics (ontology) in general. These differences were competently listed by Zawirski (cf. the paper devoted to Zawirski in this volume; see also Szaniawski, ed., 1989). Hence, the Lvov-Warsaw School cannot be included among the positivist schools, although many philosophers, especially adversaries of the Lvov-Warsaw School, called its members "positivists".

There were many adversaries of the Lvov-Warsaw School in Poland between the two world wars (and later as well). Even though they were not philosophers of science, we will list the most eminent among them: Roman Ingarden (1893-1970), the founder of the phenomenological school in Poland, Henryk Elzenberg (1887-1967), a philosopher of culture, Wincenty Lutosławski (1863-1954), a messianist, Stanisław Ignacy Witkiewicz (1885-1939), a "biological monadist", many traditional Thomists. It must be stressed that there were also Thomists who were adherents of the Lvov-Warsaw School and tried to develop the Thomist philosophy and even theology in a more precise manner, by making use of the tools of logic. This circle included Józef M. Bocheński (1902-1994), Jan Drewnowski (1896-1978) and Jan Salamucha (1903-1944). The latter contributed a logical analysis of some of St Thomas's proofs for the existence of God.

One more personage must be mentioned in this context, namely **Ludwik Fleck** (1896-1961). He was a microbiologist who also dealt with the methodology of science. His papers in this domain differed significantly from the Lvov-Warsaw School paradigm. Fleck's ideas were in fact hardly noticed by the members of the school. Many years later, his work was praised by T.S. Kuhn and inspired a new trend in the philosophy of science.

During the German occupation (1939-1945) many philosophers were murdered by the Nazis. The greatest losses were among those of Jewish origin (among the above-mentioned: Metallman and both Lindenbaums), but others, including priests (Salamucha) lost their lives, too. A few members of the Lvov-Warsaw School left Poland before or during the war. They did not return after 1945 and worked abroad (Łukasiewicz, Bocheński, Tarski). However, most remained in Poland (Kotarbiński, Ajdukiewicz, Czeżowski, Zawirski, Dąmbska, Kokoszyńska, Kotarbińska, Rohmanowa, also Ingarden and Elzenberg) and resumed their university work after the war.

3. The Postwar (Post-1945) Period in the Polish Philosophy of Science and Philosophy of Nature

In the period directly following the Second World War (1945-1949), there was a revival of all trends of Polish philosophy: analytical philosophy of the Lvov-Warsaw School, phenomenology and Thomism. There was one substantial change, though. Marxism in the form of dialectical and historical materialism appeared on the scene. Marxist philosophers were not numerous but were very active. That early period witnessed freedom of philosophical (not political!) discussion and vivid polemics took place. For example, Kłósak (a Thomist) and Łubnicki criticized dialectical materialism, while Marxists (Adam Schaff and others) responded to them.

This situation did not last long. During the period of "stalinization" (1950-1956), Marxism became the official philosophy. "Old" professors were expelled from their chairs at the universities. Some of them (Kotarbiński, Kotarbińska, Ajdukiewicz, Czeżowski, Kokoszyńska) could teach logic, others (Ossowski, Ossowska, Władysław Tatarkiewicz, Stefan Swieżawski, Dąmbska, Ingarden, Elzenberg) could not teach at universities at all (however, they took up a different task: the translation of the classics of the world philosophy into Polish which initiated an excellent book series).

The teaching of philosophy was monopolized by Marxists, although only few of them had received any proper philosophical education.[4] During that period, dialectical materialism was taught in a dogmatic way. The lectures were modeled on Stalin's works. Marxist philosophers of science promulgated the so-called "Mitchurinian" biology and attacked allegedly idealistic schools and ideas in the Western science.[5] In the philosophy of physics, the situation was better. In 1951-1952, a campaign was launched in the USSR against Einstein's "idealistic" relativity theory (it ended in 1953, after Stalin's death). Leopold Infeld, Einstein's collaborator, who returned to Poland in 1950, noted that none of the Polish philosophers of science attacked Einstein.

---

[4] An exception was the Catholic University of Lublin (KUL), whose faculty comprised no Marxists at all. Philosophy was taught exclusively by Thomists. In this respect Poland differed from all other countries of the so-called Socialist Bloc.

[5] I must admit that I was one of them (W. Krajewski).

The "Polish October" of 1956 gave birth to a revolt of the society and changed the situation once more. The "old" philosophers could teach philosophy once again. However, they were only permitted to teach students of philosophy not other students (at that time, philosophy was compulsory for all students). Until the 1980s, the mandatory lectures on representatives of other philosophical orientations were given only by Marxists. It has to be said, however, their teaching became more and more competent and objective. After 1956, philosophers again enjoyed a relative freedom of philosophical publications. The only exception was political philosophy. Different philosophical orientations emerged during that period. We shall limit our attention only to philosophers who were active in the philosophy of science, the philosophy of nature or logical methodology.

Three groups can be distinguished. First, members of the Lvov-Warsaw School and their disciples should be mentioned. Kotarbiński continued his work on *reism* and founded *praxiology* (the science of effective action). Ajdukiewicz gave up his radical conventionalism but was still deeply engaged in the study of the methodology of empirical sciences. His book *Pragmatic Logic*, published posthumously in 1965 in Polish and in 1974 in English (Ajdukiewicz 1974), was a momentous event in the life of the Polish philosophical community. Czeżowski performed further analyzes of methodological, ethical and other concepts. Dąmbska wrote valuable and inspiring books about the instruments of knowledge and about conventions. Kotarbińska analyzed the concepts of sign and definition. Kokoszyńska and Łuszczewska continued their semantic and methodological research.

The next generation of scholars included four philosophers who died prematurely: **Roman Suszko** (1919-1979) who was the author of *diachronic logic* and *non-Fregean sentential logic* (Suszko 1968), **Klemens Szaniawski** (1925-1990) who worked in the area of logical methodology using probability theory (see Szaniawski 1998), **Halina Mortimer** (1921-1984), who dealt with the logic of induction (Mortimer 1985) and Andrzej Malewski (1929-1963) who put forward a program of integrating sciences concerning human behavior. They all were disciples of Ajdukiewicz and Kotarbiński. **Jerzy Giedymin** (1925-1993), a founder of the "Poznań school", worked later in England (Brighton). He analyzed conventionalism and various problems in philosophy of science.

The most eminent disciples of Kotarbiński and Ajdukiewicz who are still active today are Marian Przełęcki, Ryszard Wójcicki, Jerzy Pelc, Andrzej Grzegorczyk, and Witold Marciszewski. We shall return to them later. Two disciples of Czeżowski must also be mentioned: Leon Gumański (in Toruń) and Bogusław Wolniewicz (now in Warsaw), both logicians and philosophers.

The second group belonging to the same generation had a different background. Its members were Marxists in the beginning of their activity. At the same time, they studied natural sciences, mainly physics or chemistry, and were interested in the philosophy of these sciences. The most active participants of this group were Helena Eilstein and Władysław Krajewski (both from Warsaw), Zdzisław Augustynek and Zdzisław Kochański (both from Cracow), Stefan Amsterdamski (from Łódź), Irena Szumilewicz (from Gdańsk), Jan Such (from

Poznań) and Wacław Mejbaum (from Wrocław). They gradually became analytical philosophers, adopting the traditions of the Lvov-Warsaw School and of the Anglo-Saxon philosophy of science, which they tried to combine with the dialectical materialism, and which consequently led to the revision of many of its theses, especially those influenced by the Hegelian tradition. As a result, they were considered to be "revisionists" by the more traditional Marxists, the so-called "dogmatists". Among the latter, was Czesław Nowiński (1907-1981), a philosopher of biology and an epistemologist, and Jarosław Ładosz (1924-1997) who worked in philosophy of mathematics and later on in social philosophy. Other and more famous "revisionists" (who were not philosophers of science), headed by Leszek Kołakowski and Bronisław Baczko, "accused" the analytic group of positivism (see Krajewski 1982, for more information on the differences between the analytic and "continental" interpretations of Marxism).

Later, almost all members of both groups of "revisionists" abandoned Marxism because Marxist historiography, "scientific socialism", etc., failed. However, we must admit that, in spite of all its shortcomings, Marxist ideas played some positive role, too. They inspired Suszko in the creation of diachronic logic, Leszek Nowak in the elaboration of the method of idealization, Jerzy Kmita in taking into account the social conditions of science, Krajewski, Izabella Nowakowa, and others in the analysis of approximate truth and its convergence to the full truth in the course of the growth of science. On the other hand, some Polish logicians of the Lvov-Warsaw School tried to formalize some Hegelian ideas. In 1948, Stanisław Jaśkowski built a system of paraconsistent logic, while in 1964 Leonard Sławomir Rogowski put forward the so-called "directional logic".

In 1968, Polish Marxism was affected by an "anti-revisionist" (and anti-Semitic) campaign of the communist regime. Some philosophers were expelled by the political authorities from the universities (including Amsterdamski and Szumilewicz); Eilstein emigrated to England and then to the USA (she returned to Poland in 1993 and is still active as a philosopher), and Kochański emigrated to the USA (where he soon died). In the 1980s, similar restrictions were imposed, this time affecting philosophers who were deeply engaged in the Solidarity movement, and who supported a democratization of the country (Amsterdamski, Nowak and others).

The third philosophical group of this period was formed by Catholic philosophers. In line with the Thomist tradition, they were mainly concerned with the philosophy of nature, as a part of metaphysics, but also with philosophy of science. Kazimierz Kłósak (1911-1982) investigated the philosophy of nature, e.g. the problem of the origins of life, trying to combine the Thomist philosophy with some achievements of contemporary science. **Stanisław Kamiński** (1919-1986) wrote about the classification of the sciences and other problems of the philosophy of science. **Stanisław Mazierski** (1915-1993) analyzed determinism. Their disciples analyzed various methodological problems, which will be dealt with below. **Bolesław Gawecki** (1889-1984) may also be included in this group; he analyzed problems of causality, of evolution, and others.

In Cracow, the phenomenological school headed by Ingarden was still active. In principle, its members did not deal with the philosophy of science with one exception. Danuta Gierulanka (1909-1995) wrote a book presenting a phenomenological approach to the epistemology of mathematics.

In the last three decades of the 20th century, philosophers of science of all groups of the post-war generation and many younger ones had substantial achievements. Let us start with the older generation.

Przełęcki published in English a small but important book on the logic of science (Przełęcki 1969). The book was later translated into Polish. In 1974, Wójcicki published a valuable book on the formal methodology of science in Polish and five years later a revised edition was published, including an English version (Wójcicki 1979). In 1974, an international conference on formal methodology was organized in Warsaw (see Przełęcki, Szaniawski, Wójcicki 1976). Pelc wrote papers on semiotics, collected in a separate volume in English (Pelc 1971). Marciszewski edited a dictionary of logic, first in Polish, and then, with some changes, in English (Marciszewski 1981). Grzegorczyk published his outline of mathematical logic in English (Grzegorczyk 1974); he has also dealt with the philosophy of mathematics and other domains of philosophy. Łubnicki (1904-1988), mentioned above, dealt with scientific epistemology, stressing the role of practice (*praxism*).

Amsterdamski wrote two books on the methodology and growth of science, which were also published in English (Amsterdamski 1975, 1992). Augustynek wrote papers and books on time and on formal ontology, taking contemporary science into account. Two of them were translated into English (Augustynek 1991 and Augustynek & Jadacki 1993). Upon her return to Poland, Eilstein published a book on fatalism taking into account contemporary physics (Eilstein 1997). Before, J. Such published books on the universality of laws, on the *experimentum crucis* and others. Szumilewicz wrote papers on various problems in the philosophy of science and a small book on Poincaré. Mejbaum wrote papers and books on various philosophical problems. Krajewski wrote, beside books in Polish, a book in English (Krajewski 1977) and edited a collection of English papers by Polish authors (Krajewski 1982).

Below, the younger generation of philosophers is introduced, and the cities and universities which are the main centers of the philosophy of science in contemporary Poland are indicated, starting with Warsaw. Only the names and primary interests are briefly listed.

The following philosophers have been active in the Institute of Philosophy and Sociology of the Polish Academy of Sciences: Michał Hempoliński (now in Szczecin) – various problems of scientific epistemology; Wojciech Gasparski – praxiology; Michał Tempczyk – philosophy of physics, especially the theory of chaos; Józef Niżnik and Stanisław Czerniak – epistemology and sociology of knowledge; Alina Motycka – relativism in the philosophy of science, scientific epistemology; Małgorzata Czarnocka – philosophy of physics, the analysis of scientific experience; Stefan Zamecki – philosophy of chemistry, problems of discovery; Włodzimierz Ługowski – philosophy of biology, especially biogenesis; Stanisław Butryn – philosophy of cosmology.

At Warsaw University: Elżbieta Pietruska-Madej (1938-2001) – the growth of science, scientific discovery, critical rationalism; Jan M. Żytkow (1944-2001) – philosophy of physics (in the 1980s, he emigrated to the USA and his interests shifted to computer science); Zbigniew Majewski and Marek Bielecki (now in the USA) – philosophy of physics; Witold Strawiński – the principle of simplicity, the unity of science; Jacek J. Jadacki – semantic and scientific ontology; Mieczysław Omyła and others – philosophy of mathematics and of logic; Andrzej Bednarczyk – history of scientific ideas; Anna Jedynak – empiricism, conventionalism; Tomasz Bigaj – philosophy of physics; Anna Wójtowicz – philosophy of logic; Krzysztof Wójtowicz – philosophy of mathematics.

At the Cardinal Wyszyński University of Warsaw: Mieczysław Lubański – philosophy of physics and of information theory; Edward Nieznański and Kordula Świętorzecka – logic of religion; Szczepan Ślaga (1934-1995), Bernard Hałaczek and Kazimierz Kloskowski (1953-1999) – philosophy of biology; Józef M. Dołęga – philosophy of nature and ecophilosophy; Anna Latawiec – methodology of the system sciences; Anna Lemańska – philosophy of mathematics.

In Cracow, at the Jagiellonian University: Jan Woleński – philosophy of logic and mathematics, history of the scientific philosophy (in 1985 he published a fundamental book on the Lvov-Warsaw School, then translated into English; see Woleński 1989); Zdzisława Piątek – philosophy of biology, now an environmental ethicist, Adam Grobler (now in Zielona Góra) – general philosophy of science; Józef Misiek, Jan Płazowski and Tomasz Placek – philosophy of physics and mathematics.

In the 1980s, an active center for the philosophy of science was organized at the Papal Theological Academy by: Michał Heller (a physicist and philosopher) and Józef Życiński (a philosopher of science and a theologian).

In Poznań, a strong school in the philosophy of science was founded by Giedymin; after his emigration to England it is led by Kmita, who works in the field of the methodology of the humanities and the philosophy of culture (Kmita 1988, 1990). Another leader of the Poznań school is Nowak, whose work is in the methodology of idealization in all sciences, especially social sciences (Nowak 1980). Other philosophers of science are mainly the disciples of Giedymin, Kmita, Nowak and Such: Jerzy Brzeziński and Krystyna Zamiara – philosophy of psychology; Anna Pałubicka and Barbara Kotowa – philosophy of the humanities; Wojciech Patryas and Tadeusz Buksiński – philosophy of the social sciences; Krzysztof Łastowski and Elżbieta Pakszys – philosophy of biology; Nowakowa and Antoni Szczuciński – philosophy of physics; Paweł Zeidler, Danuta Sobczyńska and Ewa Zielonacka-Lis – philosophy of chemistry and of experiment; Ewa Piotrowska – philosophy of mathematics; Honorata Korpikiewicz and Małgorzata Szcześniak – philosophy of astronomy; Jerzy Szymański – philosophy of technology. There is a group of logicians in Poznań combining their formal and philosophical interests: Tadeusz Batóg and Roman Murawski – philosophy of logic and mathematics; Janusz Wiśniewski – a formal theory of causality; Wojciech Buszkowski – categorial grammar; Jerzy

Pogonowski – philosophy of language; Andrzej Wiśniewski (now in Zielona Góra) – erotetic logic.

In Wrocław: Andrzej Siemianowski – conventionalism, philosophy of applied sciences; Elżbieta Kałuszyńska (now in Warsaw) – formal methodology of science, analysis of methodological structuralism; Eugeniusz Żabski – logical reconstruction of various scientific concepts; Teresa Grabińska and Mirosław Zabierowski – philosophy of physics and of astronomy; Adam Chmielewski – general philosophy of science.

In Lublin, at the Maria Curie-Skłodowska University: Zdzisław Cackowski – scientific epistemology; Leon Koj – axiology of science; Jacek Paśniczek – philosophical problems of logic; Sabina Magierska and Jadwiga Mizińska – epistemology and sociology of knowledge; Kazimierz Jodkowski and Wojciech Sady (both are now in Zielona Góra) – general philosophy of science, the rational reconstruction of scientific discoveries. At the Catholic University of Lublin: Stanisław Kiczuk – logic of changes and logic of causality; Zygmunt Hajduk, Andrzej Bronk and Józef Turek – various problems in philosophy of science.

In Łódź: Adam Nowaczyk – formal methodology of science; Sławoj Olczyk – general philosophy of science; Ryszard Kleszcz – rationalism; Elżbieta Mickiewicz-Olczyk and Aldona Pobojewska – philosophy of biology; Barbara Tuchańska – social conditions of science.

In Toruń: Jerzy Perzanowski – philosophy of logic and formal ontology; Marian Grabowski – philosophy of physics; Max Urchs (now in Konstanz) – philosophy of logic and systems theory.

In Gdańsk: Adam Synowiecki (1929-2000) – philosophy of chemistry, problems of scientism.

In Cieszyn: Adam Jonkisz – formal methodology of science.

In Opole: Urszula Wybraniec-Skardowska - formal methodology of science, theory of artificial intelligence; Zenona M. Nowak – philosophy of physics.

Many scientists, physicists and biologists also addressed, to a greater or lesser extent, methodological and philosophical problems of science. Warsaw physicists include **Leopold Infeld** (1898-1968), Andrzej Trautman, Jerzy Plebański, Stanisław Bażański, Krzysztof Maurin, Józef Werle (1923-1998), **Grzegorz Białkowski** (1932-1989), Jan J. Sławianowski; in Cracow: **Jerzy Rayski** (1916-1993), **Zygmunt Chyliński** (1930-1994), Andrzej Staruszkiewicz; in Toruń: Roman S. Ingarden; in Gdańsk: Ludwik Kostro.

Biologists writing on the philosophical problems of science in Warsaw include Jan Dembowski (1889-1963), Władysław Goldfinger-Kunicki (1916-1995), Adam Urbanek, Leszek Kuźnicki; in Cracow: Adam Łomnicki.

Numerous philosophical books and journals have always been published in Poland. Philosophy of science was present in many journals but especially those listed below. *Studia Logica* (English-language journal), edited for many years by Wójcicki, promotes high standards of formal methodology. In Poznań *Studia Metodologiczne*, edited by Jerzy Topolski (1928-1998) and Kmita, appeared twice a year since 1965; since 1976, *Poznańskie Studia z Filozofii Nauki*, edited by Nowak, appeared once a year. Also, the English-language series *Poznań*

Studies in the Philosophy of the Sciences and the Humanities, edited by Nowak, in which this volume appears, has been published since 1975 (this book series has featured some volumes devoted to the history of Polish philosophy, including Vol. 28 *Polish Scientific Philosophy: The Lvov-Warsaw School*, edited by Francesco Coniglione, Roberto Poli and Woleński, and Vol. 40 *The Heritage of Kazimierz Ajdukiewicz*, edited by Vito Sinisi and Woleński). There are also Polish book series. Since the 1980s, a series *RRR* (*Realizm, Racjonalność, Relatywizm*) on epistemology, logic and philosophy of science has been published in Lublin. Since the 1990s, a series *Cosmos & Logos*, mainly on philosophy of physics and astronomy (edited by Grabińska and Zabierowski) has appeared in Wrocław. In Warsaw, a quarterly *Filozofia Nauki* has been published since 1993. Augustynek is its initiator and the chairman of the editorial board, while Jadacki is the editor-in-chief. Some new English-language journals appeared as well. Nicolaus Copernicus University in Toruń publishes a series *Theoria et historia scientiarum*, edited by Tomasz Komendziński. In Warsaw, a journal *Foundations of Science* was founded in 1995, with Wójcicki as its editor-in-chief. It is the official journal of the Association for Foundations of Science, Language and Cognition, an international society founded by Wójcicki. He also initiated the Polish Association for the Logic and Philosophy of Science (its main interest is mathematical logic).

Many national as well as international conferences dedicated to the philosophy of science were organized in Poland. In August 1999, the 11th International Congress of Logic, Methodology and Philosophy of Science took place in Cracow, with Woleński as the Chairman of its Organizing Committee.

*

The present volume contains 27 papers devoted to eminent Polish philosophers of science and nature who worked and died in the 20th century (those who are alive were not taken into account). The papers are divided into two sections: 1. papers on professional philosophers (20 papers), 2. papers on scientists (7 papers). The papers are arranged in a chronological order. In the first part, mainly the most distinguished philosophers of the Lvov-Warsaw School are presented; the second part contains papers on 6 physicists and one physician (and a philosopher at the same time).

Three papers were added as an appendix. They are devoted to Polish philosophy, especially philosophy of science, and written in different periods:

1. *Logicist Anti-irrationalism in Poland*, by Kazimierz Ajdukiewicz. This paper was published in 1934 in Polish (*Przegląd Filozoficzny* vol. 36) and in 1935 in German (*Erkenntnis* vol. 5); the English translation was prepared especially for this volume. The author mainly presents the output of the Lvov-Warsaw School, including the philosophy of science.

2. *Philosophy of Science in Poland*, by Klemens Szaniawski. This paper was written in English for the *Handbook of World Philosophy*, edited by J. Burr, which was published in 1980 by Greenwood Press. The author mainly presents Polish analytical philosophy, to a great extent focusing on the philosophy of science.

3. *Main Orientations in the Contemporary Polish Philosophy of Science*, by Izabela Nowakowa, a paper written for this volume. The author presents five conceptions of living authors of her choice: three philosophers-logicians from Warsaw, continuing the traditions of the Lvov-Warsaw School and two philosophers from Poznań, located more in the Marxist tradition (now rare in Poland), which is naturally approached in an unorthodox manner and interpreted by each of the authors in a different way.[6]

REFERENCES

Ajdukiewicz, K. (1974). *Pragmatic Logic*, transl. by O. Wojtasiewicz. Warszawa: PWN.

Ajdukiewicz, K. (1978). *The Scientific World-Perspective and other Essays* (1938-1963), ed. by J. Giedymin. Dordrecht: Reidel.

Amsterdamski, S. (1975). *Between Experience and Metaphysics. Philosophical Problems of the Evolution of Science*, transl. by P. Michałowski (*BSPS* 35). Dordrecht: Reidel.

Amsterdamski, S. (1992). *Between History and Method. Disputes about the Rationality of Science* (*BSPS* 145). Dordrecht: Reidel.

Augustynek, Z. (1991). *Time, Past, Present, Future*. Warszawa: PWN & Dordrecht: Kluwer.

Augustynek, Z. and J.J. Jadacki (1993). *Possible Ontologies* (*Poznań Studies in the Philosophy of the Sciences and the Humanities* 29). Amsterdam–Atlanta (GA): Rodopi.

Chwistek, L. (1949). *The Limits of Science. Outline of Logic and of Methodology of the Exact Sciences*. New York: Harcourt, Brace a. Company & London: Routledge a. Kegan Pault Ltd.

Coniglione, F., R. Poli, and J. Woleński (Eds.) (1993). *Polish Scientific Philosophy. The Lvov-Warsaw School* (*Poznań Studies in the Philosophy of the Sciences and the Humanities* 28). Amsterdam–Atlanta (GA): Rodopi.

Eilstein, H. (1997). *Life Contemplative, Life Practical. An Essay on Fatalism* (*Poznań Studies in the Philosophy of the Sciences and the Humanities* 52). Amsterdam–Atlanta (GA): Rodopi.

Grzegorczyk, A. (1974). *An Outline of Mathematical Logic: Fundamental Results and Notions Explained with All Details*. Warszawa: PWN & Dordrecht: Reidel.

Gumański, L. (1999). *"To Be or Not to Be? Is that the Question?" and Other Studies in Ontology, Epistemology and Logic* (*Poznań Studies in the Philosophy of the Sciences and the Humanities* 66). Amsterdam–Atlanta (GA): Rodopi.

Ingarden, R. (1964). *Time and Modes of Being*. Springfield: Thomas.

Jadacki, J.J. (1980). On the Sources of Contemporary Polish Logic. *Dialectics and Humanism* 7, 4, 163-188.

Kotarbiński, T. (1966). *Gnosiology – The Scientific Approach to the Theory of Knowledge*, transl. by O. Wojtasiewicz. London: Pergamon.

Kmita, J. (1988). *Problems in Historical Epistemology* (*Synthese Library* 191). Dordrecht: Reidel.

Kmita, J. (1990). *Essays in the Theory of Scientific Cognition* (*Synthese Library* 210). Dordrecht: Reidel.

Krajewski, W. (1977) *Correspondence Principle and Growth of Science* (*Episteme* 4). Dordrecht: Reidel.

Krajewski, W. (Ed.) (1982). *Polish Essays in the Philosophy of the Natural Sciences* (*BSPS* 68). Dordrecht: Reidel.

Leśniewski, S. (1992). *Collected Works*. Vol. I-II, ed. by S.J. Surma, J.T. Srzednicki and O.I. Bannett. Warszawa: PWN & Dordrecht: Kluwer.

Łukasiewicz, J. (1970). *Selected Works*, ed. by L. Borkowski. Warszawa: PWN & Amsterdam: North-Holland Publishing Company.

---

[6] We are grateful to M. Czarnocka, E. Pietruska-Madej, L. Nowak and K. Brzechczyn for their remarks on the earlier version of this *Introduction*.

Marciszewski, W. (Ed.) (1981). *Dictionary of Logic as Applied in the Study of Language*. The Hague: Martinus Nijhoff.

Mortimer, H. (1985). *The Logic of Induction*. Chester: Ellis Horwood Lim.

Nowak, L. (1980). *The Structure of Idealization. Towards a Systematic Interpretation of the Marxian Idea of Science*. Dordrecht-Boston: Reidel.

Paśniczek, J. (Ed.) (1992). *Theories of Objects: Meinong and Twardowski*. Lublin: Wydawnictwo UMCS.

Pelc, J. (1971). *Studies in Functional Logical Semiotics of Natural Language*. The Hague: Mouton.

Pelc, J. (Ed.) (1979). *Semiotics in Poland 1894-1969*. Dordrecht: Reidel.

Poli R. (Ed.) (1997). *In Itinere. European Cities and the Birth of Modern Scientific Philosophy* (*Poznań Studies in the Philosophy of the Sciences and the Humanities* **54**). Amsterdam–Atlanta (GA): Rodopi.

Przełęcki, M. (1969). *The Logic of Empirical Theories*. London: Routledge and Kegan Paul.

Przełęcki, M., K. Szaniawski and R. Wójcicki (Eds.) (1976). *Formal Methods in the Methodology of Empirical Sciences*. Dordrecht: Reidel, Wrocław: Ossolineum.

Simons, P. (1992). *Philosophy and Logic in Central Europe from Bolzano to Tarski*. Dordrecht: Kluwer.

Sinisi, V., J. Woleński (Eds.) (1995). *The Heritage of Kazimierz Ajdukiewicz* (*Poznań Studies in the Philosophy of the Sciences and the Humanities* **40**). Amsterdam–Atlanta (GA): Rodopi.

Skolimowski, H. (1967). *Polish Analytical Philosophy. A Survey and a Comparison with British Analytical Philosophy*. London: Routledge and Kegan Paul.

Suszko, R. (1968). Formal Logic and Development of Knowledge, in: I. Lakatos and A. Musgrave (Eds.) (1965). *Problems in Philosophy of Science. Proceedings of the International Colloquium in the Philosophy of Science*. Vol. I. London. Amsterdam: North-Holland.

Szaniawski, K. (1998). *On Science, Inference, Information and Decision Theory*, ed. by A. Chmielewski and J. Woleński (*Synthese Library* **271**). Dordrecht: Reidel.

Twardowski, K. (1977). *On the Content and Object of Presentations. A Psychological Investigation*, transl. by R. Grossmann. The Hague: Nijhoff.

Twardowski, K. (1999). *On Actions, Products and Other Topics in Philosophy*, ed. by J.L. Brandl and J. Woleński (*Poznań Studies in the Philosophy of the Sciences and the Humanities* **67**). Amsterdam–Atlanta (GA): Rodopi.

Wójcicki, R. (1979). *Topics in the Formal Methodology of Empirical Sciences*. Dordrecht: Reidel & Wrocław: Ossolineum.

Woleński, J. (1989). *Logic and Philosophy in the Lvov-Warsaw School* (*Synthese Library* **198**). Dordrecht: Reidel.

Wolniewicz, B. (1999). *Logic and Metaphysics*. Warszawa: Polskie Towarzystwo Semiotyczne.

LIST OF THE MAJOR WORKS IN PHILOSOPHY OF SCIENCE BY POLISH AUTHORS

A. Before the 20th Century

Witelon (ca. 1230 - ca. 1314), *Perspectiva*.

Witelon, *De causa primaria poenitentiae et de natura daemonum*.

Andrzej Wężyk (ca. 1377 - 1430), *Exercitium librorum Physicorum*.

Benedykt Hesse from Cracow (ca. 1389 - 1456), *Quaestiones super libros Physicorum*.

Jan from Słupcza (1408-1488), A commentary on Aristotle's *De caelo et mundo*.

Jan Schilling from Głogów (ca. 1445-1507), *Exercitium veteris artis*.

Jakub from Gostynin (ca. 1450 - 1506), *Theoremata, seu propostitiones auctoris causarum*.

Michał Twaróg from Bystrzyków (ca. 1450 - ca. 1520), *Quaestiones in tractatus Parvorum logicalium Petri Hispani*.

Michał Falkener from Wrocław (ca. 1460 - 1534), *Epitoma figurarum in libros Physicorum et De anima Aristotelis*.

Nicolaus Copernicus (1473-1543), *De revolutionibus.*
Wojciech Nowopolczyk (1508-1559), "Oratio de laude physices".
Jakub Górski (ca. 1525 - 1585), *Commentariorum artis dialecticae libri decem.*
Adam Burski (ca. 1560 - 1611), *Dialectica Ciceronis.*
Marcin Śmiglecki (ca. 1562 - ca. 1618), *Logica.*
Bartłomiej Keckermann (1572-1609), *Systema Logicae tribus libris adornatum.*
Jan Brożek (1585-1652), *Peripateticus cracoviensis.*
Jan Jonston (1603-1675), *Naturae constantia.*
Wojciech Wijuk-Kojałowicz (1605-1677), *Oculus ratione correctus.*
Marcin Świątkowski (ca. 1620 - 1790), *Prodromus Polonus eruditae veritatis.*
Tomasz Młodzianowski (1622 - 1686), *Prelectiones philosophicae in octo libors Physicorum.*
Adam Kwiryn Krasnodębski (1628-1702), *Philosophia Aristotelis explicata.*
Adam Kochański (1631-1700) and K. Schott, *Analecta mathematica.*
Ferdynand Ohm-Januszowski (1639-1712), *Summa philosophica.*
Stanisław Konarski (1700-1777), *De arte bene cogitandi ad artem bene dicendi.*
Antoni Skorulski (1715-1789), *Commentariorum philosophiae, logicae scilicet metaphysicae, physicae generalis et particularis.*
Antoni Wiśniewski (1718-1774), *Propositiones philosophicae ex physica recentiorum.*
Benedykt Dobszewicz (1722-1794), *Placita recentiorum philosophorum explanata.*
Ignacy Włodek (1723-1780), *O naukach wyzwolonych w powszechności i szczególności księgi dwie* [Two Books on Liberal Arts in General and in Particular].
Ignacy Wilczek (1728 - after 1788), *Carmina.*
Samuel Chróścikowski (1730-1799), *Fizyka doświadczeniami potwierdzona* [Physics Confirmed by Experiments].
Jan Śniadecki (1756-1830), *Pisma rozmaite* [Miscellaneous Writings].
Józef Łęski (1760-1825), *Rozprawa o nauce przyrodzenia* [A Treatise on the Science of Nature].
Jędrzej Śniadecki (1768-1838), *Teoria jestestw organicznych* [A Theory of Organic Beings].
"Mowa o niepewności zdań i nauk na doświadczeniu fundowanych" [A Speech on the Uncertainty of Sentences and Sciences Founded on Experience].
Anioł Dowgird (1776-1835), *Wykład przyrodzonych myślenia prawideł* [An Exposition of the Natural Rules of Thinking].
Józef Hoene-Wroński (1778-1853), *L'oevre philosophique.*
Michał Wiszniewski (1794-1865), *Bacona metoda tłumaczenia natury* [Bacon's Method of Explaining Nature].
Dominik Szulc (1797-1860), *O źródle wiedzy tegoczesnej* [On the Source of Modern Knowledge].
Ferdynand Dworzaczek (1804-1876), "Rzecz dotycząca filozofii medycyny" [On the Philosophy of Medicine].
Tytus Chałubiński (1820-1889), *O metodzie wynajdowania wskazań lekarskich* [On the Method of Searching for Medical Indications].
Wojciech Urbański (1820-1889), *Pisma pomniejsze* [Minor Writings].
Władysław Kozłowski (1832-1899), *Pisma filozoficzne i psychologiczne* [Philosophical and Psychological Writings].
Henryk Hoyer (1834-1907), *O metodzie badania naukowego* [On the Method of Scientific Study].
Stefan Pawlicki (1839-1916), *Studia nad pozytywizmem* [Studies on Positivism], and *Kilka uwag o podstawie i granicach filozofii* [A Few Remarks on the Foundations and Limits of Philosophy].
Stanisław Kramsztyk (1841-1906), *Szkice przyrodnicze z dziedziny fizyki, geofizyki i astronomii* [Studies of Nature: Essays in Physics, Geophysics and Astronomy].
Marian Morawski (1845-1901), *Filozofia i jej zadanie* [Philosophy and Its Task].
Marian Morawski, *Celowość w naturze* [Teleology in Nature].
Aleksander Raciborski (1845-1920), *Pojęcie przyczynowości w "Systemie logiki dedukcyjnej i indukcyjnej" J.S. Milla* [The Concept of Causality in J.S. Mill's "System of Logic Ratiocinative and Inductive"].
Julian Ochorowicz (1850-1917), *Wstęp i pogląd ogólny na filozofię pozytywną* [An Introduction to and an Overview of Positive Philosophy].
Samuel Dickstein (1851-1939), *Matematyka i rzeczywistość* [Mathematics and Reality].

Władysław Biegański (1857-1917), *Logika medycyny, czyli zasady ogólnej metodologii nauk lekarskich* [The Logic of Medicine, or the Principles of the General Methodology of the Medical Sciences].

Władysław Biegański, *Neo-Teleologia* [Neo-teleology].

Adam Mahrburg (1860-1913), *Pisma filozoficzne* [Philosophical Writings].

Zofia Daszyńska-Golińska (1866-1934), *Szkice metodologiczne* [Methodological Essays].

B. From the Beginning of the 20th Century to 1945

Napoleon Cybulski (1854-1919), *O współczesnym witalizmie i mechanizmie* [On Contemporary Vitalism and Mechanism].

Władysław Mieczysław Kozłowski (1858-1935), *Przyrodoznawstwo i filozofia* [Natural Science and Philosophy].

Józef Nusbaum-Hilarowicz (1859-1917), *Idea ewolucji w biologii* [The Idea of Evolution in Biology].

Stanisław Kobyłecki (1864-1939), *Postulaty psychologii doświadczalnej* [Postulates of Experimental Psychology].

Józefa Kodisowa (1865-1940), *Studia filozoficzne* [Philosophical Studies].

Edmund Biernacki (1866-1911), *Zasady poznania lekarskiego* [Principles of Medical Knowledge].

Kazimierz Twardowski (1866-1938), *O czynnościach i wytworach* (English version in K. Twardowski, *On Actions, Products and Other Topics in Philosophy*, this series vol. I, 1999).

Tadeusz Garbowski (1869-1940), *Organizm a społeczeństwo* [Organism and Society].

Władysław Heinrich (1869-1957), *O metodologii nauk* [On the Methodology of the Sciences].

Władysław Szumowski (1875-1954), *Logika dla medyków* [Logic for Physicians].

Helena Konczewska (1877-1959), *Le problème de la substance.*

Leon Petrażycki (1877-1931), *Nowe podstawy logiki i klasyfikacja umiejętności* [New Foundations of Logic and Classification of Abilities].

Jan Łukasiewicz (1878-1956), *O twórczości w nauce* (English version *Creative Elements in Science*, see above: Łukasiewicz 1970).

Florian Znaniecki (1882-1958), *Przedmiot i zadania nauki o wiedzy* [The Object and Aims of the Science of Knowledge].

Leon Chwistek (1884-1944), *Granice nauki* (English version, see above: Chwistek 1949).

Bronisław Malinowski (1884-1942), *"A Scientific Theory of Culture" and Other Essays.*

Władysław Horodyski, (1885-1920), *Pojęcie stosunku przyczynowego* [The Notion of a Causal Relation].

Edward Stamm (1886-1940), *Przyczynowość a stosunek funkcjonalny* [Causality and the Functional Relation].

Tadeusz Kotarbiński (1886-1981), *Elementy teorii poznania, logiki formalnej i metodologii nauk* (English version, see above: Kotarbiński 1966).

Joachim Metallmann (1889-1942), *Determinizm nauk przyrodniczych* [Determinism of the Natural Sciences].

Tadeusz Czeżowski (1889-1981), *Klasyczna nauka o sądzie i wnioskach w świetle logiki współczesnej* [The Classic Teaching on Judgement and Inference in the Light of Contemporary Logic].

Adam Wiegner (1889-1967), *Uwagi nad indeterminizmem w fizyce* [Remarks on Indeterminism in Physics].

Kazimierz Ajdukiewicz (1890-1963), *Główne zasady metodologii nauk i logiki formalnej* [Main Principles of Methodology and Formal Logic].

Janina Hosiasson-Lindenbaumowa (1899-1942), *On Confirmation.*

Janina Kotarbińska (1901-1997), *O tak zwanej konieczności związków przyrodzonych* [On the so-called Necessity of Regularities].

Władysław Krajewski
and Jacek Juliusz Jadacki

I

PHILOSOPHERS

Poznań Studies in the Philosophy
of the Sciences and the Humanities
2001, vol. 74, pp. 27-35

Mieszko Talasiewicz

JAN ŁUKASIEWICZ – THE QUEST FOR THE FORM OF SCIENCE

I. Jan Łukasiewicz (21.12.1878, Lvov – 13.02.1956, Dublin) studied philosophy and mathematics in Lvov. He completed his PhD dissertation under Kazimierz Twardowski's supervision in 1902 and continued his studies abroad, mainly in Germany and in Belgium. Upon his return to Lvov, he started work at the university in 1906 as a lecturer; in 1911 he received the title of professor extraordinary. In 1915, Łukasiewicz was appointed head of one of philosophical chair at the newly restored Warsaw University. He had close ties with Warsaw University until the outbreak of the Second World War, except for the years 1918-1920, when he worked in the Ministry of Education (including the position of Minister in 1919). His brilliant academic and administrative career (Łukasiewicz was twice elected the president of the university) as well as his exceptional gift for teaching enabled him to inspire a spectacular movement both in logic and in philosophy – the Warsaw School of Logic and the Warsaw "branch" of the philosophical Lvov-Warsaw School. Among the members of his seminar were Stanisław Leśniewski, Alfred Tarski and many other logicians and philosophers of world-wide renown who made Warsaw an international centre of analytical research, such as Jan Salamucha, Mordchaj Wajsberg, Stanisław Jaśkowski and Bolesław Sobociński. Their activities came to an end with the war. In fact, Łukasiewicz himself did not suffer harm in the September 1939 campaign, but his library as well as all his manuscripts and notes perished in fire during the first days of the war. Until summer 1944, he stayed in the occupied Warsaw, engaged in underground teaching, then left Poland. Heinrich Scholz, a German logician, professor in Münster and Łukasiewicz's friend, helped him to move to Münster. Łukasiewicz intended to go to Switzerland but because the German-Swiss border was blocked after the attempt on Adolf Hitler's life, he was forced to stay illegally in Germany until the end of the war. He then went to Brussels and in 1946 was offered a professorship at the Royal Irish Academy in Dublin, where he spent the last decade of his life.

II. Łukasiewicz was known mainly for his original and revolutionary historical reconstructions of Aristotle's and Chrysippus' logic in modern formal language, for his research on the foundations of the classical propositional calculus and for inventing and creating systems of three- and many-valued logic. Yet, he also significantly contributed to the philosophy of science in many respects. To begin with, his ideas concerning many-valued logic and the philosophical motivations of creating many different logical systems largely belonged to that field. As he stressed in many of his writings, such systems were invented in search of a formal model of the real world, the model that should be embodied in theories of empirical science[1]. It is experiment and scientific observation that are supposed to determine which of the logical systems is the correct one, or which is useful for describing the world. Łukasiewicz was far from regarding formal logic as merely "playing with words". He started his investigations with the metaphysical presuppositions of logic (for instance, his very first idea of three-valued logic was closely connected with the problem of justifying metaphysical indeterminism), and even though he moved quite far towards pure formalism in his later works, he never abandoned the idea that any formal system of logic should be an interpreted one, and that its interpretation, and - furthermore - its relevance as a language of empirical science is the cognitive ratio of that system. Łukasiewicz revealed such opinions as late as in 1936 [2.19], long after the "philosophical period" in his scientific activity was over. Once we realize that these ideas were defended in the times of the overwhelming prominence of logical positivism, when the use of logic in philosophy of science was restricted just to investigating the rules of inference in the context of the problem of induction, it becomes clear that Łukasiewicz's concept of logic as a set of competing theories that are to be invented *a priori* but chosen and justified due to empirical contents of their interpretations was a really impressive anticipation of modern trends in analytical philosophy.

III. Łukasiewicz appeals to the methodology of science more directly in his famous classification of reasonings. Following some preliminary remarks made by his teacher Kazimierz Twardowski, Łukasiewicz formulated the classification which was widely appreciated in the Lvov-Warsaw School thanks to the clarity and elegance of its *fundamenta divisionis* (and in spite of some oddity in the final result). But it was not only clarity and elegance that made the classification so special; perhaps even more important was the strongly objective, formal characteristic of reasoning. Łukasiewicz was a declared anti-psychologist and nothing would be more against his intentions than regarding reasonings as psychological activities of particular men. Naturally, such an activity might satisfy the conditions of proper reasoning, but if it did not that

---

[1] These ideas were certainly continued. See for example Giles (1974).

would by no means influence the definition of reasoning. But again, this formal theory was supposed to fit the needs of the methodology of empirical sciences, showing what reasonings should be the proper ones. In [1.3, p. 13] he wrote:

> Every theory rises from a reasoning. Reasoning in turn comes from the fact that the desire for knowledge can hardly be satisfied only by observation. We want to know not only what a given fact looks like, but what its causes are, what effects it can produce; we want to derive it from a principle and understand it fully. In all these cases, we must refer to the factors that are not given immediately, in sensual or mental experience or in memory. And the only way to go beyond the narrow circle of immediate data is reasoning.

The classification itself is as follows. Reasoning consists in matching a sentence with a given one whereas one entails another (matching a reason with a consequence, or a consequence with a reason).

A direction – in a certain sense of the word – can be attributed to reasoning as well as to entailment: reasoning proceeds from a given sentence to the one that it shall be matched with, and entailment proceeds from reasons to consequences. Now, reasonings shall be divided primarily according to whether the direction of reasoning accords with the direction of entailment. Thus, we obtain deductive reasonings – those which proceed from a reason to its consequence (matching a consequence with a given reason); and reductive reasonings – those which proceed against the direction of entailment (matching a reason with a given consequence). Secondly, reasonings are divided according to whether or not a given sentence is an accepted (asserted) one. Crossing these divisions we hence obtain the following types of reasoning:

(deductive): INFERENCE – matching a consequence with an accepted reason;

EXAMINATION (verification) – matching an accepted consequence with a given reason;

(reductive): EXPLANATION – matching a reason with an accepted consequence;

DEMONSTRATION (proof) – matching an accepted reason with a given consequence.

This result, although obtained from very natural premises (*fundamenta divisionis*), seems somewhat unexpected. The oddity mentioned above is easily visible if we consider that, for instance, demonstration (proving), usually regarded as a perfect example of deduction, and being fundamental in the so-called deductive sciences: logic, mathematics, etc., turns out to be reductive reasoning. Again, examination (verification), one of the main tools of empirical sciences, usually regarded as being reductive, turns out to be a deductive procedure. However, all this oddity is plain to understand if we note that the main difference between reduction and deduction is in the procesual aspect of reasoning, not in the "resultative" one: to perform an inference or an examination one needs nothing more but deriving logical conclusions from given premises. On the other hand, demonstration, for instance, requires much more: although the final result of a demonstration is a perfect chain of entailments, the process of demonstrating is a highly undetermined, largely creative action. Inference and examination are always performable. It is not the case with explanation and demonstration – they ask for a play of associations, for a touch of intuition and luck.

IV. By the way, the problem of creativity in science was generally of great importance to Łukasiewicz. He devoted several papers to the subject; in [2.5, pp. 8-9] he wrote:

> If a [explanatory] generalization expresses a relationship, it introduces a factor that is alien to experience. Since Hume's times we have been permitted to say only that we perceive a coincidence or a sequence of events, but not a relationship between them. Thus, a judgement about a relationship *does not reproduce* facts that are empirically given, but again is a manifestation of man's *creative* thought [...]. Such is the role of experience in every theory of natural science: *to be a stimulus for creative ideas and to provide subjects for their verification.*

Nevertheless, Łukasiewicz is far from neglecting the differences between formal sciences, such as logic and mathematics, and empirical sciences, such as physics. But now the *fundamentum divisionis* for sciences is not the kind of appropriate fundamental reasonings, but the kind of fundamental data that the sciences deal with. In the case of formal sciences, these are general statements, e.g. they take the form of axioms which could be regarded as accepted reasons in subsequent chains of entailments. Having an accepted reason, we look for new theses by inferring some interesting consequences from it; having a theorem that we want to introduce to the system, we look for an accepted reason by making a demonstration. And again, for testing if the demonstration is correct, we try to infer our theorem from the reason found. On the other hand, the fundamental data for empirical sciences are particular statements about facts that can be regarded as accepted consequences. Thus, the main reasonings in these sciences are explanations, when we look for a hypothesis on why such-and-such event occurs (one of possible forms of explanation is induction), and examination, when we test the hypothesis, looking for its justification.

Although nowadays such a concept of science does not seem very peculiar, the underlying Łukasiewicz's philosophy was quite a novelty at the time when these ideas were first published [2.9; 1.3]. Thanks to that concept it can be plainly seen that induction and other kinds of explanation cannot serve as a means of justification; Łukasiewicz clearly stated on many occasions that a reason matched with an accepted consequence cannot be accepted just as the result of such a reasoning. Explanation of facts is the aim of science, induction may play an important role in heuristics, but it can by no means justify our theories. Justifying is a matter of a deductive procedure, i.e. examination. Although we cannot prove any hypothesis, we can test it, and justify it to some extent by applying some additional criteria of acceptance. This idea, initially defeated due to the expansion of neopositivism with its dogmas of inductive support, finally reappeared – *via* discussions with Tarski, Łukasiewicz's student – in the thought of Karl Popper, in his devotion to deduction and in his fallibilism.

However, deductionism and fallibilism are not the only Łukasiewicz's ideas later associated with Popper. So is his view that science is closer to art than it is usually concerned. Łukasiewicz compared doing science to an artistic activity, e.g., in [2.13]:

> Natural laws and theories [...] should be compared not to mechanical photography, but to a painting made by an artist. Just as the same landscape may be differently reflected in works of different artists, the same phenomena may be explained by different theories.

In [2.10; 2.5, p. 14] he wrote more emphatically:

> Poetic creativity does not differ from scientific creativity by a greater amount of fantasy. Anyone who, like Copernicus, has moved the Earth from its position and sent it revolving around the Sun, or, like Darwin, has perceived in the mists of the past in the genetic transformations of species, may vie with the greatest poet.

As has been mentioned, philosophy of science in the modern sense – in which it is closely associated with methodology – was never the main subject of Łukasiewicz's interests. If not a logician, he was more of an ontologically rather than epistemologically oriented philosopher. He never formulated any developed methodological theory, nor even attempted to. Nevertheless, we can find a lot of enlightening remarks on that subject in his writings, especially the earlier ones. In [1.1], for instance, (where also the very first ideas of his classification of reasonings are to be found) he gave some very interesting hints.

Łukasiewicz fully revealed his fallibilism and even a kind of pragmatism with respect to empirical sciences, both devoid of any touch of relativism, by writing:

> There are judgements that are logically worthless, but practically precious. Nearly all empirical laws, theories and hypotheses of nature belong here [...]. Although all hitherto known facts seem to match these laws and theories, we do not know if this conformance would not vanish some day. However, although these judgements are not equally true as, for example, theorems of mathematics, and although many of them can even reasonably be thought of as false, they have great practical value, embracing many different phenomena in one whole, putting those phenomena in order and enabling us to predict future ones. Thus, even if some day Newtonian *leges motus* totter, if the highest laws of mechanics turn to be an inexact expression of reality, they would still preserve their practical value in spite of their falsity as they order the complicated chaos of phenomena and formulas of mechanics and put it all in a systematic whole [1.1, p. 135 in the 1987 edition].

As regards the question of creativity, Łukasiewicz shows this special view of the world of abstract objects that preserves the space for creativity in formal sciences, the view akin to Gottlob Frege's Third Realm and Popper's World III – that objects, once created by us, then live their own independent lives:

> Constructive objects [of logic and mathematics – MT] only seem to be free products of mind. We can assume that definitions of all constructive objects are unrestricted; but we create many such objects, and attribute manifold properties to them. Together with such free constructions some relations among them appear – relations that are independent of our will [1.1, p. 116 in the 1987 edition].

The main Łukasiewicz's idea concerning the philosophy of science – the idea that formal systems are to be competing forms for our knowledge; that practising science is to produce, by means of logic, highly sophisticated theories and then (and only then) to "ask" the nature if they are right – was reflected in following passages:

> There are no apriorical *ergo* necessary laws of experiment [...]. Apriorical laws are based upon definitions; that is why they are certain, that is why they are dogmas. But whether our definitions fit the reality or not, that is not a dogma of science but just a hypothesis that could never be verified with all certainty [1.1, pp. 128-129 in the 1987 edition].
>
> I would like to recall a fact that is often forgotten or even unknown – that formal logic played an important role in the rise of empirical sciences. Anyone who ever read Galileo's works or Pascal's physical writings, must have noticed how exact and subtle their powerful minds were. Good observations of only few facts were sufficient for them to create magnificent theories of great generality; theories in which deductive and formal thinking was decisive [1.1, p. 182 in the 1987 edition].

According to these ideas, it is not only the (properly interpreted) language of logic that can be useful as a language of science; but metaphysics can serve as a basis of science as well. Analytical metaphysics, as a system of definitions, shall be regarded as a tool of the structural ordering of the data – neglecting metaphysics is full of bias, and is just irrational, leading straight to the poverty and banality of science. This appreciation of metaphysics, which was common among Łukasiewicz's students in the Lvov-Warsaw School, became one of the most serious differences that set the School apart from the neopositivist movement.

V. Łukasiewicz also contributed substantially to numerous specific subjects related to the philosophy of science. He paid special attention to induction. More specifically, he defended the view that induction is an inversion of deduction [1]. As we know, the classical scheme of induction was as follows:

If judgements A (evidence) are true, then judgement X (generalization) is probable.
Judgements A are true.
_____
Thus judgement X is probable.

Such a scheme confuses the true nature of this kind of reasoning by introducing a dubious major premise and representing the whole procedure in a deductive manner. Instead, Łukasiewicz defended the following:

If judgement X (generalization) is true, then judgements A (evidence) are also true.
Judgements A are true.

Thus judgement X may be true.

Now, the major premise is unquestionable – judgements A are logical consequences of X – and the form of the reasoning reveals fully the point of the problem: this reasoning is reductive.

Łukasiewicz formulated a classification of inductive procedures [2.2] into generalising induction ("a_1 is B", "a_2 is B", "a_3 is B", ... then "All a are B") and complementary induction ("a is b_1", "a is b_2", "a is b_3", ... then "a is B") and maintained that a sense of probability changes can be associated only with the latter – probability can increase only if some hypotheses are rejected (finding out that e.g. "a is b_4" excludes the possibility of a hypothesis that "a is B'", where B' consists of b_1, b_2, b_3, c, ...); increasing the number of positive generalising instances is irrelevant in this case.

The theory of probability itself (being closely connected with the problem of induction) was another subject of special interest for Łukasiewicz. He created a very sophisticated theory, the so-called logical theory of probability, based on his own theory of indefinite propositions [1.2]. The theory "is objective in so far as it interprets probability as a property of propositions [not events!] which is characterized by its relationship to the *objective* world" [1.2; 1.5, p. 47]. The ideas contained in that theory turned out to be very fertile and were being developed and applied in the philosophy of science long after Łukasiewicz's death.

Łukasiewicz was also known for supplying a distinctive classification of sciences [2.11]. He divided them into empirical sciences and formal sciences and then divided empirical ones into natural (physics, chemistry, biology etc.) and human sciences (sociology, history, law, etc. – these were empirical since their subject – the results of human mental activity such as language, customs, institutions, etc. – could, and had to, be empirically experienced). Philosophy in one sense was for him a heterogeneous set of particular sciences (logic, psychology, etc.); in another sense it was regarded not as a pure science, but rather as a domain located between science, literature and religion.

VI. Finally, it should be stressed that Łukasiewicz – being a prominent logician – was also a philosopher of formal sciences. He was also a philosopher who initiated many metalogical investigations and conducted profound research in the foundations of logic and mathematics[2].

---

[2] See for instance [23] where Łukasiewicz discusses the relations among classical, intuitionistic and many-valued logics and analyses their place and role in the foundations of mathematics.

REFERENCES

Giles, R. (1974). A Non-classical Logic for Physics. *Studia Logica* **33**, 4, 397-415.

SELECTED BIBLIOGRAPHY

1. **Books:**
    1. (1910). *O zasadzie sprzeczności u Arystotelesa* [On the Principle of Contradiction in Aristotle]. Kraków: Akademia Umiejętności. Re-edited: Warsaw: PWN 1987.
    2. (1913). *Die logischen Grundlagen der Wahrscheinlichkeitsrechnung.* Kraków: Spółka Wydawnicza.
    3. Słupecki, J. (Ed.) (1961). *Z zagadnień logiki i filozofii. Pisma wybrane* [From the Problems of Logic and Philosophy. Selected Writings]. Warszawa: PWN.
    4. McCall, S. (Ed.) (1967). *Polish Logic 1920-1939.* Oxford: Oxford University Press.
    5. Borkowski, L. (Ed.) (1970). *Selected Works.* Amsterdam: North-Holland, Warsaw: PWN.
    6. Jadacki, J.J. (Ed.) (1998). *Logika i metafizyka* [Logic and Metaphysics]. *Miscellanea.* Warszawa: Wydział Filozofii i Socjologii Uniwersytetu Warszawskiego.

2. **Papers:**
    1. (1903). O indukcji jako inwersji dedukcji [On Induction as the Inversion of Deduction], *Przegląd Filozoficzny* **5**, 9-24, 138-152. Reprinted in [1.6], 203-227.
    2. (1906). O dwóch rodzajach wniosków indukcyjnych [On Two Kinds of Inductive Conclusions]. *Przegląd Filozoficzny* **9**, 83-84. Reprinted in [1.6], 227-228.
    3. (1906). Analiza i konstrukcja pojęcia przyczyny [An Analysis and Construction of the Concept of Cause]. *Przegląd Filozoficzny* **9**, 105-179. Reprinted in [1.3], 9-62.
    4. (1907). Co począć z pojęciem nieskończoności? [What to Do with the Concept of Infinity?]. *Przegląd Filozoficzny* **10**, 135-137. Reprinted in [1.6], 48-49.
    5. (1907). O wnioskowaniu indukcyjnym [On Inductive Reasoning]. *Przegląd Filozoficzny* **10**, 474-479. Reprinted in [1.6], 229-230.
    6. (1908). Pragmatyzm, nowa nazwa pewnych starych kierunków myślenia [Pragmatism, a New Name of Certain Old Trends in Thinking]. *Przegląd Filozoficzny* **11**, 341-342. Reprinted in [1.6], 389.
    7. (1909). O prawdopodobieństwie wniosków indukcyjnych [On the Probability of Inductive Conclusions]. *Przegląd Filozoficzny* **12**, 209-210. Reprinted in [1.6], 231-232.
    8. (1910). Über den Satz von Widerspruch bei Aristoteles. *Bulletin international de Académie des Sciences de Cracovie, Classe de Philosophie*, Cl I-II, 15-38. Published in English: On the Principle of Contradiction in Aristotle. *Review of Metaphysics* **24** (1971), 485-509.
    9. (1911). O rodzajach rozumowania. Wstęp do teorii stosunków [On the Types of Reasoning. An Introduction to the Theory of Relations]. *Ruch Filozoficzny* **1**, 78. Reprinted in [1.6], 232.
   10. (1912). O twórczości w nauce (Creative Elements in Science). *Księga pamiątkowa ku uczczeniu 250 rocznicy założenia Uniwersytetu Lwowskiego*, Lwów: Uniwersytet Lwowski, pp. 1-19. Published in English in [1.5], 1-15.
   11. (1915). O nauce i filozofii [On Science and Philosophy]. *Przegląd Filozoficzny* **28**, 190-196. Reprinted in [1.6], 33-38.
   12. (1916). O pojęciu wielkości [On the Concept of Magnitude]. *Przegląd Filozoficzny* **19**, 1-70. Reprinted in [1.5], 64-83.
   13. (1918). Wykład pożegnalny wygłoszony w auli Uniwersytetu Warszawskiego [Farewell Lecture Delivered in the Warsaw University Lecture Hall]. *Pro arte et studio* **3**, 3-4. Published in English in [1.5], 84-86.

14. (1928). "O metodę w filozofii [Towards a Method in Philosophy]. *Przegląd Filozoficzny* **31**, 3-9. Reprinted in [1.6], 41-42.
15. (1928/1929). Rola definicji w systemach dedukcyjnych [The Role of Definition in Deductive Systems]. *Ruch Filozoficzny* **11**, 134. Reprinted in [1.6], 129-130.
16. (1928/1929). O definicjach w teorii dedukcji [On Definitions in Deduction Theory]. *Ruch Filozoficzny* **11**, 177-178. Reprinted in [1.6], 130-131.
17. (1930). Philosophishe Bemerkungen zu mehrwertigen Systemen des Aussagenkalkül, *Comptes Rendus de la Société des Sciences et des Lettres de Varsovie*, Cl III. **23**, 51-77. Published in English: Philosophical Remarks on Many Valued Systems of Prepositional Logic, [1.4], 40-65. Reprinted in [1.5], 153-178.
18. (1934). Znaczenie analizy logicznej dla poznania [The Importance of Logical Analysis for Cognition]. *Przegląd Filozoficzny* **37**, 369-377. Reprinted in [1.6], 60-67.
19. (1936). Co dała filozofii współczesna logika matematyczna? [What Has Contemporary Mathematical Logic Contributed to Philosophy?]. *Przegląd Filozoficzny* **39**, 329-326. Reprinted in [1.6], 68-69.
20. (1941). Die Logik und das Grundlagenproblem. *Les entretiens de Zürich sur les fondements et la méthode des sciences mathématiques*. Zürich: Editeurs S.A. Leemann Peres & Cie, 82-100.
21. (1950). O zasadzie najmniejszej liczby [On the Principle of the Least Number]. *Rocznik Polskiego Towarzystwa Matematycznego* **21** (1948-1949), 28-29. Reprinted in [1.6], 322-324.
22. (1952). On the Intuitionistic Theory of Deduction. *Indagationes Mathematicae*, Series A, No 3, 202-212.
23. (1953). Sur la formalisation des théories mathématiques. *Colloques internationaux du Centre National de la Recherche Scientifique* **36** (*Les méthodes formelles en axiomatique*), 11-19. Published in English: Formalization of Mathematical Theories, [1.5], 341-351.
24. (1953). The Principle of Individuation, *Proceedings of the Aristotelian Society*, supplementary volume **27**, 69-82.
25. (1954). Arithmetic and Modal Logic. *The Journal of Computing System* **1**, 4, 213-219. Reprinted in [1.5], 391-400.

Poznań Studies in the Philosophy
of the Sciences and the Humanities
2001, vol. 74, pp. 37-46

Irena Szumilewicz-Lachman

ZYGMUNT ZAWIRSKI – THE NOTION OF TIME

I. Zygmunt Michał Zawirski (29.09.1882, Berezowica Mała near Tarnopol – 2.04.1948, Końskie) was born in in the part of Poland annexed by Austrians and called by them "Eastern Galicia" (now a part of the Ukraine). He completed his studies at Lvov University and received a PhD there in 1910. He was fortunate enough to have Twardowski, the founder of the Lvov-Warsaw School (LWS), as his academic mentor. Zawirski was an eminent member of the school. In 1928, he became professor at the University of Poznań, in 1937 at the Jagiellonian University in Cracow. During World War II he took part in the clandestine educational work at the university level. After the war, he was again a professor of the Jagiellonian University. He took part in the 7th and 8th World Congresses of Philosophy. He sent his paper to the 9th Congress in Amsterdam (August 1948). The proofs arrived at Cracow one day after his death.

II. When the Vienna Circle (VC) came into being, a lively and fruitful co-operation between LWS and VC developed. Both schools belonged to the analytical philosophy (in the broader sense), both esteemed logic and the natural sciences, advocated methodological empiricism, and rejected speculative philosophy. Nevertheless, there were some essential differences between both schools and the members of LWS were sensitive to them. Let me give some examples.

In 1931, R. Carnap and O. Neurath put forward the idea of physicalism. According to it, the most perfect language was developed in physics and the languages of all empirical sciences should be reduced to it; then, the unity of science will be achieved. Zawirski, like many other members of the LWS, was very critical of the idea. He pointed out differences between sciences; for instance, it is impossible to apply the physical space-time language to psychology; one would come across all difficulties of mind-body problems, which are far from being solved.

The differences between LWS and VC were clearly perceptible in the attitude to metaphysics. The condemnation of the traditional speculative metaphysics was shared by both schools. However, this condemnation was understood in Poland differently than in Vienna. Polish philosophers, whilst opposing sterile speculations, nonetheless respected philosophy.

> Polish philosophers were treated by the representatives of logical positivism as coming close to their standpoint. This was right to a some extent, but not very much, because Polish scientific philosophy did not share the most important point of the old and new positivism. Namely, Polish philosophy did not preclude the possibility that at least some issues of traditional metaphysics should be treated in a scientific manner [2.2].

On the other hand, science cannot do without metaphysical assumptions. Even speculative considerations might play, or have played, an enormous role in the advancement of science. Let us take, for instance, the fruitless search for solutions of some problems. Fruitless – says Zawirski – was the search for *perpetuum mobile* or for the squaring of the circle. Yet, these apparently fruitless efforts had enriched science enormously. The search for the inertial frame helped in the formulation of the general theory of relativity, the pursuit of a *perpetuum mobile* provided foundations for the principle of the conservation of energy, studies of the squaring of circle made it possible to understand the importance of irrational (transcendental) numbers as such. The role played by metaphysics is a *sine qua non* condition for the development of science.

III. Zawirski, who met Popper in Vienna in 1935, had a great respect for him. Nevertheless, he subjected Popper's views to a detailed critical analysis (see Szumilewicz-Lachman 1994, pp. 52-55). Below are his arguments in a brief summary.

1. Induction and deduction are equally important in the natural sciences. Both are indispensable for science. Induction is more important when a theory or law are formulated, *in statu nascendi*, while deduction plays its part during the process of testing.

2. Falsification, like verification, is never of a final character, since one does not test one singular theory, but always a group of theories or laws. One never knows, for certain, on the basis of a negative outcome of an experiment, which theory exactly has been falsified.

3. Falsification (as well as verification) is based on protocol sentences. But protocol sentences are not rock solid. They can be undermined as well. They depend on a number of factors: technical, theoretical and linguistic.

4. The development and changes of science are of infinite nature. One never knows whether truth had been already achieved. Although, thanks to Tarski, we have a definition of a true sentence, we do lack a criterion of truth.

A definition of a true sentence does not provide a criterion of truth. Zawirski concludes:

> A falsification by way of an experiment or an observation is of the same relative nature as verification; hence, if we look more closely at the verification and falsification processes, we shall find that the apparent asymmetry does not exist.

Zawirski emphasized that in the case of a negative outcome of an experiment a simple application of *modus tollendo tollens* would not be appropriate. In every experiment or observation, more than one hypothesis is involved. Let us assume that the hypothesis is a logical conjunction of three propositions p, q and r, and that it is rejected in view of the falsity of its consequences. One obtains a negation of this conjunction. However, the negation $\neg(p \wedge q \wedge r)$ can be transformed in accordance with the laws of logic in such a way as to yield a seven-member alternative: $\neg(p \wedge q \wedge r) \equiv [(\neg p \wedge q \wedge r) \vee (p \wedge \neg q \wedge r) \vee (p \wedge q \wedge \neg r) \vee (\neg p \wedge \neg q \wedge r) \vee (\neg p \wedge q \wedge \neg r) \vee (p \wedge \neg q \wedge \neg r) \vee (\neg p \wedge \neg q \wedge \neg r)]$.

Having rejected the conjunction $(p \wedge q \wedge r)$, we face the necessity of choosing one of the seven possibilities. Nevertheless, some reservations are indispensable at this point. The total refutation of the hypothesis takes place only when no change has to be introduced into the ideas entering the composition of the hypothesis and playing a role in the description of the experiment. The point is that if one changes the meaning of the terms applied, even to a very small degree, the refuted hypothesis can be taken up again. Thus, for example, the corpuscular theory of light, which was rejected after Foucault's experiments, was reintroduced into physics owing to the quantum theory when, along with various changes in the theory of light, the meaning of the term "corpuscle" was changed. The same applies to the case of *experimentum crucis* and *instantia crucis,* i.e. to the case when, out of the two competing hypotheses or theories, equally suitable for being adopted, the one selected is that which predicts a newly discovered fact. Out of the two hypotheses one will be verified and the other falsified. There exists an analogy between the testing of a singular hypothesis and the solution of a conflict between two competing hypotheses. In either case, we have to choose one of several possibilities, which form an alternative. In the latter case, the alternative is only more complex. In the former case, one expects to obtain an affirmative or negative answer to one question only. In the latter case, the positive answer with regard to one hypothesis is, at the same time, a negative answer with regard to the other hypothesis.

However, science sometimes comes across situations where the whole edifice of physics has to be changed. This is a revolutionary situation, and no small changes could solve the difficulties. There are similar cases, says Zawirski, when two originally inconsistent theories are later found to be capable of co-existence:

IV. Zawirski's lifelong passion was undoubtedly time. His most important achievements belonged to this field. In addition to many papers dealing with time, he also wrote two books. The first, *Wieczne powroty światów* [The Eternal Returns of the Universes] was published in Cracow in 1927 [2.1]. As he admitted, the work on the subject of the eternal return, involving a rather odd theory, awoke in his mind a profound interest in the problem of time.

The time enigma had been a focus of investigation almost since human curiosity had been awakened at the earliest stage of culture. No wonder! We only have to think of the mystery connected with existence, and the flow of time. The past has ceased to exist, the future has not yet arrived, and the present is only a point of contact between the past and the future. We cannot grasp any part of time, and yet its relentless flow is so closely connected with human fate, with the irreversibility of its direction, with the pilgrimage of humankind from childhood to old age and the unavoidable ultimate end. The irreversibility of time, the threat of death which, like the Sword of Damocles, hangs over everyone's head, has attracted the attention of philosophers and scientists for ages. We find this problem, again and again, in the ancient books, even in the Bible.

The most distinguished thinkers, from the Greek philosophers to the modern times, were fascinated by time. To cite just some names, they included Anaximander, Heraclitus, Plato, Aristotle, St Augustine, St Thomas Aquinas, Pascal, Leibniz, and Bergson.

At the turn of the centuries, the problem of time could be found at the centre of scientific and philosophical discussions.

In his book *Wieczne powroty światów*, Zawirski deals with the idea of cyclic time. According to this theory, the universe exists eternally, but this is by no means tantamount to saying that no changes take place. On the contrary, the universe is subject to variations, both of evolutionary and revolutionary nature; however, after a relatively long period, it returns to the initial point, in order to begin a new cycle identical with the previous one.

Zawirski's book consists of two parts. The first part is of a historical nature; the other is a critical analysis of the theory of cycles, in the light of contemporary science.

The historical part is highly instructive. It presents the first systematic exposition of the idea of returns that are chronologically and objectively arranged. The reader is often struck by the fact that, throughout history, the idea of eternal returns had been brought forward frequently and from many standpoints.

The whole history of the theory of eternal returns cannot be presented in this article. Therefore, I will limit the presentation to the ending. Zawirski cites Einstein's opinion on this subject: "The external returns cannot be denied with complete certainty". Zawirski ends his book by stating:

The scientific odds for the theory are, on the whole, not unfavourable, in the case both of inductive and deductive argumentation [...]. If the Universe is an unending melody which is played by Divine Eternity, then our existence seems to be those tones which are repeated now and then in this melody.

It appears to us that the conclusions drawn by Zawirski when appraising the probability of the theory of returns are doubtful. To accept a theory only because it "cannot be denied with complete certainty" seems to be risky and against scientific standards. Zawirski's approach can be explained only by his emotional attachment to the idea of returns.

V. Zawirski's opus magnum appeared in French: *L'évolution de la notion du temps* [1.2]. Selected parts of this book are now published in English (in: Szumilewicz-Lachman 1994). Zawirski was aware of the difficulties posed for the understanding of time by the twofold origin: in private "subjective" and in public "objective" experience. In his book, he discussed the conceptual puzzles of continuity, infinity and irreversibility of time. He claimed that empirical science will decide whether time and space are infinite.

No other publication known at that time was as up-to-date as Zawirski's. Duhem's *Le temps selon les philosophes hellenes* was devoted solely to the ancient period; Baumann's *Die Lehren von Raum, Zeit und Mathematik* ends with Hume; and Werner Gent's *Die Philosophie des Raumes und der Zeit* begins with Aristotle and ends at 1768. Hence, one great merit of Zawirski's book consists in that it carries the development of the idea of time down to the latest period, and takes into account problems related to relativistic physics as well as to quantum mechanics. The book is written in a clear and lucid style. Zawirski introduced a convincing development classification, with a division into historical periods and the isolation of the main problem to facilitate orientation in this vast material. We recall that conceptual clarity and stylistic lucidity are characteristic of the mental formation to which Zawirski belongs as the faithful Twardowski's disciple.

The book consists of two parts. The first contains the theories of time beginning with the Pythagorean period and reaching Zawirski's times. The second is critical, devoted to the analysis of some *aporias* related to the notion of time.

From among the ancient doctrines, the author devotes the most attention to Aristotle, whose conception he believes to be the closest to his own. From the pre-Aristotelian period, he takes into account mainly the doctrines of the Pythagoreans, Eleatics and Plato; and from the thinkers after Aristotle, the theories of the Stoics, Atomists, Sceptics, Plotinus and his successors, such as Proclus and Simplicius.

From the Christian philosophy of the medieval period Zawirski chooses only the most important items, the theories of St Augustine and of St Thomas Aquinas.

In terms of the evolution of the concepts of time, Zawirski divides modern history into four periods. The first one, which extends to Newton, still remains under the influence of scholastic doctrines, which is exemplified especially by Descartes and Spinoza. The second period begins with Newton and ends with Kant. Kant opens up the third period which spans almost the entire 19th century. At this stage, psychological studies develop and the critique of Newton's conception become more profound. This results in opening the way to new conceptions of time in the fourth period which lasts throughout the 20th century. Those new conceptions encompass Bergson's theory and the phenomenological investigation of Husserl's school. The main set of problems is related to the theory of relativity, which is discussed at length: Zawirski devotes a lot of time to Poincaré's conventionalism and to the conception of Enriques who is against radical conventionalism.

In order to allow readers easier orientation, Zawirski divides this set of problems into groups. According to him, the basic problems are: 1) the reality of time, and 2) the psychological genesis of the idea of time.

From the point of view of the reality of time, all of the theories are divided into realistic and idealistic ones. The latter comprise theories which either degrade time to the role of delusion (as in the case of Eleatics), with which reality has nothing in common, or else to the a priori form of our sensuality (as in Kant). The realistic views are, in turn, divided into extreme (substantial) and moderate realism. The extreme realism occurred in antiquity (some neo-Platonists), and, in modern times, in Newton. This view attributes to time a reality which is independent of external objects. Time flows "by virtue of its own nature", and it would exist even if the world ceased to exist.

Extreme realism is opposed by moderate realism, according to which time, while not being an *ens per se*, still denotes something real that co-exists with the changes and would disappear if the world would set and become motionless, just as in the fable of the sleeping princess (after the princess has been pricked with a needle). Moderate realism, which the author himself supports, can assume various forms. Moderate realism proclaims the thesis that time does not exist, but the sentence has a different meaning in this case than it had for Kant or the Eleatics. "Time does not exist" because there is no such substance or thing with which this name could be co-ordinated, but events do exist and the temporal order in nature would exist even in the absence of consciousness and of an observer. Throughout history, the dominant opinion was that of moderate realism, whose adherents included Aristotle, the scholastics, Leibniz and, what is more important, it owed its scientific support to relativistic physics.

Another basic problem is the genesis of the idea of time. Zawirski distinguishes two possible answers here: the source of the idea of time is found either in the facts of mental life or in those of the physical world. In addition to these two positions, there also exists an intermediate position which takes both sources into account. Depending on the answer to the questions related to the genesis of the idea of time, the above-mentioned position of moderate realism may assume the form of either psychological or naturalistic or even universal doctrine. The psychological theories stress the qualitative properties of time, while the naturalistic ones emphasize the metric properties.

The contrast between these trends runs across the entire history of philosophy; the followers of Plato, as well as St Augustine and Locke, form the stages of the evolution which has led to Bergson. The positions adopted of Aristotle and St Thomas Aquinas may be associated with modern physics.

Zawirski thinks that, since time has both qualitative and quantitative properties, one has to reconcile both the aforementioned tendencies.

In the critical part of his book, the author suggested solutions of problems which have appeared in the course of history. These are successively: the presuppositions taken over by scientists, the correctness of reasoning, and terminology. Zawirski concludes that, although the relation between generalizations and the experimental data leaves much to be desired, on the whole the reasonings are correct. As far as terminology is concerned, the situation is much worse: there appear to be many sins related to the ambiguity of terms, faults in their definitions, etc. Zawirski thinks that most of the doubts can be applied to the presuppositions accepted as the starting point. It is best in this case to apply Ockham's razor: *Entia non sunt multiplicanda praeter necessitatem.* "Thus the presuppositions ought to be both as few as possible, as well as obvious and intuitively understandable and certain". Ockham's razor is, however, found to be insufficiently sharp. Zawirski quotes examples of moderate and extreme realism, in order to demonstrate that intuition and obviousness do not provide a sufficiently acute criterion. It follows from the examples that, sometimes, contradictory presuppositions may appear to be equally obvious.

The problem of the verifiability of presuppositions produces insuperable difficulties: "none of the three possible theories are verifiable". For instance, we cannot verify what would happen with time if the world ceased to exist, or if it suddenly became motionless and changes ceased to exist. In this situation, Zawirski opts for the position of moderate realism, as the "least adventurous". The word "time" ought to be considered an "apparent name". Zawirski explains Kant's position at length. According to Zawirski, the proposition that "time" is an apparent name means that it is neither a substance nor a thing. As understood by Kant, time is only a form of sensuality, and its existence is related to the existence of consciousness.

Zawirski's objection to Kant is that he does not take into account the anisotropic character of time. He writes:

> Kant failed to take into account all the properties of time [...]. No one-dimensional continuum constitutes time by itself, as long as the difference between present, past and future is not associated with the relevant ideas. It is precisely this fundamental property of time that Kant passes over in silence in all of his works [1.2, p. 46].

Now, Zawirski is wrong here, for Kant does discuss the anisotropy of time. Here are the relevant quotations:

> Second Analogy, Principle of Succession in Time in Accordance with the Law of Causality. All alterations take place in conformity with the law of the connection of cause and effect [Kant 1933, p. 218]. But the concept which carries with it a necessity of synthetic unity can only be a pure concept that lies in the understanding, not in perception; and in this case it is the concept of the relation of cause and effect, the former of which determines the latter in time, as its consequence, not as in a sequence that may occur solely in the imagination (or that may not be perceived at all). Experience itself – in other words, empirical knowledge of appearances – is thus possible only in so far as we subject the succession of appearances, and therefore all alteration, to the law of causality (Kant 1933, p. 219).

Kant explains that by "the necessity of the causal relation" he means that the succession of the causally connected events cannot be reversed. On the contrary, the succession of events which are not connected causally depends on our will. He illustrates this idea with two examples: when we see a house, the succession of our perceptions depends solely on ourselves; we may begin with the roof and end with the gate, but we can reverse the order. The situation is quite different when, for instance, we see a ship move down stream.

> My perception of its lower position follows upon the perception of its position higher up in the stream [...]. The order in which the perceptions succeed one another in apprehension is in this instance determined, and to this order is bound down [...]. But in the perception of an event there is always a rule that makes the order in which the perceptions (in the apprehension of this appearance) follow upon one another in a necessary order (Kant 1933, p. 221).

I think that the quotations adduced suffice to quash the objection relating to Kant's allegedly not taking into account the anisotropic character of time.

As regards the other main problem, i.e. that of the origin of the idea of time, Zawirski does reject both apriorism in the Kantian sense and the epistemological apriorism. Bergson's conception (with his psychological concept of time) is also subjected to criticism. Zawirski thinks that the main difficulty in trying to solve this question is connected with insufficient standards of psychological investigations. The author blames Bergson for exaggerating the role of intuition and memory in the creation of the idea of time. The same applies to the qualitative characteristics of time, which had been exaggerated to the detriment of the metric properties of time.

Time can have metric properties without being measured by any observer: this does not change the fact that time is measurable.

The last question to be discussed is the oldest and most general one of the various ideas of time. Zawirski thinks that this is the notion of the time interval. He distinguishes three kinds of time relations: simultaneity, succession and duration. Duration is a term involving the greatest difficulties for historical reasons. According to Kant, duration means constancy and immutability of the given existence as opposed to existences which undergo changes. The Scholastics considered that the permanent duration of the immutable being was something different from time, since in such a duration there is no succession. Descartes was the first to protest against such a radical difference. The last link in this development is Bergson, who proceeded even further. According to him, not only is there no difference between duration and the time understood as a succession of states, but there does not exist any permanent substance either. Everything undergoes changes in a "creative evolution", something new always appears.

The third and the most important group of aporias deals with the structure of the notion of time. On this matter, Zawirski refers to the book of Reichenbach, "one of the most eminent philosophers among the relativists", as he calls him. Zawirski accepts Reichenbach's division of the time properties into the metric (quantitative) and topological (qualitative) notions. The qualitative notions include the relations of succession, simultaneity and the relation "in between". The metric notions include primarily the equality of two time intervals.

On the whole, Zawirski accepts Reichenbach's suggestions; he also accepts the proposed definitions. In the basic epistemological field, he begins the appropriate discussion with Reichenbach's statement that the definitions involved are allegedly of conventional character and do not express any cognition (Reichenbach 1958).

When starting a discussion with Reichenbach, Zawirski emphasizes, quite rightly, that as far as the definitions of objects and real relations are concerned the scientist's choice is limited by reality itself. As an example, he gives the history of the abolition of the classical definition of simultaneity.

Fifty years after his death, Zawirski deserves to be given particular attention for his fusion of analytic and historical scholarship. Strikingly versatile and contributing original work in all his fields of competence, Zawirski addressed the philosophical aspects of the relativity theory, the claims of the intuitionistic foundations of mathematics, the nature of usefulness of many-valued logics, the calculus of probability, the axiomatic method in science and the philosophy of science, and – in his crowning achievement – the conceptual history of the notions of time.

His work has been little known in English. It is my hope that this is going to change. As one of the most impressive and original Polish philosophers, Zawirski certainly deserves it.

REFERENCES

Kant, I. (1933). *Critique of Pure Reason*. London: MacMillan and Co. Ltd.
Reichenbach, H. (1958). *The Philosophy of Space and Time*. London: Dover Publications Inc., Constable and Co. Ltd.
Szumilewicz-Lachman, I. (1994). *Zygmunt Zawirski, His Life and Work*. Dordrecht: Kluwer.

SELECTED BIBLIOGRAPHY

1. **Books:**
   1. (1921). *Relatywizm filozoficzny a fizykalna teoria względności* [Philosophical Relativism and Physical Theory of Relativity]. Lwów: Drukarnia Słowa Polskiego.
   2. (1936). *L'evolution de la notion du temps*. Cracovie: Librairie Gebethner et Wolf.

2. **Papers:**
   1. (1927). Wieczne powroty światów [The Eternal Returns of the Worlds]. *Kwartalnik Filozoficzny* **5**, 3, 328-384; 4, 421-426; **6**, 1, 1-25.
   2. (1947). *O współczesnych kierunkach filozofii* [Contemporary Trends in Philosophy]. Kraków: Wiedza, Zawód, Kultura.

Poznań Studies in the Philosophy
of the Sciences and the Humanities
2001, vol. 74, pp. 47-51

Jan Woleński

TADEUSZ KOTARBIŃSKI – REISM AND SCIENCE

I. Tadeusz Kotarbiński (31.03.1886, Warsaw – 23.10.1981, Warsaw) studied philosophy in Lvov with Kazimierz Twardowski in 1907-1912. In 1913, he received his PhD upon submitting a dissertation on utilitarianism in Mill's and Spencer's ethics. In 1912-1919, he taught in secondary schools in Warsaw. He was appointed professor of philosophy at the University of Warsaw in 1919, where he lectured until his retirement in 1961 (in 1945-1949, he also taught at the University of Łódź). Kotarbiński was one of the most respected public figures in the Polish intellectual community, and a great teacher. In 1957, he was elected President of the Polish Academy of Arts and Sciences.

II. Kotarbiński belonged to the generation of Twardowski's students who formed the Lvov philosophical circle about 1911; it was from that circle that the later Lvov-Warsaw School originated. The Lvov circle accepted Twardowski's metaphilosophical programme of practising philosophy by legitimate methodological tools. Philosophy should be formulated in a clear and unambiguous language, proceed by correct inferences and consist of justified theses. There was a price for satisfying those demands: philosophy must limit its metaphysical ambitions (Twardowski did not abandon all metaphysics, but merely rejected "bad" speculative thoughts) and be sharply distinguished from world-views.

Twardowski's students were also influenced by the "new" mathematical logic which had been taught in Lvov by Jan Łukasiewicz since 1906. In fact, many Twardowski's students worked in logic, not only in mathematical or formal logic, but also in semiotics and the philosophy of science. For Kotarbiński, the close friendship with Stanisław Leśniewski, a great Polish logician, was extremely important. It was Leśniewski who pushed Kotarbiński in the direction of nominalism. As a professor in Warsaw, Kotarbiński became one of the main representatives of the Lvov-Warsaw School which flourished in the inter-war period.

III. In spite of his strong analytical attitude, Kotarbiński produced a philosophical system. However, it was not a speculative construction, but rather a collection of views interconnected by some leading general ideas.

Following the ancient tradition, Kotarbiński worked extensively in theoretical as well as practical philosophy. Reism, praxiology and independent ethics were the most important of his views. In my further remarks, I will not address ethics, as it is not particularly relevant for Kotarbiński's philosophy of science.

IV. Reism is a view that there are (exist) only concrete material things; *a contrario*, there are no abstract objects such as classes, events, properties or relations. This is reism in its ontological formulation. Moreover, there is reism as a semantic theory. It is a view on languages, more specifically about the conditions of the meaningfulness of sentences. According to Kotarbiński, only those sentences which, apart from logical constants, have names of material concreta as their constituents (or sentences translated into such sentences) are meaningful. The names of abstracta are apparent terms and should be eliminated from a language which is intended as a descriptive tool for speaking about the world. For example, the sentence "Alexander is wise" is a correct reistic sentence because it contains only names of concreta as its constituents. Now, the sentence "Wisdom is a property of Alexander" can be translated into the sentence "Alexander is wise". Thus, the sentence "Wisdom is a property of Alexander" is also meaningful, not directly, but *via* a suitable translation. On the other hand, a philosophical statement "Properties are irreducible abstract objects" is, by assumption, not translatable into the reistic language. At first, Kotarbiński did not distinguish ontological and semantic reism. Later, he was inclined to stress the importance of reism as a semantic theory. Naturally, he also believed in ontological reism, but semantic matters were of primary importance for him. He used several labels in order to explain his reistic intentions. "Reism" was the earliest one. Then, he used the terms "concretism" (there are only particulars or concreta) and "pansomatism" (every object is an extended corporal body) more often. Pansomatism, which is radical materialism, was a constant element of Kotarbiński's philosophy. Finally, Kotarbiński considered reism a project rather than a theory. He argued that it was a sound project, helping to achieve clarity and contributing in essential ways to the elimination of various pseudo-problems of philosophy. Such problems often appear when apparent names are used; they result in "hyposthases", that is suppositions that abstract object exist. In an impressive wording, reism is simply a battle against hyposthases.

V. Kotarbiński was always very keenly interested in practical matters (his first book was entitled *Eseje praktyczne* [Practical Essays] [1]). He decided to build a general theory of efficient action which he called "praxiology", and which had affinities with the general theory of action in the sense of Ludwig von Mises. He intended it to encompass all kinds of action: theoretical and practical. Not surprisingly, for Kotarbiński praxiology was simply the general methodology of any conduct governed by assumed goals, and in particular the methodology of scientific work. Praxiology was designed by Kotarbiński as the science of analyzing general concepts which are necessary for investigating special kinds of action. Goal, method, co-operation, efficiency, cause, effect, result, plan, invention, profit, method, etc., are some of the variety of praxiological notions. Kotarbiński's understanding of efficiency was closely related to certain general ideas: methodological rationalism (accept only statements that are sufficiently justified) and practical realism (assume goals which are possible to realize). The main results of praxiology were formulated by Kotarbiński in his famous book *Traktat o dobrej robocie* [A Treatise on Good Work] [3].

Kotarbiński's practical interests resulted in his general thinking concerning science. He was not only interested in the philosophy of science in its traditional scope, but also approached science as a social phenomenon. In particular, he stressed the importance of the proper (rational) organization of scientific research and of educating young scientists. Kotarbiński proposed a "science of science" as a general theory of science which would investigate the very complexity of science rather than merely its methodological aspects. The science of science, in Kotarbiński's view, should investigate all aspects of science: methodological, economic, organizational, psychological, political, sociological, etc. This project continues to be partly realized in Poland.

Some traditional problems of the philosophy of science were addressed by Kotarbiński in praxiology. He analyzed several others in his textbooks of which *Elementy teorii poznania, logiki formalnej i metodologii nauk* [Elements of the Theory of Knowledge, Formal Logic and Methodology of Sciences] [2] is the most famous. This book contains a detailed analysis of many basic problems and concepts of the philosophy of science (inference, deductive system, science, kinds of science and their methodological peculiarities, causality, space, time, method, induction, etc.). In general, Kotarbiński's analysis of those matters was perhaps not original, but always very precise, lucid and well documented by examples. *Elementy* (which is the popular short name of the book) became a very influential textbook in Poland and certainly very substantially contributed to the development of the philosophy of science in Poland.

VI. A very special problem concerns general consequences of reism for the philosophy of science. In *Elementy*, Kotarbiński tried to analyze several problems, for example the concepts of space and time, in the reistic language, that is under the assumption that only material things exist. He also tried to provide a consistently materialistic interpretation of psychological phenomena. Later, he considered problems which became the most difficult: the reistic interpretation of mathematics and humanities. It is well known that every nominalism (and reism is, of course, a kind of nominalism) runs against serious difficulties with respect to mathematics and the humanities. Without going into details, let me note that reism has special apparatuses for dealing with mathematics, namely Leśniewski's logical systems, which were intentionally constructed from the nominalistic point of view. In particular, Leśniewski's calculus of names (commonly called "ontology") is a system which is stronger than the standard predicate calculus. Although Kotarbiński succeeded in the "reification" of an elementary algebra of classes, the reistic interpretation of the whole mathematics is still an open problem. Let me also recall that another Leśniewski's system, namely mereology (a theory of classes in the collective meaning) also contributes to reism in an essential way, because it enables us to deal with collections of individuals.

The humanities raise two problems for reism. The first problem concerns values while the second is related psychological aspects of human activities. As I mentioned earlier, reism (in the more recent stages of Kotarbiński's views) was mostly a project. Thus, Kotarbiński aimed to demonstrate step by step that several problems of humanities and social sciences can be expressed in a language which satisfies reistic standards. What is perhaps particularly interesting is that Kotarbiński did not share behaviourism and naturalism that were popular among philosophers seeking to provide a nominalistic interpretation of psychology, humanities and social sciences. In general, an experiencing body (a treatment of human beings from the point of view of pansomatism) related to other experiencing bodies by relations of co-operation or conflict is the main category used in the reistic interpretation of society. Values are results of intentions which are states of experiencing bodies. This is continued by the analysis of groups which are also concreta, according to mereology. Independently of reism, Kotarbiński pointed to yet another methodological problem of the humanities, namely "the evolutionary point of view", which was very often neglected in investigations conducted in the field of the humanities and social sciences. He insisted that humanities that fail to take into account the evolutionary point of view must be defective.

V. Kotarbiński's position in the philosophy of science was untypical. His main interest was in topics usually neglected by leading philosophers of science, such as Popper or Carnap. In some respects, he was a precursor of quite recent developments, focusing on a more complex picture of science than that given by logical empiricists.

REFERENCES

Woleński, J. (Ed.) (1990). *Kotarbiński: Logic, Semantics and Ontology.* Dordrecht: Kluwer Academic Publishers.

SELECTED BIBLIOGRAPHY

1. (1913). *Szkice praktyczne* [Practical Essays], Warszawa: Wydawnictwo Kasy im. Mianowskiego.
2. (1929). *Elementy teorii poznania, logiki formalnej i metodologii nauki* [Elements of Theory of Knowledge, Formal Logic and Methodology of Science]. Lwów: Zakład Narodowy im. Ossolińskich. English ed. as *Gnosiology. The Scientific Approach to the Theory of Knowledge*, transl. by O. Wojtasiewicz. Oxford: Pergamon Press & Wrocław: Ossolineum, 1966.
3. (1955). *Traktat o dobrej robocie* [Treatise on Good Work]. Łódź: Łódzkie Towarzystwo Naukowe. English ed. as *Praxiology: An Introduction to the Science of Efficient Action*, transl. by O. Wojtasiewicz. Oxford: Pergamon Press & Warszawa: Państwowe Wydawnictwo Naukowe, 1965.

Poznań Studies in the Philosophy of the Sciences and the Humanities
2001, vol. 74, pp. 53-63

Tomasz Bigaj

JOACHIM METALLMANN – CAUSALITY, DETERMINISM AND SCIENCE

I. Joachim Metallmann (24.06.1889, Cracow – 1942?) studied natural sciences and philosophy at the Jagiellonian University. In 1912, he received a doctorate in philosophy from this University. In the years 1915-1917, he did military service in the Austrian Army. After the First World War, in 1918, Metallmann passed the examination qualifying him to teach in secondary schools. From 1928 to 1929 he travelled abroad, studying at different European universities, where he met Emil Meyerson, among others. In 1933, he submitted his postdoctoral dissertation and was appointed associate professor, and in 1939 was promoted to the position of professor at the Jagiellonian University. He had close academic ties with **Leon Chwistek, Władysław Natanson, and especially with Zygmunt Zawirski.** After the outbreak of the Second World War, in November 1939, Metallmann was arrested with other professors of the Jagiellonian University by the occupying Nazi forces. The arrest was the so-called "Special Action" (*Sonderaktion*), aimed at destroying Polish cultural and scientific elite. Some of the professors arrested during this action were subsequently released, but Metallmann, because of his Jewish origin, was imprisoned in the concentration camp, and after suffering continuous physical and mental torture was finally murdered, probably in 1942.

Metallmann was one of relatively less known Polish analytical philosophers of the inter-war period, probably because he did not have his own students and followers. However, his works were distinguished because of his extensive knowledge of the subject and profound yet clear thought. Metallmann's field of interest comprised mainly philosophical problems of natural sciences, and epistemology including general methodology of sciences. His main work is an extensive (over 400 pages long) book *Determinizm nauk przyrodniczych* [Determinism of Natural Sciences] [1.1], summing up seven years of studies on the problem of causality and determinism in natural sciences. This book covers almost the whole range of questions connected with the problem of determinism, chance, probability, indeterminism in quantum mechanics, in biology, etc.

Metallmann also published several articles and reviews in Polish, French and German on various themes connected with the philosophy of science. In 1939 he wrote *Wprowadzenie do zagadnień filozoficznych. Część I* [An Introduction to Philosophical Problems, Part 1] [1.2]. It was one of the best introductory textbooks of philosophy ever published in Poland, and probably abroad too. It addresses such problems as the analysis of the scientific method, observations and experiments, predicting and verifying, causal relation and causal laws, deductive systems, and the relationship of logic with science and philosophy. It is a serious loss that the second part of this book has never been finished, and that Metallman's written notes on the subject were destroyed during the war.

II. Metallmann believed that there were three types of laws in science, which were not reducible to each other. They are: causal laws, statistical laws and morphological laws. Accordingly, three types of determinism can be distinguished: causal, statistical and qualitative (morphological) determinism. In his book *Determinizm nauk przyrodniczych* he thoroughly analyzed all the three types of determinism, making many digressions and remarks concerning the epistemology and ontology of natural sciences. At the beginning, he focused his attention on the problem of causality and causal determinism.

Metallmann began his analysis with terminological remarks concerning the notion of "causality". First of all, he insisted on distinguishing between the causal relation and the principle of causation. It is obvious that the former notion refers to a certain objective connection, whereas the latter refers to a general thesis concerning the structure of the real world. In order to reveal the proper meaning of the principle of causation, Metallmann analyzed typical uses of this principle in the works of both philosophers and scientists. He noted that the concept of "the principle of causation" was ambiguous. Some writers would equate it with determinism of the Laplacean type, some would hold that this principle amounts to a thesis that there is a universal regularity in the universe, and some would say that it is identical with the schema of prediction. By the "schema of prediction" Metallmann understood the general thesis that, if x, y are certain objects (states, processes, events or properties) and R – is a relationship between x and y, then provided the knowledge of x and R, it is possible to predict the occurrence of y.

Metallmann's own position was that the principle of causality could not be identified with the schema of prediction. It is necessary to introduce the second element, an ontological one. According to Metallmann, the principle of causation consists of two parts: one is the schema of prediction and the second is the principle of the repetition of events, stating, roughly speaking, that under the same conditions the same happens.

It is interesting to see what relationship holds, according to Metallmann, between the principle of causality and experience. He claimed that this principle was not a simple generalization of individual facts. However, it does not mean that the principle of causation is immune from change. Metallmann noted that the exact meaning of it had undergone some changes during the history of science. This was possible because the term "the same conditions", which can be understood differently in different periods of scientific development, was mentioned in the second part of the principle. Thus, for instance, the principle of causation would have a different meaning in the context of the action-by-contact theory, and a different one in the context of the field theory of force. The most radical change in meaning of this principle took place when quantum theory emerged. We will return to this issue later.

Metallmann's opinion on the dependence of the causal principle on experience, is a consequence of his epistemological conception, which he named "correlativism". He opposed both empiricism and transcendentalism (apriorism). He criticized empiricism for introducing the ideal of "pure experience", or "pure sensing" which can never be realized. On the other hand, transcendentalists are unable to explain why certain principles and concepts, supposedly independent of experience, are being overthrown in science. Metallmann's correlativism states that there is mutual interdependence (or correlation) between general principles of science and experience. It means that these principles, contrary to transcendentalism, are subject to experimental justification and possibly revision, but, contrary to empiricism, experimental data are organized with the aid of already existing theoretical notions and laws.

Metallmann also discussed the notion of causality. His general characterization of the causal relation is as follows. We can say that a is a cause of b if the following conditions are met:

(1) a and b are four-dimensional events (i.e. events in four-dimensional space-time);

(2) a differs from b not only with respect to space-time coordinates (for example a and b cannot be the same thing taken at two different moments);

(3) a and b can occur more than once;

(4) the causal relationship R between a and b has the following features:

    (a) R is unchangeable (it means that whenever a occurs, it is followed by b),

    (b) R is asymmetric,

    (c) R is transitive,

    (d) R is a many-to-one relationship (R allows for the prediction of b),

    (e) R can be used in causal explanations of events.

Metallmann was especially interested in the asymmetry of the causal relation. He formulated the question of whether the asymmetry of the causal relation can be accounted for by the asymmetry of time order, or *vice versa*. He critically analyzed existing attempts of reducing one order to the second. Among others, he criticized H. Reichenbach's theory. Reichenbach sought to define time order in terms of a causal relation. According to this definition, if x is a cause of y, then x is earlier than y. But Metallmann noted that Reichenbach's theory had some defects. Consider the case of cyclic processes, such as the swinging of a pendulum. Let s_1 be a state in which the pendulum reaches the highest point, s_2 be the highest point on the opposite side and s_3 be the state, in which the pendulum comes back to the previous location. Now the question of identity of s_1 and s_3 arises. If s_1 and s_3 are identical states, then it is impossible to define time order on the circular set of states. But if s_1 is different from s_3, we may ask what differs one from another? Metallmann answered that the difference was in time sequence: s_3 had occurred later than s_1. But if so, then Reichenbach's definition of time order is circular.

Metallmann concluded from this that it is impossible to reduce time order to causal order, and *vice versa*. He maintained that both orders are results of the more fundamental feature of the world, namely of continuous becoming, or the current of events. Hence, Metallmann expressed his belief in the objective existence of the flow of time. However, his remarks on this subject are enigmatic and therefore not fully satisfactory.

III. Metallmann was aware of the fact that the main threat for the principle of causality came from quantum physics. He was very interested in new developments in physics and their philosophical consequences. He distinguished three aspects of quantum physics, which were of great philosophical importance for him. They were: the requirement of "observability", the wave-particle dualism, and the alleged indeterminism of quantum mechanics. His main goal was to analyze the third aspect of quantum physics and its connection with the two others.

When scientists state that quantum physics rejects determinism, by "determinism" they usually mean "the schema of prediction". Metallmann noted that such "indeterminism" was present already in earlier physical theories, such as statistical physics. We must remember that for a successful prediction conforming with the above schema (cf. **II.**), two requirements should be satisfied: knowledge of a general law and exact knowledge of an initial state. In statistical physics, the second is impossible to achieve because of the excessive level of the complexity of the initial state. In quantum mechanics, the situation is similar, but the reason for this is different. We cannot know the exact initial state because of the indispensable restriction expressed by the Heisenberg uncertainty principle.

However, Metallmann maintained that the uncertainty principle, and hence the wave-particle dualism, do not imply indeterminism. His argument was as follows. Because it is impossible to determine simultaneously the location and momentum of an electron over a certain limit of accuracy, it is inappropriate to talk of an electron existing independently as a particle over this limit. But if the electron in the classical sense no longer exists, we cannot ask for predicting its future states. According to Metallmann, quantum physics did not overthrow determinism, but merely restricted its applicability. In quantum physics, determinism is still accepted, but in a statistical form.

Metallmann's argument against the indeterminism of quantum mechanics to a certain degree displays an analogy with another argument, put forward by Zawirski (1932). Zawirski noted that if the formulation of determinism took the form "if we know initial conditions, then we can predict the future states of a given system", then in quantum mechanics this sentence should be treated not as false, but as vacuously true, because the antecedent of the conditional is false. Although Metallmann himself expressed his doubts on whether Zawirski correctly interpreted the thesis of indeterminism, his own argument against indeterminism was not very far from Zawirski's. However, it was Zawirski who criticized Metallmann's method of defending "generalized" determinism (Zawirski 1935). First of all, Zawirski remarked that whether we decide to name quantum physics "indeterministic" or "deterministic in a broader sense" was a question of terminology. But more important is the remark that Metallmann was probably wrong claiming that over a certain limit of accuracy the notion of an "electron" ceases to be meaningful. For example, physicists continue to speak about individual electrons, even when their states are in a sense "uncertain".

Be as it may, Metallmann claimed that within quantum domain the schema of prediction continues to be valid, although in a slightly modified form: if we know *statistically* the state x, and the connection R, then we can predict the state y with the accuracy no greater then the accuracy of the state x. In other words, it is not causal determinism, but statistical determinism which retains its validity within the microphysical domain. But now the question arises of whether statistical laws could be in some sense reduced to causal laws. After analyzing various attempts of reducing one type of laws to the other, Metallmann concluded that this reduction is impossible. Indeed, statistical correlation refers always to certain classes of individuals, whereas causal relation connects individual objects. And if we state statistical correlation, we implicitly assume that the behaviour of individuals is not determined. However, Metallmann strongly insisted that causal determinism and statistical determinism are different forms of one generalized determinism.

IV. Metallmann did not restrict himself to only two types of determinism: causal and statistical determinism. He noted that there is one more type of scientific laws, namely morphological, or qualitative, laws. Whereas causal laws state causal connections between events, or processes, morphological laws express the coexistence of properties. According to Metallmann, coexistence of properties cannot in general be reduced to causal connections. Therefore, it is inappropriate to claim, as some philosophers of science do, that morphological laws can be, and should be, reduced to causal laws.

Morphological laws appear in all sciences, but are most common in such disciplines as biology, geology, palaeontology, and crystallography. Searching for morphological laws is often connected with searching for the most "natural" classification of objects of a given type. Metallmann strongly emphasized that we should distinguish between natural classification and artificial classification. Natural kinds consist not only of similar objects, but of objects that share essential properties, and whose properties are connected together. Using numerous examples taken from different branches of knowledge, Metallmann showed that creating an appropriate hierarchy of classifications is one of the most important aims of science.

But there are some arguments against the thesis that causal laws and morphological laws are on a par. For instance, J.S. Mill pointed out that for causal laws there exists a general principle, namely the principle of causation, which states that each event has its cause. However, we cannot formulate such a counterpart for morphological laws – that for each property there is a different property, which coexists with the former. Metallmann replied that Mill had used a wrong analogy. According to him, the correct form of the "coexistential principle" is: each object of nature can be included into a class which displays firm coexistence of properties and which belongs to a hierarchy of increasing firmness of coexistence.

In concluding the presentation of Metallmann's analysis of determinism, it should be noted that he distinguished scientific determinism from the so-called naive determinism. According to Metallmann, naive determinism, or fatalism, claims that every single event in the history of the world is already "determined", or that everything is ruled by fate. He rightly pointed out that such "determinism", stating also that the human mind cannot predict its future, is surprisingly not very far from the non-scientific view that everything happens by chance. It is so because naive determinism lacks the crucial part of each proper determinism – namely the schema of prediction.

In his later paper on the notion of emergence in biology, Metallmann mentioned another type of determinism, which can be named "emergent determinism" [2.2]. This form of determinism, applicable to the domain of living creatures, has two special features. Firstly, it emphasizes the role of internal conditions of a giving system in determining life processes. Secondly, emergent determinism allows for the "emergence" or creation of new laws. According to the emergence hypothesis, a given schema of prediction is valid only within a finite period of time, and is then replaced by another. Unfortunately, this interesting conception seems to be unfinished. Metallmann himself had not been given a chance to elaborate on these ideas.

V. The main question, stated by Metallmann in his article "O budowie i własnościach nauki" [On the Structure and Properties of Science] [2.1] is: What is science? Some people try to reformulate this question as: What is to be a scientific statement? What sentences can count as scientific? The simplest answer is that in order to be scientific, a sentence needs to be true. But is it a sufficient and necessary condition for being scientific? Some (including Jan Lukasiewicz) would say that this condition is not necessary, because there are plenty of irrelevant, albeit true sentences, which will never be a part of any science. Banal truths cannot count as scientific. But Metallmann rightly noted that "banality", or "irrelevance", are highly subjective and relative notions. There are truths which seem to be irrelevant at a particular time, and become very important after certain discoveries. This is the case of, for example, facts about certain minute differences between individuals belonging to the same species (such as species of plants or animals). Until the phenomenon of mutation was discovered, these facts seemed to be highly irrelevant. Metallmann concluded from this that we cannot exclude from science any sentence just because it is irrelevant at the current stage of scientific development.

Nevertheless, Metallmann did not agree that truthfulness is a sufficient and necessary condition for being scientific. First of all, the notion of "truth" is ambiguous. Metallmann suggested that the classical definition of truth should be accepted: a sentence is true always and only if it conforms with reality. More precisely, what should be conformed is a certain relationship in the reality and its formal counterpart expressed in the sentence. Metallmann maintained that in this sense of "truth", at least some statements of logic and mathematics do not count as true – namely those which derive from definitions only. If it is the case, then we would have examples of scientific sentences which are not true (and, of course, not false).

Moreover, Metallmann pointed to other counterexamples. Consider for instance general empirical hypotheses. It is well known that we cannot be sure that they are true. General statements can only be falsified, not verified. Nevertheless, they constitute a very important part of science. Therefore, truthfulness cannot be taken as a necessary condition for being scientific. But Metallmann also insisted that there are true, non-scientific sentences. They are subjective, psychological sentences, reporting our feelings, sensing, emotions, etc. As they are not given intersubjectively, they cannot count as scientific. Hence, the conclusion is: truthfulness is neither a sufficient nor a necessary condition for being scientific.

The other criterion of demarcation between scientific and non-scientific is the famous criterion of verifiability, put forward by the Vienna Circle. Metallmann rejected this criterion for two reasons. Firstly, he noted that it is impossible to verify one single sentence. According to Metallmann, we always verify a sentence with the aid of certain different sentences. For example, if we deduce a consequence of a general empirical statement, then what this consequence is compared with is not an observation, but an observational sentence. Metallmann insisted that it is impossible to compare any sentence directly with observation. He concluded that "no sentence is verifiable 'on its own', hence none is scientific independently of others; verifiability is a property of a system of sentences" [2.1, p. 115].

Regardless of the acceptability of this conclusion, we must say that Metallmann's argument does not seem to be plausible. If we compare a consequence of a general law with an observational sentence, then we can say that the former sentence is verified when two conditions are fulfilled: (1) the two sentences being compared are identical (or at least their meanings are identical), and (2) the latter is true. Thus, in fact, after verification we are left with only one sentence (up to synonimity) confronted directly with experience.

However, the second argument against verificationism, put forward by Metallmann, seems to be more convincing. He pointed out that a single šentence can be verified only as a part of a greater whole, i.e. a scientific theory. For example, the statement that ether does not exist, verified in the famous Michelson-Morley experiment, could be confronted with experience only together with many other statements and theorems from optics, theory of electricity, magnetism, and so on. Without the whole theoretical equipage, this single sentence would certainly count as non-verifiable, and hence as non-scientific. At this point, Metallmann had anticipated results that were to be achieved later by Quine and many others.

As a result of these considerations Metallmann declared that the initial formulation of our problem was inappropriate. We should not ask which sentences, considered separately, are scientific. Instead, "we must ask what set of conditions needs to be satisfied for a certain system of sentences to be science". At the beginning of his analysis of this question, Metallmann noted that science should be a system of sentences that are connected, both logically and factually. Logical connectedness is reduced to the requirement that sentences within given science should be logically deducible one from another. But each science must also have its special "viewpoint", in Metallmann's terminology. In other words, each science is characterized by its method, which is in turn determined by two factors: typical forms of reasoning and type of objects under investigations.

In connection with the last issue, Metallmann subdivided objects into a set of real objects and ideal objects. The former set constitutes the domain of empirical sciences, and the latter the domain of deductive (formal) sciences. He provided an original characterization of real objects and ideal objects. Real objects display the following features: (1) they are spatio-temporal (or at least temporal); (2) they are concrete, i.e. they have infinitely many independent properties; (3) it is impossible to formulate their definition, i.e. such a characteristic that logically implies all true sentences about these objects; (4) they are individual, i.e. two distinct real objects always differ with respect to some property; and (5) they are "given", which means that they can be empirically investigated. On the contrary, ideal objects fulfil quite opposite conditions: (1) they are neither spatial nor temporal; (2) they are "detached", i.e. their unique definition can be given; (3) they are universal, i.e. distinct objects can have the same properties (for example two congruent triangles), and (4) they are constructed.

Subsequently, Metallmann characterized methods of deductive sciences and empirical sciences. Several rules are observed within deductive sciences. First of all, Metallmann held that before considering a certain formal science we must explicitly name a science that was logically prior to it. What he had in mind was probably that each formal theory logically presupposes another more basic theory (e.g. analysis presupposes arithmetic and arithmetic presupposes first-order logic). The second rule states that each term of deductive science should be either explicitly defined, or be a primitive term whose sense is given in the axiomatic way. And according to the third rule, every statement must be deductively derivable from the complete, mutually independent and consistent set of axioms.

On the other hand, methodological rules of empirical sciences concern the ways of stating facts, formulating hypotheses and laws, binding laws together into one theoretical system, predicting, and verifying. It is interesting that Metallmann accepted intuition as an additional method applicable both for deductive and empirical sciences. By "intuition" he means "direct *a priori* knowledge". What is more surprising, he maintained that "direct *a priori* knowledge" is no less justified than other "ultimate" cognitive methods, such as observation and justification of axioms.

Besides the above-mentioned conditions of being science, there is one more aspect which distinguishes science from other activities. This aspect is rationality. Scientific attitude is characterized primarily by its criticism. For Metallmann, the main sign of scientific criticism was that in science there are no sentences that are absolutely protected against revision. Each scientific statement is subject to rejection. But it does not mean that a scientist makes no basic assumptions. One such assumption, as Metallmann pointed out, is that there exists regularity in nature, or that every phenomenon is in principle determined by some, possibly unknown, conditions.

Metallmann also made very profound remarks on the issue of changes in science. The standard inductivist account of scientific development emphasizes gaining facts and extending the scope of laws as a main criterion of progress. Even most recent changes in physics seem to conform to this pattern. For example, laws of relativity may be said to generalize the laws of Newton's mechanics, because the latter are approximate consequences of the former. However, Metallmann noted that this picture is an oversimplification of real processes. He suggested that science develop according to the inductivist schema only to a certain limit, i.e. up to the moment when its basic concepts could no longer be useful. Metallmann called this moment "a crisis". As a result of the crisis, new concepts must be developed and basic assumptions changed. At this point, Metallmann's conception exhibits a striking resemblance (even in terminology!) to Kuhn's theory of scientific revolutions, which became famous some thirty years later.

REFERENCES

Zawirski, Z. (1932). W sprawie indeterminizmu fizyki kwantowej [On the Matter of the Indeterminism of Quantum Physics]. In: *Księga pamiątkowa Polskiego Towarzystwa Filozoficznego we Lwowie*. Lwów, 456-483.

Zawirski, Z., (1935). J. Metallmann: Determinizm nauk przyrodniczych [a review of J. Metallmann's Determinism of Natural Sciences]. *Przegląd Filozoficzny* **38**, 145-156.

SELECTED BIBLIOGRAPHY

1. **Books:**
    1. (1934). *Determinizm nauk przyrodniczych* [Determinism of Natural Sciences]. Kraków: PAU.

2. (1939). *Wprowadzenie do zagadnień filozoficznych.* Część I [An Introduction to Philosophical Problems. Part I]. Kraków: Księgarnia D.E. Friedleina.

2. Papers:
   1. (1937). O budowie i własnościach nauki [On the Structure and Properties of Science]. *Wiedza i Życie* **4**, 356-67. Reprinted in: J.J. Jadacki, B. Markiewicz (Eds.). *A mądrości zlo nie przemoże.* Warszawa: PTF, 1993, 109-126.
   2. (1938). Determinizm i pojęcie emergencji w biologii [Determinism and the Notion of Emergence in Biology]. *Przegląd Filozoficzny* **41**, 45-53.

Poznań Studies in the Philosophy of the Sciences and the Humanities
2001, vol. 74, pp. 65-73

Leon Gumański

TADEUSZ CZEŻOWSKI – OUR KNOWLEDGE, THOUGH UNCERTAIN, IS PROBABLE

I. Tadeusz Czeżowski (26.07.1889, Vienna – 28.02.1981, Toruń) graduated from the Jan Kazimierz University in Lvov, where he studied philosophy, physics and mathematics. His two university teachers, Kazimierz Twardowski, the founder of the Lvov-Warsaw Philosophical School, and Jan Łukasiewicz, an outstanding logician, exerted dominant influence on his scientific interests and partly on his opinions in the domain of philosophy and logic. However, being an independent thinker and inquiring scholar, Czeżowski not only developed some of their ideas but also elaborated a number of interesting new conceptions. It is therefore only natural that he is usually named among the most prominent representatives of the Lvov-Warsaw School.

A comparatively short period of administrative work, as counsellor and then director of the Department of Science and Schools of Higher Education at the Ministry of Religious Denominations and Public Education, was a prelude to his long academic career that started in 1923, when he was given the Chair of Philosophy at the Stefan Batory University in Vilna. After the distressful experience of the Second World War, he left Vilna and came to Toruń in 1945, where he took part in founding the Nicolas Copernicus University and headed the Chair of Logic until his retirement in 1960.

Czeżowski's scholarly output includes more than 180 publications, most of which date from the post-war period and belong to the realm of logic, methodology and the philosophy of science. His style of writing is a manifestation of the tendency to ensure precision, consistency and appropriate justification of statements on the one hand, and didactic intelligibility on the other. As a rule, the starting point of his investigations were some historical studies thanks to which interesting problems could be found and knowledge of various solutions that had previously been proposed could be gained. At first, he usually presented the results of his explorations in the form of lectures that he frequently delivered in several academic centres. The lectures provided opportunities for a comprehensive discussion. Only then the revised version of the text went to press. Most of his concise papers were later collected in two volumes[1].

Czeżowski's research field was very extensive. In the present brief overview, we shall restrict ourselves merely to outlining a few selected results of his investigations in the areas of methodology and the philosophy of science.

II. Different types of reasoning are applied in every scientific discipline. Thus, it is small wonder that Czeżowski devoted special attention to defining and classifying various forms of simple reasoning, i.e. such in which the relation of premise to conclusion occurs only once. Continuing the studies of Jevons, Sigwart and Łukasiewicz, he put forward a new conception, according to which all simple reasonings have the same inferential scheme

$$\frac{Cab}{a}$$
$$\frac{}{b}$$

where *Cab*, *a* are called "inferential premises" and *b* – "conclusion". The reasoning is deductive since *Cab* represents a law of logic and the rules of substitution and detachment are applied. However, two kinds of such reasoning can be discerned: if the law belongs to the propositional or functional calculus, the reasoning is apodeictic; when *Cab* is a theorem of the logical theory of probability, the reasoning is probabilistic. Both the apodeictic and probabilistic types of reasoning can take the discovering (heuristic) or justifying form. In the former case, the premises serve as the starting point, whereas in the latter it is the conclusion that plays this part[2]. As could be expected, simple induction turns out to be a special variant of probabilistic reasoning.

III. Czeżowski devoted a separate paper to the problem of induction[3]. He argued that Hume's analysis is unjust because the reasoning of the kind involved refers to the whole set of instances comprised by a certain inductive generalization, and not to a particular unknown instance. The more elements of the set we know, the better we are acquainted with the whole set. However, the traditional concept of induction deals with categorical sentences only, therefore it is too narrow and should be extended. Namely, in accordance with Czeżowski's suggestion, the name "induction" encompasses any reasoning that fulfils the following two conditions: (1) the conclusion is a general law, whereas the premises state

---

[1] See item [1.3] and [1.4] of the selected bibliography.
[2] We refer here to the paper [2.5] and [1.4].
[3] Cf. [1.4], pp. 78-83.

particular instances of the law, and (2) the premises logically follow from the conclusion. As to induction by elimination it is a complex reasoning that consists of two steps: The first one is probabilistic, for it applies simple induction which leads to a certain number of alternative generalizations. The second step is apodeictic and employs the law of *modus tollendo ponens, CKApqNpq,* in order to eliminate some of the generalizations. The so-called method of residues, which Mill treated as a kind of induction by elimination, is not at all induction, because it does not result in general assertions but only in hypotheses about some unknown facts. General assertions obtained by simple induction are justified as being statements which have a certain degree of probability that is different from 0. On the contrary, induction by elimination turns out to be nothing but a mere – though useful – tool of scientific heuristics.

IV. Tadeusz Czeżowski supported Jevons' standpoint that argument by analogy is a complex reasoning. Namely, it consists of two phases: a generalization by simple induction precedes either the classical deduction executed according to the *Barbara* (or *Darii*) mode, or some explanatory reduction. Arguments by analogy may be used in two ways: in order to anticipate something or to test an observational sentence. In the former case, the scheme of the reasoning may be presented in the form:

every a is b
x is a
---
x is b

where the minor premise is an observational sentence or a sentence which Czeżowski calls 'hypothesis in the methodological narrow sense'. In the second case the scheme is transformed into:

every a is b
x is b
---
x is a

where the minor premise is again an observational sentence but the conclusion is a hypothesis that is probabilistically justified. In view of the initial inductive phase, both forms of the argument must be classified as probabilistic reasonings. The degree of probability of the conclusion increases as the analogy is more precise and the range of the inductive generalization is more limited[4].

V. Observational sentences are never certain. They may be false when the observations are wrong. In order to be justified, they should be put to test by further observations. If a new observation is discordant with the

---

[4] Cf. [1.4], pp. 82-96.

former ones, the conditions of observation have to be examined. Nevertheless, it is impossible to reproduce all the conditions (as time changes, some factors decline), hence there is no point in trying to refute observational sentences decisively; they always remain only probable.

It may be worth adding that inner observations have to be justified in the same way as sensory observations. Introspection – as Czeżowski puts it – is neither a better nor a worse source of knowledge than observation through one's senses and is likewise intersubjectively testable[5].

In the light of what has been said thus far, it should be clear that sentences about single facts must not be treated as the basic premises of empirical sciences – as some theoreticians maintained – for such sentences have to be tested and the process of testing is never terminated.

VI. The word "hypothesis" is often used as tantamount to "conjecture" or "guesswork". In Czeżowski's terminology it has a special, more technical meaning. According to him, hypotheses are general, particular or singular sentences with an existential import and serve to subordinate some observational sentence to a general law. They can be divided into two groups: naturalistic and historical. The former usually take the form of a general sentence about some presumable properties of facts stated in observational sentences. The latter are singular (or particular) sentences which are most often about the supposed past cause of certain observed facts or about other events of the past. In either case, a hypothesis p has to explain some implied sentences $q_1, q_2, ..., q_n$ each of which describes an observation. On the other hand, the sentences $q_1, q_2, ..., q_n$ reciprocally justify the hypothesis p. Usually, the set of sentences $q_1, q_2, ..., q_n$ may be explained by different hypotheses $p_1, p_2, ..., p_m$. In order to refute some of competing alternative hypotheses, further premises $q_{n+1}, q_{n+2}, ...$ are to be taken into account. However, it may turn out that none of the hypotheses $p_1, p_2, ..., p_m$ is reliable and all of them have to be abandoned. A hypothesis is better justified if it explains more facts more exactly, and allows to foresee more events. Nonetheless, it is important to realize and to keep in mind that no hypothesis is certain. They have at most a certain degree of probability which increases as the number n of the sentences $q_1, q_2, ..., q_n$ rises.

VII. Let us now turn to the problem of scientific laws. A sentence S of a theory T may be called "independent" if and only if S is accepted on the ground of the theory T. When such an independent sentence S_i is simple, i.e. not composed of other sentences, then S_i plays the role of a supposition of the theory T. General assertions of natural science (laws of nature) occur in theories as their dependent components only, i.e. as antecedents of

---

[5] Cf. [1.4], p. 77.

independent implications having a sentence about facts as the consequent. The antecedent yields explanation to the consequent. As a matter of fact, it is usual in practice to formulate the laws of nature in such a way as if they were independent. But this should be treated merely as an application of a serviceable shortening. It follows from this view on the nature of general assertions that the truth of an empirical law is indifferent in relation to the truth of the theory in which the law occurs, because an implication remains true independently of the value of its antecedent. That is why the requirement of the truth of laws ought to be replaced by the postulate of usefulness or expediency. A law L is useful for explanation as long as no false consequence of L has been discovered. Therefore, the laws of nature have to be constantly inspected by means of evaluating their consequences. Each positive result increases the probability of the law; however one can never be sure whether the next result will not be negative. And so the laws remain merely probable. This approach, in Czeżowski's opinion, is conformable with Aristotle's realism (*universalia in rebus*) as well as with the way of thinking of contemporary physicists[6].

VIII. Francis Bacon and other empiricists supposed that inductive generalization is the sole heuristic method in sciences. And yet a quite different method was invented and applied by Galileo in his kinematics, namely the method of analytical description (or briefly: analytic method) which has a very broad range of applications. The use of this method was a distinctive feature of Kazimierz Twardowski's school. Therefore, it is only natural that Czeżowski devoted a separate paper to a detailed discussion of the method, all the more so that he considered it advisable to be fully conscious of the methodological means which he used[7]. The gist of the method is as follows. The description of one or at most a few specimens of a certain empirical entirety (object, event) constitutes the first step of the method. The described specimen is taken as a representative of the entirety. Passing from the initial to the analytical description, the investigator transforms the initial description, composed of observational sentences, into a definition. The initial description may be checked by further observations and even modified. Nonetheless, when it gets transformed into a certain definition, it refers only to such objects that fall under the definition, for a definition is always taken as a true sentence. An important point to note is that the above-mentioned transformation changes the investigated object. Namely, the definition does not describe an individual real thing but instead an abstract object: a kind, species or genus under which the described things come. The transition from initial

---

[6] Czeżowski expressed such an opinion in [2.1]. Compare [1.3], p. 73.
[7] The method is discussed in [2.7]. Cf. [1.3], pp. 136-142, and [2.11], pp. 27*f*.

description to definition is possible thanks to an intuitive act of generalization (cf. Plato's *noesis*). It is a specific act of cognition based on the analysis of features of the described object and consisting in a selection of some of the features. The selection is not always right nor the generalization appropriate. An adequate result can be achieved by way of several trials and errors. The definition obtained is analytic and real at the same time, since it tries to copy the empirical reality. Moreover, the definition excludes all cases that are discordant with its contents. This elucidates why analytical descriptions are not only general but also apodeictic.

The method of analytical description is applicable in any branch that makes use of empirical materials and is in fact a common basis of empirical inquiry. It is particularly necessary in view of two goals: first, to determine the meaning of scientific terms, and secondly, to classify or arrange objects of a given investigated domain. The method thus characterized may likewise have valuable applications beyond the limits of natural sciences, i.e. in the humanities, and even in philosophy and logic.

IX. There is an important point here that must not be overlooked. Czeżowski emphasizes the assertion that the method of analytical description is *par excellence* empirical. Then again he claims that the generalization which is involved is an act of intuitive cognition different from sensory and introspective perception. Moreover, the generalization results in an abstract construction. On that account it might be objected that there is a discordance among those statements. However, such an objection misses the point, for in Czeżowski's opinion the positivist understanding of empiricism is too narrow. The scope of empiricism should embrace all kinds of intuitive cognition, i.e. any cognition that grasps the object as an individual, directly and as a wholeness. Thus, one can speak not only about empiricism of natural science but also about ethical, aesthetic and other forms of empirical cognition. Assuming an attitude called "attention", one can apprehend sensory qualities and taking other appropriate attitudes one can recognize moral, aesthetic, and other values. There is no chasm between singular propositions concerning sensory qualities and singular propositions concerning values. Both are not true but only probable, both must be confronted with further experiences formulated in analogous singular propositions[8].

X. A system of sentences deduced from definitions makes a deductive theory which portrays a realm of abstract objects. The ascription of truth to definitions is nothing but a supposition. This merely hypothetical truth is inherited by all other statements of the theory. Truth of this kind should

---

[8] Czeżowski summarized his views on empiricism in [1.4]. See pp. 171*ff* there.

not be confused with material truth ascribed to observational sentences. In order to connect a deductive theory T with observational sentences describing real objects of a certain domain D, the theory T should be interpreted by means of expressions having reference to objects of D. This can be accomplished with the aid of two different methods: metric or ametric description. The former is practicable when the events under investigation are to be expressed quantitatively, as is usual in natural science, the latter has a good standing in psychology, theory of literature, praxiology, etc.[9].

A construction composed of an explanatory law, an explanatory hypothesis and an observational sentence being explained can be considered the elementary structural model used in the course of building an empirical theory, since it contains a sufficient reason of the observational sentence. When a theory T_1 is to be subordinated to a more general theory T_2, an analogy must be detected between them at first, an analogy which allows the subordination of all theses of T_1 to more general laws of T_2.

No thesis of an interpreted theory T is distinguished as an absolute independent element, for each depends of those sentences by means of which the thesis has been tested. In view of its further applications the interpreted theory T must be subjected to constant examinations and confronted with new observations.

XI. For all those reasons, science is a dynamic rather than static composition that undergoes permanent changes. As to the unity of science, Czeżowski maintained that it is logic which unites all possible sorts of theories, because each scientific theory, both axiomatic and empirical, has a logical structure. Accordingly, he conceived logic as a science of the structure of science or structure of knowledge. He regarded logic as the most universal discipline which, though it does not yield premises to other branches of knowledge, nevertheless indicates for them appropriate forms of statements as well as principles of their acceptance and combination into a theoretical system.

Czeżowski was convinced that his view on scientific knowledge was neither relativistic nor sceptical. It differs from relativism, because it does not subject the notion of truth to conditions. It differs from scepticism since it takes advantage of the fact that human knowledge has a certain justification and moreover a degree of probability which may increase[10].

XII. The scientific legacy of Czeżowski's research includes noteworthy works addressing yet other problems of methodology. For instance, in a

---

[9] Czeżowski writes about the interpretation of deductive theories as well as on metric and ametric description in [2.11], especially on pp. 28*f*.

[10] Cf. [1.4], p. 171.

separate paper he discussed the concept of proof, distinguished several meanings of the term "proof" and characterized diverse types of proofs. In some periods of his studies he was concerned with the methods of defining and classifying various kinds of definitions. Moreover, he analyzed the operation of generalizing, abstracting, formalizing, etc.[11]. He was also interested in the history of methodology, conducted research on the development of the problem of causality in ancient times[12], reviewed the Cartesian methodological postulates, compared the methodologies of Aristotle, Galileo and Bacon. Needless to say, as a logician and philosopher, he scrutinized penetratingly a great number of problems and subject matters that play an essential if not pivotal role in these disciplines.

SELECTED BIBLIOGRAPHY

1. **Books:**
    1. (1933). *Jak powstało zagadnienie przyczynowości. Zarys jego rozwoju w filozofii starożytnej* [How Did the Problem of Causality Arise. An Outline of its Development in Ancient Philosophy]. Wilno: Wydawnictwa Wileńskiego Towarzystwa Filozoficznego.
    2. (1946). *O naukach humanistycznych* [On the Humanities]. Toruń: T. Szczęsny.
    3. (1958). *Odczyty filozoficzne* [Philosophical Lectures]. Toruń: Towarzystwo Naukowe.
    4. (1965). *Filozofia na rozdrożu. Analizy metodologiczne* [Philosophy at the Cross-Roads. Methodological Analyses]. Warszawa: PWN.
    5. (2000). Ed. by Gumański. *Knowledge, Science, and Values. A Program for Scientific Philosophy* (*Poznań Studies in the Philosophy of the Sciences and the Humanities* **68**). Amsterdam-Atlanta (GA): Rodopi.

2. **Papers:**
    1. (1947). Twierdzenia ogólne w teorii naukowej [General Assertions in Scientific Theory]. *Życie Nauki* **3**, 302-308. Reprinted in [1.3], 70-74.
    2. (1951). De la vérification dans les sciences empiriques. *Revue International de la Philosophie* **5**, 347-366. English translation in: M. Przełęcki, R. Wójcicki (Eds.). *Twenty-five Years of Logical Methodology in Poland*, Warszawa: PWN & Dordrecht: Reidel 1977, 93-109.
    3. (1953). On Certainty in Empirical Sciences. *Proceedings of the 10th International Congress of Philosophy* **6**. Bruxelles, 126-129. Polish version in [1.3], 75-77.
    4. (1953). Ethics as an Empirical Science. *Philosophy and Phenomenological Research* **14**, 2, 163-171. Polish version in [1.3], 40-45.
    5. (1955). Klasyfikacja rozumowań [Classification of Reasoning]. *Studia Logica* **2**, 254-262. An extended version in [1.3], 128-135.
    6. (1955). On Certain Pecularities of Singular Propositions, *Mind* **64**, 392-395.
    7. (1956). O metodzie opisu analitycznego [On the Method of Analytical Description]. *Sprawozdania Towarzystwa Naukowego w Toruniu*, 114-118; enlarged version in [1.3], 136-142.
    8. (1960). Indukcja a rozumowanie przez analogię [Induction and Reasoning by Analogy]. *Ruch Filozoficzny* **20**, 297-299. Enlarged version in [1.4], 82-96.

---

[11] Cf. the papers in [1.4], pp. 19*ff*, 29*ff*, 97*ff*, 106*ff*.

[12] The opinions of ancient philosophers are presented in [2.1].

9. (1967). O jedności nauki [On the Unity of Science]. *Fragmenty filozoficzne* [Philosophical Fragments], Series 3. Warszawa: PWN, 19-29.

10. (1970). Prawda w nauce [Truth in Science]. *Studia Filozoficzne* 3, 19-25; English translation in *Sprawozdania Towarzystwa Naukowego w Toruniu* 8, 114-118; *Dialectics and Humanism* 1973, 1, 165-171.

11. (1973). Empiria i teoria. Rozważania metodologiczne [Empiricism and Theory. Methodological Considerations]. *Studia Filozoficzne* 6, 29-33.

12. (1975). Definitions in Science. *Poznań Studies in the Philosophy of the Sciences and the Humanities* 1, no 4, 9-17.

Poznań Studies in the Philosophy
of the Sciences and the Humanities
2001, vol. 74, pp. 75-78

Józef M. Dołęga

BOLESŁAW JÓZEF GAWECKI – A PHILOSOPHER OF THE NATURAL SCIENCES

I. Bolesław Józef Gawecki (15.10.1889, Spirov near Tver – 13.04.1984, Warsaw) was born in Russia. In 1897, the Gawecki family migrated to Silesia and settled in Sosnowiec. In 1908-09 he took up studies at Munich, specialising in mathematics, physics and philosophy. He continued his academic education at the Jagiellonian University in Cracow, his tutors being, among others, Władysław Heinrich (1869-1957) and Władysław Natanson (1864-1937). Having passed final examinations in 1914, he was awarded a doctoral degree in philosophy after submitting a doctoral dissertation entitled "Przyczynowość i funkcjonalizm w fizyce" [Causality and Functionalism in Physics]. In 1929, he qualified as an assistant professor at the Jagiellonian University in Cracow, presenting a dissertation entitled "Stosunek czasowy przyczyny i skutku" [The Temporal Relation of Cause and Effect]. Since 1947, he worked for the Committee of the History of Polish Philosophy of the Polish Academy of Arts and Sciences, and was a member of the Ossolineum Friends' Society. At the same time, he began lecturing in physics at the Warsaw University of Technology and in the Evening Engineering School. However, due to lack of time he had to give up lecturing and devoted himself to creative work and the translation of philosophical works. In 1954, he accepted the post of an independent scientific editor of the Library of Philosophy Classics publications and became a member of its editorial staff.

In 1955, Gawecki started lecturing at the Chair of Philosophy of Nature in the Department of Christian Philosophy of the Academy of Christian Theology in Warsaw. Gawecki continued lecturing at the Academy of Catholic Theology until 1967 and retired at the age of 77. Even after his retirement he was still active in creative work and willingly undertook reviews of doctoral and postdoctoral dissertations.

II. Gawecki dealt with the theory and philosophy of natural sciences. His definition of science has been much in use by several contemporary scientists specializing in theory of scientific knowledge. The definition is as follows:

> Science is a product of a methodical pursuit of a logically ordered system of intersubjectively provable true statements about associations of facts and relations between constructions of reasoning, that have a universal meaning, thereby performing a certain social function [2.11, pp. 415-422].

The method was defined by Gawecki as the most effective way of acting, leading to a certain definite goal that is concordant with experiences of a given epoch or a given period in the development of science. There is a great number of practically applied methods and a single, universal scientific method does not exist. Methods should be distinguished from directives in the same way as the methodology of sciences was separated from the methodology of scientific work. Methodology forms a part of the theory of science.

Three stages may be distinguished in scientific cognition, namely: a) the discovery, description and gathering of scientific facts; b) the discovery of scientific principles and laws that capture relations between scientific facts; c) the incorporation of these principles into the body of scientific theories. Scientific theories, logically ordered scientific facts, as well as scientific principles and laws form a scientific system.

Scientific methods were grouped by Gawecki as follows: methods of natural sciences, axiomatic methods, methods of the humanities. The characteristic traits of methods of natural sciences include the following stages of action: formulation of hypotheses that aim at determining relations between the accumulated scientific facts; concluding – inferring of consequences from hypotheses; checking of consequences by referring to results and effects of experiments.

The axiomatic method that is applied in formal sciences is a deductive system where axioms are of cardinal significance. Thus, two types of scientific statements are found in this system, namely: axioms, definitions and scientific theorems.

With regard to methods of the humanities, a historical method in particular, three stages may be distinguished: a) gathering of material, description and classification of sources; b) analysis of the material that leads to determining historical facts; c) synthesis – elaboration of the material consisting in establishment of relations between historical facts.

The basis for dividing sciences into formal and real sciences (the latter including natural sciences and the humanities) is the study object and method of research as well as the manner of checking truthfulness of scientific statements.

According to Gawecki, the characteristic features of contemporary sciences are specialisation, decentralisation and dispersion, posing a threat to the progress in the development of science. However, a tendency may be observed to acquire more general attitudes that may lead to the restoration of the unity of science despite its intrinsic diversity. Attempts to arrive at a scientific synthesis is the basis which facilitates philosophic synthesis.

III. Gawecki focused especially on the philosophy of natural sciences. A distinct division between natural sciences, the philosophy of natural sciences and the philosophy of nature, emphasized in his works, allowed defining epistemological and methodological status of these scientific branches.

Gawecki perceived the objectives of the contemporary philosophy of nature as follows: philosophers of science start from the results of natural sciences that are subsequently elaborated by them according to general principles of the theory of knowledge; then they attempt to combine them into a (relatively) finite uniform entity and interprets them metaphysically, elucidating the very substance of phenomena described by naturalists. This conception of philosophy of science is marked, according to Gawecki, by its extra-scientific character and comes as part of particular ontology. The conception itself as well as the very methodological status of the philosophy of nature have been widely discussed in philosophical literature.

Gawecki made a distinction between general and particular epistemology, the latter including the philosophy of natural sciences, which, in turn, comprised philosophy of particular natural sciences, such as, for instance, the philosophy of physics or the philosophy of biology. The study object of the philosophy of physics is the critical analysis of assumptions and basic physical conceptions as well as the critical analysis of methods applied in physics. In particular, Gawecki was interested in the following issues: the principle of causality, the time relation between cause and effect, interrelations of phenomena, the unity of nature, the anticipation scheme, causal laws, determinism and indeterminism. The question of causality and related problems should be stressed at this point. Gawecki paid special attention to the methodical directive connected with the progressing scientific specialisation in physics and chemistry – the scientist should not lose sight of the unity of science and its relation with his/her specialisation. Just as the real world is a unity, also the knowledge of this world should form an unity, notwithstanding all the artificial and temporary divisions made for practical purposes.

SELECTED BIBLIOGRAPHY

1. Books:
1. (1918). *O wartości nauki* [On the Value of Science]. Warszawa.
2. (1938). *Propedeutyka filozofii* [Propaedeutic Philosophy]. Warszawa: Biblioteka Polska.
3. (1964). *Przygotowanie do filozofii* [Preparation to Philosophy]. Warszawa: Pax.

4. (1967). *Filozofia rozwoju. Zarys stanowiska filozoficznego* [Philosophy of Development. An Outline of Philosophical Attitude]. Warszawa: Pax.
5. (1969). *Zagadnienie przyczynowości w fizyce* [A Question of Causality in Physics]. Warszawa: Pax.
6. (1975). *Myślenie i postępowanie* [Thinking and Acting]. Warszawa: Pax.

2. Papers:

1. (1921-22). Z powodu uwag prof. T. Kotarbińskiego o potrzebie zaniechania wyrazu "filozofia" itp. [On the Occasion Professor T. Kotarbiński's Remarks on the Need to Abandon the Word "Philosophy" etc.). *Ruch Filozoficzny* **6**, 8-10, 113-116.
2. (1923). Przyczynowość i funkcjonalizm w fizyce [Causality and Functionality in Physics]. *Kwartalnik Filozoficzny* **1**, 2, 204-232; 3, 336-361; 4, 487-507.
3. (1924). Makrokosmos i mikrokosmos [Macrocosm and Microcosm]. *Astrea* **1**, 2, 140-151.
4. (1927) Co to jest filozofia przyrody? [What is Philosophy of Nature?]. In: *Księga pamiątkowa ku czci prof. W. Heinricha*. Kraków, 31-42.
5. (1927). Zagadnienie istoty filozofii w polskiej literaturze filozoficznej ostatniego dwudziestolecia [The Question of the Essence of Philosophy in Polish Philosophical Literature of the Last Twenty Years]. *Przegląd Filozoficzny* **30**, 4, 334-346.
6. (1928). *O stosunku czasowym do przyczyny i skutku* (On the Temporal Relation of the Cause to Effect]. *Kwartalnik Filozoficzny* **6**, 3, 336-384; 4, 401-418.
7. (1931). Konsekwencje filozoficzne indeterminizmu w fizyce współczesnej [Philosophical Consequences of Indeterminism in Contemporary Physics]. *Przegląd Filozoficzny* **34**, 1, 3-14.
8. (1933). Indukcja i dedukcja w badaniu zjawisk przyrody [Induction and Deduction in Studies on Natural Phenomena]. *Mathesis Polska* **8**, 1-2, 7-25.
9. (1936). Nauki ścisłe a metafizyka [The Sciences and Metaphysics]. *Przegląd Współczesny* **15**, 10, 103-112.
10. (1947). Kryteria prawdziwości [Criteria of Truthfulness]. *Wiedza i Życie* **16**, 1-2, 67-74.
11. (1947). Pojęcie nauki [The Notion of Science]. *Wiedza i Życie* **16**, 5, 415-422.
12. (1947). Metoda nauk przyrodniczych [Methods of Natural Sciences]. *Wiedza i Życie* **16**, 6, 511-519.
13. (1947). Metoda aksjomatyczna i metody nauk humanistycznych [The Axiomatic Method and Methods of the Humanities]. *Wiedza i Życie* **16**, 9, 705-712.
14. (1958). Hipotezy w fizyce [Hypotheses in Physics]. *Roczniki Filozoficzne* **6**, 3, 147-156.

*Poznań Studies in the Philosophy
of the Sciences and the Humanities
2001, vol. 74, pp. 79-87*

Izabella Nowakowa

ADAM WIEGNER – NONSTANDARD EMPIRICISM

I. Adam Wiegner (16.12.1889, Poznań – 28.09.1967, Poznań) studied philosophy at the Jagiellonian University in the years 1908-1914, where he obtained a PhD degree in 1923. Since 1928, he worked at Poznań University, where he completed his postdoctoral (*habilitacja*) dissertation (1934). His work and lectures were concentrated mainly on epistemology and logic. Wiegner headed the chair of logic at the Adam Mickiewicz University in Poznań. He died on 28th September, 1967, in Poznań. His main philosophical conceptions were: the original epistemology of "holistic empiricism", a nonstandard analytic philosophy of science, and an analytic reconstruction of Marxist epistemology.

II. Adam Wiegner's philosophical work was close to the Polish brand of analytical philosophy – the Lvov-Warsaw School. The mainstream of the Lvov-Warsaw School, was parallel to the German neopositivism and British empiricism, all of which were under the spell of Hume's epistemology. By contrast, Wiegner was inspired by the tradition of neo-Kantism and proposed a theory which in many ways preceded the views developed later by K.R. Popper.

III. Below is a list of the main epistemological ideas that were the starting point for A. Wiegner:

1. *Analytic-synthetic dualism.* All beliefs are based either on the analysis of its component concepts or on the association of experiential data. The former can be certain if the analysis is correct, the latter can be at best probable. The content of the former is limited to the content of the concepts involved (analytic beliefs), the content of the latter is enriched through experience (synthetic beliefs).

2. *A priori synthetic beliefs are impossible.* The combination of certainty and informativeness is excluded. Contentful (synthetic) beliefs are uncertain, certain (analytic) beliefs are empty. Metaphysics aspires to discover certain and informative truths. Its allegedly synthetic *a priori* claims are only illusions. The only disciplines whose claims have cognitive value are mathematics and logic, on the one hand, and empirical sciences on the other.

3. *Atomistic nature of experience.* All experience — whether through extraspection or introspection — is based on particulars (sense-data), which are the atoms of the mind. Sense-data represent particular simple features of the world. They are combined (in association) to represent more complex features. Various mental procedures lead to the creation of ever more complex ideas.

4. *Atheoretical nature of experience.* If sense-data constitute the ultimate basis for all our synthetic knowledge, they themselves cannot presuppose any knowledge of this kind. Sense-data are atheoretical and neither they nor our experience of them rely on any further presuppositions. On the contrary, any assumptions that may appear to be independent of experience can be ultimately reduced to configurations of sense-data.

Theses 1-4 are ultimately based on the British empiricist tradition (J. Locke, D. Hume) in epistemology. Adam Wiegner calls into question this epistemological paradigm by revising, or at least reinterpreting, theses 3-4.

IV. Wiegner inherits the critique of the empiricist's epistemology of extraspection from the Fries/Nelson version of neo-Kantianism. Two arguments are of particular importance:

Hume reduces the notion of a necessary connection to the recollection of a succession of impressions, i.e. to the association of one impression with another. As a result, the occurrence of one impression leads to the expectation that another impression will occur. But an expectation is something over and above a recollection. After the occurrence of one event we may expect another, which we did not have a chance to associate with the former – expectations go beyond an individual's experience.

The fact of associating a relation is something else than a relation between associations. There is a relation between the experience of high heat and the experience of pain, but it is dubious whether it is identical to the experience of a relation between these two impressions. The former is an objective linking experiential facts, the latter is a part of the conscious experience just as the facts just mentioned above. According to Wiegner, the neo-Kantian tradition does not reach beyond Hume when the understanding of the basic structure of experience is concerned. It holds that sense-data ("impressions", "feelings") are the elements of experience, and the relations between them (succession, association, recollection, etc.) are not part of the experience, but rather something that organizes experience from the outside, as it were. But this is inadequate:

Each element of consciousness (even the most basic but concrete one) already implicitly contains a relation of itself to others. To say "An object is directly present to me" amounts to saying "I am conscious of a certain relation between that object and another". To a [proper] empiricist "relations" are just as directly given as individuals between which they hold [2.1, p. 50].

An impression is not a basic sense-datum but can be analyzed into components. Wiegner believed in the physicalistic interpretation of sense-data as states of affairs appearing, in the last resort, on the things. It is always possible to distinguish in an impression, representations of at least two objects x and y and a relation R between them. Moreover, not only the two objects but also the relation between them are directly experienced. More precisely, neither x nor y are the objects of our direct experience. Rather, what is directly experienced is a certain structure (R, x, y), such that $R(x,y)$. In other words, we experience x's being R-related to y. This is the basic structure. Ordinarily, however, what we take for sense-data are structures of a higher order. Let 'C' stand for 'x being R-related to y' and 'D' stand for 'u being S-related to w'. Among the sense-data there are structures of a second-order: we may experience structure $C\varphi D$: xRy is φ-related to uSw. And so on.

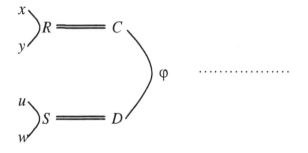

Thus understood

[e]mpiricism becomes relativistic in the sense that the concept of an object and the concept of a relation depend on one another. It is impossible to make sense of one without the other. Both are found in what is "directly given" [2.1, pp. 50-51].

We experience not elementary data but complex structures. In particular, we do not experience only the things being the components of those structures, but also the component relations (of the first order) holding between the things, component relations (of the second order) holding between the things related one to another by the relations of first order, etc. The question arises whether all relations constituting such structures can be experienced by the senses.

At this point, we encounter the problem of the theoretical involvement of sense-data. Wiegner raises this problem and is convinced that theory ("mind") does influence experience ("senses"):

The notion of a sense-organ belongs [...] to physiological psychology and designates a peripheral bodily organ, whose stimulation is a condition of our claims about objects. Empiricism is quite compatible with the claim that each concrete holistic datum that is

directly given to us in experience depends equally on physiological changes in the
peripheries (in the sense-organs) as in the centre (in the mind) [2.1, p. 52].

In details, however, this answer is not quite clear. It is not clear what it
means to say that we experience relations constitutive of structures given
in experience. It is not clear how "empiricism unequivocally takes
'relations' to be equally given in direct experience as 'objects'" [2.1,
p. 52]? Does this mean that we experience them in the same way as things?
Or does it mean that our mind projects an image of the relations on the
experienced things, thereby organizing them into a structure? And does
this concern all relations of the experiential structures or only some? And
if some, then which ones?

Extending Wiegner's terminology, let us call the relation of the highest
order in a given experiential structure the organizing relation for that
structure. In the above example of the second-order structure, φ is the
organizing relation for that structure, while R and S are subordinate
relations. It seems natural to distinguish such a relation in every
experiential structure. If someone were to hold that only some relations are
projected onto experience by the mind, he would most likely have at least
the organizing relations in mind.

Whenever one encounters an unclearness of this sort, it is best to begin
with a whole spectrum of possibilities. There are three possible views:

(1) when we experience a structure, we experience the component
     things in the same way as all component relations;
(2) when we experience a structure, we experience only the
     component things, all the component relations being derived from
     the presupposed theory;
(3) when we experience a structure, we experience the component
     things in the same way as some component relations, in particular
     the subordinate relations, but the organizing relation for that
     structure derives from the presupposed theory.

The question is which of these positions Wiegner holds. This is not
entirely clear. It seems, however, that position (1) can be excluded.
Otherwise, what would the joining of "sensory" and "cognitive" elements
in experience consist in? Everything would be derived from the senses: the
elements among which certain relations hold and the relations themselves.

Position (2) can also be eliminated. Under that view, the role of the
senses is reduced to registering the things. What could it mean, however,
to "sense-register" a thing without noting that it is somehow related to
another? Here is how Wiegner describes the determination that two objects
are identical:

It reduces to the claim that object x conceived by means of the relation R (i.e. relative to
individual y) is identical to object u conceived by means of the relation S (i.e. relative to
individual w) [2.1, p. 50, symbols changed].

Things are thus identified by means of relations in the act of perception. If the latter were derived from reason, it would not only be impossible for the relations but also for the things to be objects of perception. The perceptual content of experience would be thereby reduced to zero. Wiegner, however, emphatically identifies himself as an empiricist.

Position (3) is thus most likely to be Wiegner's. It reconciles his empiricism with his view concerning the theoretical presuppositions of the perceptual act. This also seems to be the most adequate view. Some relations are undoubtedly observable (e.g. "is higher than") but others are not (e.g. "dominates over"). Moreover, relations organizing experience are usually theoretically laden, while lower relations are observable. When one person says to another "Look, C clearly dominates D" referring to two people walking in front of them, he has in mind an unobservable relation between two persons. He says "look", however, because that relation becomes apparent in observable relations, such as that D follows in C's footsteps, while C walks in an uninterrupted rhythm without paying any particular attention to the attentive partner.

In sum, on the standard view, the perceptual act consists in a simple discerning of some observable feature in the observed object. On Wiegner's view, the perceptual act consists in the discerning of a structure of relations, the lower of which (themselves observable) hold among observed objects, while the higher involve more and more non-observational relations and it is the latter that constitute the organizing relations.

V. Let the following be the structure of the perceptual act that p: (i) $C\varphi D$, which on the basis of the fact that C and D are objects of a higher (second, say) order and defined by means of relations R and S, respectively, and (ii) the adopted theoretical assumptions T leads to (iii) xRy and uSw, where (iv$_1$) relations R, S are (let us assume) observational while (iv$_2$) φ is an non-observational relation. Given this, which as we saw was admitted by Wiegner, it is clear that the stereotype of an one-sided control of theory by experience must be revised. According to the stereotype, if theoretical claim t leads to an observational consequence o, and observation demonstrates that *not-o*, then t must be rejected. However, assume that the conjunction of (i) and (ii) holds and that it has as its consequence observation (iii): *if* (i) & (ii), *then* (iii). Then, the failure of observations (iii) may be explained not by the falsity of (i) but of the theoretical assumptions (ii), so that the experiential datum (i) may be preserved as holding good. For the falsity of (iii) prejudges merely the truth of the disjunction non-(i) or non-(ii). This is the reason, why the falsification of prediction (iii) need not lead to the revision of the view concerning the occurrence of the observed fact (i).

Obviously, from the contemporary point of view, it would be necessary to add a few more requirements: for example, requiring the postulation of relations (i) with other perceptions but forbidding an *ad hoc* postulation of such relations, i.e. the postulation justified merely by the will to save perceptual structure (i) from falsification. I will not dwell on this point, however, since there is not a sufficient textual basis for it. After all, we are talking about writings from the 1920s and 1930s, when the dominating trend in the philosophy of science – neopositivism – was silent about these issues and earlier discoveries of Duhem were marginalized. Nonetheless, Wiegner analyzed these problems deeply enough independently of the writings of Duhem to have discovered Quine's famous thesis, known today as the Duhem-Quine thesis. This is clearly implied by Wiegner's holistic conception of experiential data which he himself anchored in the philosophical assumptions of the *Gestalttheorie* and Avenarius' idea of experience [2.7; 2.11].

VI. There are good reasons to suppose that Adam Wiegner proposed the holistic conception of experience, among the consequences of which are the views concerning the differences between theory and experience and the theoretical component of experience. Let us remember that we are talking about 1925, when the neopositivist position was only beginning to take shape in the empiricist circles. According to it, sense-data were to constitute the foundation of all our knowledge – they were to be theory-independent; on the contrary, all theoretical knowledge was merely an abbreviation for potential impressions. This view was dominant also in the Polish literature at that time. The greatest Polish analytical philosopher, K. Ajdukiewicz, claimed then that observational sentences constitute the "fundamental basis" of the sciences. The problem is that, according to Ajdukiewicz, sentences belonging to a science are either "deduced from other sentences" or "not deduced from already accepted sentences". The latter constitute the fundamental basis of a given science (Ajdukiewicz 1938a, pp. 287-289). Those sentences do not presuppose any others, any theoretical knowledge — they are derived from experience alone. It is this that Wiegner denies.

Ajdukiewicz pictures the acceptance of an observational sentence thus:

(A)   observation that o ⎯⎯⎯⎯→ acceptance of sentence "o"

According to Wiegner, the mere observation is not sufficient for the acceptance of the corresponding observational sentences — it is necessary that certain non-observational knowledge is accepted as well. Wiegner's picture is more complete and closer to real perceptual acts:

(W) observation that o
theoretical knowledge T $\Big\}$ ⟶ acceptance of sentence "o"

This difference stems from the different views on the nature of the perceptual act (among others). For Ajdukiewicz, perception consists in registering some observable feature of the observed object. For Wiegner, perception consists in registering a structure organized by a non-observational relation with subordinate observational relations holding among observational objects. The organizing relation (and possibly other higher-order relations), which is (are) unobservable can be known only by means of the accepted theory. It is thus that the structuralist or holistic conception of the structure of experience leads to the thesis that experience is dependent on theory — an extraordinarily original thesis in Polish analytical philosophy if one considers the time and place in which it was first proposed.

As is well known, Popper has called the thesis about the independence of experience from theory into question, but that was a decade later than Wiegner (Kmita 1967).

VI. One more point is worth emphasizing and it may have escaped notice in the above discussion. Although Wiegner's views were different in many points from the dominating in his time outlook of the Lvov-Warsaw School, his style of philosophizing follows in the main "Twardowski's style". This is particularly clear in his seminal epistemic work, *Zagadnienie poznawcze w oświetleniu L. Nelsona* [The Problem of Knowledge in L. Nelson's Illumination] [2.1]. He distinguishes a variety of senses of the terms at stake ("knowledge", "thought", "reasoning", "proof", "deduction", "regression", etc.) penetrating deeply into the sometimes exceedingly subtle shades of meaning. His analyses were rather valuable for the task of constructing a "philosophical dictionary" hoped for by the first generation of the Lvov-Warsaw School. Wiegner also undertakes a detailed reconstruction of Nelson's reasoning, identifying and critically analyzing his reasoning patterns. This was exactly the work style of Twardowski – used in 1925, that is in a rather early stage of the Lvov-Warsaw School.

But Wiegner's views are in many points definitively different. Aside from some methodological innovations [1.1; 1.2; 2.2; 2.9], which cannot be discussed here at all such as:

• clear and deep respect for the metaphysical tradition,
• realism (anti-nominalism) in the understanding of definition,
• raising the problem of correspondence in methodology,

- putting forward the idea of "bridging rules," enabling the discussion of the problem of reduction, etc.,

Wiegner's main contribution to epistemology is his rejection of the theory of knowledge 1-4 commonly accepted at the time of his writing in the Polish analytical philosophy. To put it very briefly, Wiegner undermined two of its component ideas: the thesis about the one-sided dependence of theory on experience and the thesis about the duality of experience. He replaced them with two new ideas:

3*. *Holistic nature of experience*. All experience is based not only on physiological conditioning of our senses by the physical things related to one another, but also on some theoretical presuppositions; it is the theoretico-empirical holes that constitute the units of experience.

4*. *Theoretical nature of experience*. Experiential data are subject to theoretical interpretation; the adopted theory gives us the basic relations that organize the structure of material reaching our senses and thus enables us to recognize the meaning of it.

Indeed, Wiegner's view is empiricist but non-Humean.

REFERENCES

Ajdukiewicz, K. (1938a). *Metodologiczne typy nauk* [Methodological Types of Sciences]. In: Ajdukiewicz (1960), 287-313.

Ajdukiewicz, K. (1938b). *Postępowanie człowieka* [Human Conduct]. In: Ajdukiewicz (1960), 317-364.

Ajdukiewicz, K. (1960). *Język i poznanie* [Language and Knowledge], vol. I. Warszawa: PWN.

Kmita, J. (1967). *Koncepcja empiryzmu całościowego Adama Wiegnera* [Adam Wiegner's Conception of Holistic Empiricism]. *Studia Filozoficzne* 3(50), 41-51.

SELECTED BIBLIOGRAPHY

1. **Books:**

    1. (1932). *Uwagi nad indeterminizmem w fizyce* [Remarks on Indeterminism in Physics]. Poznań: Drukarnia *Dziennika Poznańskiego*.
    2. (1934) O istocie zjawisk psychicznych [On the Nature of Mental Phenomena]. *Prace Komisji Filozoficznej PTPN* IV, 2. Poznań: PTPN.
    3. (1948) *Elementy logiki formalnej* [Elements of Formal Logic]. Poznań: Księgarnia Akademicka.
    4. (1952). *Zarys logiki formalnej* [An Outline of Formal Logic]. Warszawa: PWN.
    5. (in print). *Observation, Hypothesis and Introspection* [translation of the selected writings of A. Wiegner] (*Poznań Studies in the Philosophy of the Sciences and the Humanities* 81). Amsterdam-Atlanta: Rodopi.

2. **Papers:**

    1. (1925). Zagadnienie poznawcze w oświetleniu L. Nelsona [The Problem of Knowledge in L. Nelson's Illumination]. Prace Komisji Filozoficznej PTPN I, 5. Poznań.

2. (1932). W sprawie wyobrażeń imaginatywnych [Concerning the Imaginative Mental Representations]. *Przegląd Filozoficzny* **XXXV**, 131-134.
3. (1934). Idea logiki poznania [An Idea of the Logic of Knowledge]. *Życie Literackie* **10**, p. 5.
4. (1935). Na marginesie *Logicznych podstaw nauczania* Kazimierza Ajdukiewicza [Some Comments on Kazimierz Ajdukiewicz's *Logical Foundations of Education*]. *Kwartalnik Filozoficzny* **13**, 327-330.
5. (1936) Über Universalismus, Reismus und Anti-Irrationalismus. *Actes du Congres International de Philosophie Scientifique, Paris 1935.* No 388. Paris: Hermann, pp. 74-48.
6. (1948). Filozoficzne znaczenie teorii postaci [The Philosophical Sense of the *Gestalttheorie*]. *Sprawozdania PTPN za I i II kwartał 1948*, 26-28.
7. (1959). O subiektywnej i obiektywnej jasności myœli i słowa [On the Subjective and Objective Clarity of Thought and Word]. *Zeszyty Naukowe UAM. Filozofia, Psychologia, Pedagogika* **3**, 3-15.
8. (1963). W sprawie tak zwanej "prawdy względnej" [Concerning the so-called Relative Truth]. *Studia Filozoficzne* **1**(32), 113-128.
9. (1964). O empiryzmie całościowym [On the Holistic Empiricism]. *Sprawozdania PTPN za II półrocze 1964*, 199-201.

Translated by Katarzyna Paprzycka

Poznań Studies in the Philosophy
of the Sciences and the Humanities
2001, vol. 74, pp. 89-96

Anna Jedynak

KAZIMIERZ AJDUKIEWICZ – FROM RADICAL
CONVENTIONALISM TO RADICAL EMPIRICISM

I. Kazimierz Ajdukiewicz (12.12.1890, Tarnopol – 12.04.1963, Warsaw) studied philosophy, mathematics and physics at Lvov University – the centre of Polish scientific philosophy of that time. He attended lectures by Twardowski, Łukasiewicz, Sierpiński and Smoluchowski. Having received his PhD degree in 1912, he continued his studies in Göttingen, attending Husserl's and Hilbert's lectures. After the First World War, he was appointed professor at Warsaw University and then at Lvov University. After World War II, Ajdukiewicz continued his academic career as a professor of the philosophy of science and logic in Poznań and Warsaw. He was very active as a scholar, teacher, organizer of conferences and symposia, and editor of philosophical and logical journals. He acted as a member or head of many scientific institutions (e.g. he was the rector of Poznań University in 1948-52). He educated many eminent students who took up some philosophical problems which he had initiated.

II. Ajdukiewicz was interested in various philosophical problems, but his first concern was the philosophy of cognition, especially the philosophy of scientific cognition. He differentiated between methodology and metascience. The former deals with such problems as how to construct languages that are suitable for developing science, how to justify scientific theorems, or what the goal of scientists seems to be. The resolution of these problems should provide the basis for developing metascience, which is the theory of scientific theories, and answers such questions as: is the language of a given theory properly formed, are its theorems properly justified, or does the theory accomplish scientists' goal.

It was clear to Ajdukiewicz that the results of reflection on deductive sciences were much more advanced and more justified than in the case of inductive ones. In the methodology of deductive sciences generally accepted results were attained, so the theory of deductive systems could develop. Contrary to that, methodological problems of empirical sciences were still heatedly discussed, without the chance of reaching general agreement and, consequently, without the possibility of developing a metascience.

III. Ajdukiewicz's idea was to use, to the extent that it was possible, the results of the methodology of deductive sciences to develop the methodology of empirical sciences. He tended to make the latter itself as scientific as possible and to make as its tools modern mathematical logic and critical analysis of language. Thus, he noted that natural sciences use measurement and contain the language of mathematics as their part. He derived two conclusions from that: 1) Given the classical definition of truth, languages of natural sciences are incomplete, for according to Gödel's theorem, the language of mathematics is incomplete itself. 2) While measuring physical objects, we ascribe numbers to their physical features, with isomorphism holding between arithmetical relations and relations between the features in question. Due to that isomorphism and to our knowledge of mathematics, we can deduce from some simple physical relations of objects, given in the results of measuring, some much more complicated physical relations of objects, which sometimes transcend all possible experience.

Ajdukiewicz claimed that empirical theories could be regarded as axiomatic systems as well as mathematical theories. The only difference is that the latter are deductive axiomatic systems, i.e. theses are actually deduced from axioms according to the rules of transformation, while the former are reductive ones, i.e. although theses are deducible from axioms, the actual order of formulating the theory is the opposite: the scientist infers axioms (which are theoretical principles) from theses (which are results of experiments or general empirical laws) in a non-deductive way. Thus, the difference between deductive and reductive axiomatic systems is a pragmatic one; intratheoretical relations seem to be similar in both cases.

Non-deductive types of inference, common in empirical sciences, can provide false conclusions, even though the premises are true. Ajdukiewicz dealt with the problem of rationality of such types of inference. He treated methods of inference as gambling systems, for inference procedures usually justify decisions on how to act when both profit and loss are possible. Then the mathematical theory of probability can be useful in establishing a degree of infallibility of a given scheme of inference. Ajdukiewicz claimed that the logic of induction, based on mathematical probability, should be supplemented with the statement that a rational method of inference allows the acceptance of conclusions with a degree of certainty not greater than its own degree of infallibility[1].

Empirical sciences are based on direct experience, for their final premises are observational sentences. Ajdukiewicz noticed that direct experience is

---

[1] The problem of rationality of non-deductive types of inference was widely dealt with by Ajdukiewicz's followers: K. Szaniawski and H. Mortimer.

always subjective and unrepeatable, while only intersubjective and repeatable methods can be recognized as scientific ones. Nevertheless, methods justifying general empirical laws are scientific; although the laws are actually inferred inductively from observational sentences, they could also be inferred from other observational sentences that describe someone else's experience.

According to Ajdukiewicz, the aim of developing science is not to copy the way we perceive the world, but to construct a conceptual scheme, copying the structure of the world free from all sensual qualities. The question of how to form languages suitable for expressing natural knowledge was the most important methodological problem for him. His views on that evolved from radical conventionalism to radical empiricism.

At first, he was highly impressed by the fact that rapid changes sometimes occurred in science, and then both the basic notions and the theorems changed. In the early thirties, he undertook the task of explaining the nature of disruptive changes in science. He was inspired by the French conventionalists, who claimed that one and the same reality can be expressed in many different conceptual structures. His aim was to develop that philosophical intuition and to put it into a formal framework. As a result, he created his radical conventionalism. It was based on a special conception of meaning which should be presented here first.

Ajdukiewicz regarded natural language as a deductive system, i.e. a system containing axioms, rules of admitting sentences, and theses or sentences admitted under these rules. He differentiated between axiomatic, deductive and empirical meaning rules. Axiomatic meaning rules demand from each user of the given language unconditional acceptance of certain sentences; deductive rules demand the acceptance of certain sentences conditioned by previous acceptance of other sentences; empirical ones demand the acceptance of certain sentences conditioned by some particular empirical data (simple empirical meaning rules) or by empirical data and previous acceptance of other sentences (compound empirical meaning rules). The scope of an axiomatic meaning rule is a set of sentences which must be accepted under this rule. The scope of a deductive meaning rule is a set of ordered pairs of the type: set of sentences – sentence, such that the latter must be accepted under the rule if the former were previously accepted. The scope of an empirical meaning rule is a set of ordered pairs of the type: empirical data – sentence, or empirical data, set of sentences – sentence, such that the latter must be accepted in the face of the former. According to the three types of meaning rules, there are axiomatic, deductive and empirical theses of language, i.e. sentences which are to be accepted under the respective rules.

The meaning of expressions and the total scope of the meaning rules determine each other reciprocally. Therefore, Ajdukiewicz defined the

meaning of an expression as, speaking informally, its position in the total
scope of the meaning rules. Such a conception of meaning implies that
bare empirical data do not force us to assent to any articulated proposition;
meaning rules as the second factor are necessary to enable us to assume
propositions. The meaning rules, however, may be different, and so may be
the theses admitted under them in the face of the same empirical data.

The thesis of radical conventionalism states that a set of theses of the
given language depends on the meaning rules of that language. This thesis,
as referred to any language which is really used, seems to be quite obvious.
However, in the inter-war period Ajdukiewicz referred it to a special type
of languages, whereby the content of the thesis was stronger and more
original. He differentiated between the four following types of languages:

A language is connected if any two of its expressions are directly or
indirectly meaning-related. Two expressions are directly meaning-related
if they both occur in the same element of the scope of any meaning rule,
while they are indirectly meaning-related if they are linked by a finite
chain of expressions which are directly meaning-related among
themselves. A language which is not connected is disconnected. It can be
divided into at least two parts which have no mutual meaning relations.
The language J is open if there exists a language J' which contains all the
expressions of J and ascribes to them the same meanings as J does, and if
it also contains some extra expressions which are not synonymous with the
expressions of J and at least one of them is directly meaning-related to an
expression of J. A language is closed if it is not open. A conceptual
apparatus is a set of meanings of a closed and connected language and a
world-picture is a set of theses of the same type of language.

When a language is open, the set of meaning rules and, consequently,
the set of meanings can be expanded. It is not so when a language is closed
and connected: the total scope of the meaning rules is then complete, and
thereby the characteristic of the meaning of the expressions is full. The
conceptual apparatus is complete and no new meaning can be added to it.
A closed and connected language cannot be enriched unless either it
becomes disconnected or a new expression is synonymous with an old one.
Two conceptual apparatuses are either identical or completely separate,
without any points in common. Correspondingly, two closed and connected
languages are either fully mutually translatable or completely untranslatable.

Ajdukiewicz took into account only languages which are closed and
connected. The stronger radical conventionalism thesis is then as follows:
the world picture depends on the conceptual apparatus and, moreover, there
is a possibility of choosing between alternative conceptual apparatuses, for
there exist at least two different ones.

Choosing the conceptual apparatus, we choose the world-picture. That means that the same empirical data can be found in different scientific theories which are neither consistent nor inconsistent: they are incomparable, for they are not mutually meaning-related. It is not a matter of the poverty or wealth of languages: note that a closed and connected language is full and complete and cannot be enriched. Two mutually untranslatable closed and connected languages do not refer to different empirical data, for as closed ones they can both be used to speak of anything. Rather, they represent two different viewpoints, from which the same reality appears differently, but in both cases – completely.

Radical conventionalism provides the grounds for the conception of a discontinuous development of science. The change of only one meaning within the conceptual apparatus involves changes in all the other meanings, i.e. the change of the whole apparatus. When this happens, a scientific theory is replaced by another one. For example, Ajdukiewicz regarded the languages of classical and relativistic physics as two mutually untranslatable closed and connected languages; the change of the whole conceptual apparatus was a result of redefining the notion of simultaneity.

The assumption that there are at least two different conceptual apparatuses makes the stronger radical conventionalism thesis quite original, but on the other hand shows the possibility of questioning it. In the post-war years, Ajdukiewicz neither upheld nor denied it: he simply gave up the notions that were necessary to express it. They had been based on the notion of a closed and connected language, which Ajdukiewicz recognized as an empty one. Nevertheless, the weak formulation of the thesis, claiming that a set of theses of an open language depends on the set of its meanings, was still valid.

IV. In the post-war period, Ajdukiewicz leaned toward empiricism. His attitude was to minimize the role of conventions in sciences, contrary to his earlier approach, in which he stressed it. His empiricist views were manifested in two ways.

Firstly, he questioned the common conviction that analytical sentences (i.e. meaning postulates and their deductive consequences) are always true in virtue of the very meaning of expressions. He pointed out that postulates are not sufficient to found any sentence unless it is a law of logic. It is possible that an analytical sentence, founded only on terminological conventions, implies a false existential consequence, whereby it proves itself false. Suppose a sentence represented by the scheme $F(\lambda)$ is an analytic one, founded on the terminological convention establishing that the term λ should designate an object fulfilling the function $F(x)$. The sentence can still be false, for it entails an existential consequence $\exists x\, F(x)$ which is false if nothing fulfils the function $F(x)$. Therefore, terminological conventions are only useful to found sentences if we establish these conventions with regard to what exists. Empirical knowledge is necessary to prove analytic sentences as well as synthetic ones[2].

Secondly, Ajdukiewicz examined the thesis of radical empiricism. According to his interpretation, this thesis claims that only empirical knowledge can be expressed in a scientific language; that is, a scientific language should not be governed by axiomatic or deductive meaning rules and only empirical sentences can occur in it. This thesis, as applied to languages that are really used, is false, but according to Ajdukiewicz that does not imply the impossibility of constructing a language containing empirical sentences only. Providing that purely empirical knowledge is possible, Ajdukiewicz had to formulate a new conception of language and meaning. The meaning of expressions should be defined in such a manner that it would not involve any axiomatic or deductive meaning rule.

In radical empirical language, all traditionally conceived logic, as based on the rules in question, should be cut off. A problem then arose of how it would be possible for language users to reason and what the status of logical laws is. Ajdukiewicz's idea was to interpret logical laws as sentences indirectly based on experience, verified in conjunction with empirical hypotheses. Positive results of tests should confirm both: the hypotheses from which the sentences describing the results in question had been derived as well as the logical laws which had made the derivation possible. Negative results could be interpreted as falsifying either the hypotheses or the laws of logic, or both. Ajdukiewicz claimed that there is no vicious circle in the fact that people reason before logical laws tested, because in order to reason people do not need to know them; they only need to use them, which is not the same.

In order to limit all knowledge to a purely empirical one it is also necessary to cut off all axioms based on the meaning relations of descriptive terms. The meaning of the expressions should be suitably meagre: it should be defined in such a manner that no meaning relation expressible as a language axiom could occur, for example, all verbal definitions should be excluded. Ajdukiewicz started to formulate a new theory of meaning, but he did not finish the task before his death. He assumed that the meaning of a compound expression was a correspondence which associates a denotation of each constituent expression with its syntactical position within the whole compound expression. He did not

---

[2] According to some of Ajdukiewicz's followers, the notion of analyticity should only be applied to sentences, the justification of which does not have to be founded on experience. Thus, given the results of Ajdukiewicz's investigation, they presented various proposals of redefining the notion of analyticity. A short outline of the discussion on this subject was given by M. Przełęcki and R. Wójcicki in (1968/1969). See also: M. Przełęcki (1969).

solve the problem of meaning of simple, one-word expressions. However, if we apply this concept of meaning of compound expressions to simple ones as well, the meaning of a term will prove to be a correspondence between its denotation and its position within itself, whereby the meaning will be practically identical with the denotation. And then there is no possibility of establishing different, alternative conceptual apparatuses assigned to speak of the same things, whereby the purely empirical character of knowledge is guaranteed.

V. Thus, Ajdukiewicz formulated two different conceptions of science, language and meaning: the radically conventionalist and the radically empiricist one. They are both based on empty notions, for they refer to languages which do not exist. However, the two types of languages in question approximate some of those types which are really used. While the language characterized in the radical conventionalism conception approximates the one of scientific lectures, the language of radical empiricism rather approximates the one of scientific research. Ajdukiewicz rightly noticed that languages tend toward rationalization: the rougher a theory, the less fixed the meanings of its terms are. While the research goes on, empirical data suggest which notions could be fruitful and how to enrich the meaning of fundamental terms, establishing suitable axioms based on terminological conventions. The terms gradually take on more and more definite meaning, whereby more and more problems are solvable independently of experience, and simultaneously the theory becomes finally fixed. As Ajdukiewicz noted, the more rationalized the language of a theory is, the more probable it is for the principles or axioms of the theory to fall into mutual contradiction or into contradiction with experience. Some axioms should then be revoked. One might add that if fundamental notions are mutually strongly meaning related, the change of even one notion or one axiom can cause the change of main parts of the whole conceptual apparatus (the origin of relativistic physics or non-Euclidean geometries can serve as an example). Changes of language lead to rapid changes in science and this was what Ajdukiewicz was interested in. The notion of a closed language seems to be unnecessary to deal with that problem.

Jerzy Giedymin claims that Ajdukiewicz's radical conventionalism, questioned and replaced by radical empiricism at the object level, moved to the metalevel[3]. At the object level, the choice of a language influences the truth of *a priori* propositions, while at the metalevel it influences the truth of epistemological theses, e.g. the thesis of radical empiricism. It is not the content of logical laws that depend on the language, but the view of their status; and not the set of meanings of expressions, but the view of what

---

[3] See: Giedymin (1978, p. XLVIII).

meaning is. Instead of a discussion on different conceptual apparatuses there is a discussion on different conceptions of language and epistemological programmes. Briefly speaking, matters which are more basic depend on conventions.

REFERENCES

Giedymin, J. (1978). Radical Conventionalism. Its Backward and Evolution: Poincaré, LeRoy and Ajdukiewicz. Editor's Introduction. In: K. Ajdukiewicz, *The Scientific World-Perspective and other Essays*. 1931-1963. Dordrecht: Reidel, pp. XIX-LIII.
Przełęcki, M. (1969). *The Logic of Empirical Theories*. London: Routledge and K. Paul.
Przełęcki, M. and R. Wójcicki (1968/1969). A Problem of Analyticity. *Synthese* **19**, 372-399.

SELECTED BIBLIOGRAPHY

1. **Books:**
   1. (1960/1965). *Język i poznanie* T. I-II [Language and Knowledge. Vol. I-II]. Warszawa: PWN.
   2. (1974). *Pragmatic Logic (Synthese Library* **62**). Dordrecht: Reidel & Warszawa: PWN.
   3. (1978). *The Scientific World–Perspective and Other Essays 1931-1936*. Dordrecht: Reidel.

2. **Papers:**
   1. (1934). Sprache und Sinn. *Erkenntnis* **IV**, 100-138.
   2. (1935). Das Weltbild und die Begriffsapparatur, *Erkenntnis* **V**, 259-282.
   3. (1958). Le problème du fondement des propositions analytiques. *Studia Logica* **VIII**, 259-282.
   4. (1995). My Philosophical Ideas. In: (Eds.) V. Sinisi and J. Woleński, *The Heritage of Kazimierz Ajdukiewicz (Poznań Studies in the Philosophy of the Sciences and the Humanities* **40**). Amsterdam-Atlanta: Rodopi, 13-33.

Poznań Studies in the Philosophy
of the Sciences and the Humanities
2001, vol. 74, pp. 97-101

Anna Jedynak

JANINA HOSIASSON-LINDENBAUM – THE LOGIC OF INDUCTION

I. Janina Hosiasson-Lindenbaum (6.12.1899, Warsaw – 04.1942, Vilna) studied philosophy at Warsaw University, attending Kotarbiński's, Łukasiewicz's and Ajdukiewicz's lectures. She attained her PhD degree after completing her dissertation *Justification of Inductive Resoning*. She combined her scientific research with work in a secondary school as a teacher of philosophy. In 1929/30 she supplemented her studies in Cambridge. When the Second World War broke out, she moved to Vilna, where she took part in the activities of the Vilnius Philosophical Society. In 1942, she was arrested and, in April, shot dead by the Nazis in Vilna, falling victim to the Holocaust.

The logic of induction was in fact her only interest. She knew the contemporary literature on the subject quite well and some of her views were inspired by works of Western authors, especially Keynes and Nicod. Although her intellectual background was the philosophical Lvov-Warsaw School, she did not take up any traditional philosophical problems. Her papers deal with the following problems:

I. Types of induction.

II. Applying the theory of probability to the logic of induction.

III. Justification of induction.

IV. Psychological analysis of inductive reasoning.

Let us present her views in each of these areas.

II. Types of induction. She defined four of them (having differentiated them with respect to the problem of justification):

1. In the case of generalizing induction, we suppose or strengthen our belief that *every A is B*, on the basis that some different *A* were *B*.

2. Hypothetical (*resp.* approximative) induction is not a generalizing one and in this case we suppose or strengthen our belief that *b*, on the basis that *a*, and that "*a*" is a consequence (resp. approximative consequence) of "*b*".

3. In the case of generalizing subinduction, we suppose or strengthen our belief that a series of *A* are *B*, on the basis that a series of other *A* were

B. Reasoning by analogy is a special case of generalizing subinduction: since *X* and *Y* share a series of features, we suppose or strengthen our belief that another series of features characteristic of *X* is characteristic of *Y* as well.

4. Hypothetical (resp. approximative) subinduction is not a generalizing one and in this case we suppose or strengthen our belief that *c*, on the basis that *a*, and that both "*a*" and "*c*" are consequences (resp. approximative consequences) of the same thesis "*b*".

III. Applying the theory of probability to the logic of induction. Keynes interpreted the probability of conclusions of fallible reasonings as an objective one. Hosiasson-Lindenbaum contradicted this, stating that this would be possible only if we could define it, but since we cannot, we should interpret it as a subjective one, i.e. as a degree of strength of a justified belief. She found both notions of probability (subjective and objective) to be worthy of explication: first, subjective probability should be based on suitable axioms, and second, it is quite possible that it depends on objective probability, so if only possible the latter should be defined. Her own efforts were invested in both of these directions.

She examined the statistical interpretation of probability. She rejected the objection that the interpretation contradicted some theorems of the theory of probability; there is a contradiction only on the grounds of some additional premises which do not need to go together with the statistical interpretation. She also examined the possibility of applying the statistical interpretation of probability to empirical problems, seeking an answer to the question of whether this interpretation is useful for predicting exactly the frequency of an event, given its probability, and *vice versa*. The answer was affirmative, unless a finite number of elementary events was concerned. In the latter case, the prediction would be fallible.

In order to axiomatize a theory of confirmation of a sentence by other sentences (i.e. a theory of justified belief in what a sentence says on the ground of some other sentences) she took Mazurkiewicz's (1932) axiomatization of the theory of probability as a model. The reason was that the function of confirmation should refer to sentences, and Mazurkiewicz's theory of probability was the only one referring probability to sentences, not to events. Hosiasson-Lindenbaum supplemented the three axioms stated by Mazurkiewicz with a fourth one, stated by herself. These axioms express some essential intuitions dealing with the notion of confirmation or justified belief, but they do not show how to calculate the confirmation of a given sentence on the grounds of given knowledge. Although the axioms do not enable us to express confirmation as a number, they do allows us, in some circumstances, to compare confirmations of different sentences in order to decide which one is greater, and to formulate some important theorems concerning such comparisons. However, Hosiasson-Lindenbaum noticed that there was an urgent need to find a method of calculating confirmations, because otherwise even if we sometimes knew that we could strengthen our belief, we would not know to what degree it could be justified.

IV. Justification of induction. Hosiasson-Lindenbaum used the axioms to prove many interesting theorems, especially those dealing with the problem of the justification of induction. As she wrote:

> [The axioms] imply many facts about the confirmation, solve different questions in a definite and precise way and simplify some statements. They enable us to avoid occasional appeals to intuition, since if they are once accepted, all further facts may be deduced in quite a formal way [9, p. 148].

Using the axioms, she proved the following statements:

1. The confirmation of a sentence h on the ground of knowledge s increases when s is supplemented by f, which is a consequence of h.

2. The confirmation of a generalization increases with the number of its observed instances; the confirmation of a hypothesis increases with the number of its observed outcomes.

3. The confirmation of a hypothesis depends on the *a priori* confirmation of its observational results: the lower the latter, the higher the former. (Hosiasson-Lindenbaum's point is similar to Popper's. Moreover, she denied the view that metaphysical sentences do not entail any observational sentences; she claimed that they could entail such sentences, but only those which are quite probable independently of them.)

4. Given a hypothesis h and its two logically independent consequences f_1 and f_2, although f_1 increases the confirmation of h, it does not always increase the confirmation of f_2 as well. If it were so, any sentence could increase the confirmation of any other sentence for every two sentences f_1 and f_2 have at least one logical premise in common, *viz.* their logical product. Hosiasson-Lindenbaum pointed out what condition is sufficient to increase the confirmation of f_2 by f_1, *viz.* f_1 should not diminish the confirmation of f_2 on the ground of any hypothesis concurring with h. (Concurring hypotheses are mutually exclusive and their logical sum is *a priori* certain.)

5. The more consequences of h confirm it, the higher the confirmation of other consequences of h.

6. The less probable *a priori* the consequences of h which confirm it, the higher the confirmation of other consequences of h.

7. The confirmation of h tends to certainty when the number of its observed consequences tends to infinity (if h is not excluded *a priori,* and its observed consequences are mutually logically independent, and it is not true that all the consequences in question follow from a hypothesis concurring with h).

8. It is not always true that the nearer the given premises of a probable conclusion are to a set of premises from which the conclusion follows, the greater the confirmation of this conclusion on the grounds of these premises. She showed this statement to be true only under some specified conditions.

Hosiasson-Lindenbaum resolved the so-called paradox of confirmation, stated by C. G. Hempel. The paradox is as follows: a sentence a: *All kitchen salt is soluble in water* is confirmed by its instance a_1: *This is kitchen salt, and it is soluble in water*. A sentence b: *No substance which is insoluble in water is kitchen salt* is confirmed by its instance b_1: *This is insoluble in water and it is not kitchen salt*. Now, since b is equivalent to a as its contrapositive, b_1 should also confirm a. This seems paradoxical: no one would try to test whether all salt is soluble in water by testing substances insoluble in water, and looking whether they are salt or not. According to Hosiasson-Lindenbaum, both sentences a_1 and b_1 confirm both a and b however a_1 confirms them higher than b_1. There are two reasons for that: 1) She proved, using axioms, that if the extent of non-B is greater than the extent of A, then the confirmation of an instance of non-B being non-A is greater than the confirmation of an instance of A being B. The number of substances that are insoluble in water is immense in comparison with the number of substances which are salt. So a_1, whose confirmation is lower than that of b_1, provides stronger confirmation to both a and b than b_1. 2) It is much more probable that *if one sample of salt is soluble in water, every sample of salt is soluble in water,* than *if one substance insoluble in water is not salt, every substance insoluble in water is not salt.* On the grounds of that knowledge, a_1 provides stronger confirmation to both a and b than b_1.

Hosiasson-Lindenbaum also dealt with the following problem: what is going on with the probability of a given A being B, when we successively enrich our knowledge of A, and why it is reasonable to try to know more and more about A. She showed some difficulties arising with respect to answers which seem convenient at first glance. Finally, she applied some theorems of the theory of probability to resolve the problem.

V. Psychological analysis of inductive reasoning. According to Hosiasson-Lindenbaum, the psychology of inductive reasoning encounters two serious obstacles. We do not know: 1) how to recognize the basis of reasoning, i.e. the whole knowledge on which a conclusion is based; 2) how to express the strength of a belief that the conclusion is true. If we could answer the latter, we could also answer the former, for the strength of belief is a result of all previous knowledge.

Hosiasson-Lindenbaum found a behaviouristic method proposed by F. Ramsey to be the only one that was useful for measuring the strength of a belief. It treats making decisions as gambling systems and is based on the assumption that people act in accordance with the mathematical expectation. Thus, Hosiasson-Lindenbaum claimed, the method cannot be applied when people do not act rationally.

Her programme was to carry on empirical research in order to discover psychological laws governing inductive reasonings, to reveal logical forms of the latter and to consider their justification. Actually, she did perform some experiments of that sort on her secondary-school students.

REFERENCES

Mazurkiewicz, S. (1932). Zur Axiomatik der Wahrscheinlichkeitsrechnung. *Comptes Rendus des Séances de la Société des Sciences et des Lettres de Varsovie* 25, pp. 1-4.

SELECTED BIBLIOGRAPHY

1. (1928). Definicje rozumowania indukcyjnego [Definitions of Inductive Reasoning]. *Przegląd Filozoficzny* 31, 352-367.
2. (1931). Why do We Prefer Probabilities Referred to Many Data? *Mind* 40, 23-36.
3. (1932). Uwagi w sprawie pojęcia prawdopodobieństwa jako granicy częstości [Comments on the Notion of Probability as a Limit of Frequency]. *Przegląd Filozoficzny* 35, 194-208.
4. (1934). O prawomocności indukcji hipotetycznej [On the Validity of Hypothetic Induction]. In *Fragmenty filozoficzne, Księga Pamiątkowa ku czci Profesora Tadeusza Kotarbińskiego*. Warszawa, 11-34.
5. (1935). Wahrscheinlichkeit und Schluss aus Teilprämissen, *Erkenntnis* 5, 44-45.
6. (1935). Uwagi w sprawie psychologii wnioskowań indukcyjnych [Comments on the Psychology of Inductive Reasonings]. In: Księga Pamiątkowa ku czci Profesora Władysława Witwickiego. *Kwartalnik Psychologiczny* 7, 275-300.
7. (1935). Przyczynek do pojęcia prawdopodobieństwa jako granicy częstości [Contribution to the Notion of Probability as a Limit of Frequency]. *Wiadomości Matematyczne* 39, 135-145.
8. (1936). O prawdopodobieństwie hipotez [On the Probability of Hypotheses]. *Przegląd Filozoficzny* 39, 416-417.
9. (1940). On Confirmation. *The Journal of Symbolic Logic* 5, 133-148.
10. (1941). Induction et analogie. *Mind* 50, 351-365.
11. (1948). Postępy wiedzy z punktu widzenia poznawczego [The Progress of Knowledge from the Cognitive Point of View]. *Przegląd Filozoficzny* 44, 59-65.

Poznań Studies in the Philosophy
of the Sciences and the Humanities
2001, vol. 74, pp. 103-106

Władysław Krajewski

JANINA KOTARBIŃSKA – LOGICAL METHODOLOGY AND SEMANTICS

I. Janina Kotarbińska's (19.10.1901, Warsaw – 2.01.1997, Warsaw) maiden name was Dina Sztejnbarg. In the 1930s, she was an assistant reader at Tadeusz Kotarbiński's chair of philosophy. During the Second World War she assumed the name Janina Kamińska and participated in underground university teaching. For those activities, she was sent to Auschwitz (however, the Nazis did not discover her Jewish origin). After the war, she resumed university teaching, initially in Łódź and then in Warsaw. When T. Kotarbiński was left a widower, she became his wife. After her husband retired, she became head of one of two chairs of logic at Warsaw University. In 1972, she retired but was still active, almost until her death.

II. Kotarbińska's scientific output covers various fields of philosophy, logic and semantics. In the first period of her activity (before the Second World War), she dealt mainly with the logic and philosophy of science, in the second period (after the war) mainly with semantics. All her papers exhibit scrupulousness of analysis, and the striving for clarity and precision. In an essay on Kotarbińska (on which the present is based to a great extent), her eminent disciple, Marian Przełęcki (1971), rightly said that she provided a paradigm of the critical-analytical style. We shall briefly present some essential problems which she investigated.

In the first period, Kotarbińska investigated concepts of the law of nature, of necessity, and of chance. She analyzed views of other philosophers and gave her own interpretation.

In an early paper [2.1], Kotarbińska analyzed the concept of the law of nature proposed by her favourite philosopher, J. St. Mill. In a larger paper published in the same year [2.2], she gave a critical analysis of the concepts of necessity elaborated by E. Meyerson, J. St. Mill, B. Russell, C. J. Ducasse, C. D. Broad. She was especially critical of Meyerson, who was fashionable at the time.

Kotarbińska analyzed thoroughly the dispute between determinism and indeterminism in three papers in which this problem was discussed, in physics, then in biology and finally in the human sciences [2.3; 2,4; 2,5].

She did not attempt to solve the controversy. On the contrary, she showed that the problem had not yet been solved, in spite of claims about the victory of indeterminism.

In a very valuable paper on the concept of chance [2.5], Kotarbińska distinguished 26 meanings of the term. It is worth noting that she quoted many Marxist authors in that paper (Engels, Plekhanov, Bukharin, Deborin), which was unusual at the time. Conversely, after the war, when it became very fashionable, she did not quote Marxists.

III. Kotarbińska dealt extensively with the analysis of works written by R. Carnap and by K. R. Popper. She wrote a critical essay about Carnapian physicalism [2.7] and an analytic review of Popper's *Logik der Forschung* [2.8]. Continuing these topics after the war, she wrote an essay about the evolution of the Vienna Circle [2.9] and a paper on the evolution of physicalism read in French at the Congress of Philosophy in Amsterdam [2.10]. Later on, she wrote an essay about the controversy between deductionism and inductionism [2.14], in which she tried to show that the controversy was to a great extent spurious, for example that Mill's canons of the eliminative induction are very close to the Popperian method of the criticism of hypotheses.

However, Kotarbińska's main interest after the war was in semiotics. Her monograph on the concept of sign [2.12] consists of two parts: a review of main theories of sign and an analysis of the concept of sign. The analysis is crowned by a definition of sign and definitions of its kinds: iconic sign, symbolic sign, and verbal sign.

IV. Another subject of Kotarbińska's analyses was definition as such. In her paper on it [2.11], various kinds of the definition are analyzed (real and nominal ones, analytic and synthetic ones) and a question is posed of whether the definition is a proposition from a logical point of view (various solutions are discussed). The next paper is devoted to a "deictic" (or "ostensive") definition which is not a definition in the strict sense [2.13]. Related problems are discussed in a paper on the "occasional statements" [2.17].

In a paper on the "strife" about the limits of the use of logical methods [2.15], Kotabińska presents a controversy between two schools of the analytic philosophy: reconstructionism and descriptionism. The issue is as follows: whether the methods of the formal logic are applicable beyond the mathematics and logic. The reconstructionists answer this question in the affirmative, and the descriptionists in the negative. The latter claim that the formal methods are applicable only to artificial languages; their use in a natural language distorts it. Kotarbińska criticizes this view and defends "moderate reconstructionism". According to it, formal methods should be used in an analysis of the empirical sciences, but only where it is useful for enhancing the precision of a problem, and not for the sake of formalization for itself. **V.** Kotarbińska also wrote about the philosophical conception elaborated by her husband: reism. She noticed the difficulties involved in this conception in an important paper on the difficulties with existence [2.16]. The main difficulties are connected with mathematics, which deals only with abstracts. Kotarbińska proposes a modification of reism. We may distinguish various kinds of existence: the things exist in the basic sense, the abstracts also exist but in another sense, a "non-basic" one.

After Tadeusz Kotarbiński's death in 1981, Kotarbińska was actively involved in popularizing his views and especially in refuting mistaken interpretations of those views. Kotarbińska was intellectually active until her death at the age of 96.

REFERENCES

Przełęcki M. (1971). O twórczości Janiny Kotarbińskiej [On Janina Kotarbińska's Output], *Studia Filozoficzne* 5(72). Reprinted as a foreword in the book: J. Kotarbińska [1.1], 5-13.

SELECTED BIBLIOGRAPHY

1. Books:
1. (1990). *Z zagadnień teorii nauki i teorii języka* [Problems of the Theory of Science and the Theory of Language]. Warszawa: PWN.

2. Papers:
1. (1931). Pojęcie prawa przyrodniczego u J. St. Milla [The Concept of the Law of Nature in J. St. Mill's Writings]. *Przegląd Filozoficzny* **34**, 15-38.
2. (1931). O tzw. konieczności związków przyrodnych [On the so-called Necessity of Natural Connections]. In: *Sprawozdania z Posiedzeń Towarzystwa Naukowego Warszawskiego* **XXIV**, 67-108. Reprinted in [1.1], 19-58.
3. (1932). Zagadnienie indeterminizmu na terenie współczesnej fizyki [The Problem of Indeterminism in Contemporary Physics]. *Przegląd Filozoficzny* **35**, 34-69.
4. (1932). Zagadnienie indeterminizmu na terenie biologii [The Problem of Indeterminism in Biology]. *Przegląd Filozoficzny* **35**, 245-272.
5. (1933). Zagadnienie indeterminizmu na terenie nauk humanistycznych [The Problem of Indeterminism in Human Sciences]. *Przegląd Filozoficzny* **36**, 77-106.
6. (1934). Analiza pojęcia przypadku [An Analysis of the Concept of Chance] In: *Księga Pamiątkowa ku uczczeniu 15-lecia pracy nauczycielskiej w Uniwersytecie Warszawskim Prof. Tadeusza Kotarbińskiego*. Warszawa, 161-179. English translation in: *Dialectics and Humanism* 1977, 1, 81-93.
7. (1934). Fizykalizm [Physicalism]. *Przegląd Filozoficzny* **37**, 91-95.
8. (1935). Recenzja z książki K. Poppera *Logik der Forschung* [Review of the book by K. Popper *Logik der Forschung*]. *Przegląd Filozoficzny* **38**, 269-278.
9. (1947). Ewolucja Koła Wiedeńskiego [The Evolution of the Vienna Circle]. *Myśl Współczesna* **2**, 145-160. Reprinted in [1.1], 106-127.

10. (1948). Le physicalisme et les etapes de son evolution. *Proceedings of the 10th International Congress of Philosophy* I. Amsterdam: North-Holland Publ., 693-696.

11. (1955). Definicja [Definition]. *Studia Logica* 2, 301-321. Reprinted in [1.1], 128-151.

12. (1957). Pojęcie znaku [The Concept of Sign]. *Studia Logica* 6, 57-134. Reprinted in [1.1], 152-244.

13. (1959). Tzw. definicja dejktyczna [The So-called Deictic Definition]. *Fragmenty filozoficzne. Seria druga*, English transl. as: On Ostensive Definition. *Philosophy of Science* 27(1960), 1, 1-22.

14. (1961). Kontrowersja: dedukcjonizm–indukcjonizm [The Controversy: Deductivism versus Inductivism]. *Studia Filozoficzne* 1(22), 25-42. English transl. in: Th. Nagel, P. Suppes and A. Tarski (Eds.). *Logic Methodology and Philosophy of Science* I. Stanford: Stanford University Press 1962, 25-42.

15. (1964). Spór o granice stosowalności metod logicznych [Controversy on the Applicability Limits of Logical Methods]. *Studia Filozoficzne* 3(38), 25-48. Reprinted in [1.1], 301-331. English transl.: *Logique et Analyse, Nouvelle serie*, 29 (1965), 3-29.

16. (1967). Kłopoty z istnieniem. Rozważania z zakresu semantyki [The Difficulties with Existence. Semantic Considerations]. In: *Fragmenty filozoficzne. Seria trzecia*. Warszawa: PWN, 129-146. Reprinted in [1.1], 332-350. English transl. as *Puzzles of Existence*. J. Pelc (Ed.), *Semiotics in Poland. 1894-1969*. Warszawa: PWN & Dordrecht: D. Reidel Publishing Company 1979, 208-226.

17. (1971). Wyrażenia okazjonalne [Occasional Expressions]. *Studia Filozoficzne* 1(68), 3-16. Reprinted in [1.1], pp.351-369. English transl. in: *Logic, Language and Probablity*. Dordrecht: Reidel Publishing Company1972, 3-16.

*Poznań Studies in the Philosophy
of the Sciences and the Humanities
2001, vol. 74, pp. 107-111*

Jan Woleński

IZYDORA DĄMBSKA – BETWEEN CONVENTIONALISM AND REALISM

I. Izydora Dąmbska (3.01.1904, Lvov – 18.06.1983, Cracow) studied philosophy with Kazimierz Twardowski and Kazimierz Ajdukiewicz in Lvov in 1922-1927. In 1926-1930, Dąmbska worked as an assistant in Twardowski's philosophical seminar; she obtained her PhD in 1927. She then taught in secondary schools in Lvov. After 1945, Dąmbska received her postdoctoral degree and was appointed as associate professor at Warsaw University; she also lectured at Poznań University. In 1950, she was dismissed from her university position by the communist regime. In 1957, Dąmbska became professor of philosophy at the Jagiellonian University in Cracow. She taught there until 1964, when she was moved to the Polish Academy of Arts and Sciences. Political authorities did not like to allow her to teach students because of her independent philosophical views.

Dąmbska was strongly influenced by her teachers, Twardowski and Ajdukiewicz. She accepted Twardowski's general metaphilosophical program which required a clear philosophy, free of metaphysical speculation and consisting of correctly justified theses. This programme favoured ... (in its broad sense, that is including formal logic, semiotics and the methodology of science). Dąmbska worked principally in those fields, but her interests were much wider. In particular, she was a distinguished historian of philosophy. Among her historical studies, works about Plato, scepticism (mainly French scepticism), conventionalism and ancient semiotics should be mentioned. One point concerning her analysis of conventionalism deserves special attention. It is the relation of conventionalism and relativism. A prevailing opinion is that conventionalism is a kind of relativism: since we can adopt arbitrary conventions, relativism is an unavoidable consequence of conventionalism. Dąmbska did not agree with this view. She believed that one should distinguish justified and arbitrary conventions. If a convention is justified, for example by empirical data, it does not lead to relativism. In particular, conventionalism does not entail the relativity of truth. In this way, Dąmbska tried to achieve a compromise between the views of her teachers: Twardowski's epistemological absolutism and the Ajdukiewicz's radical conventionalism.

II. Dąmbska located her views in the philosophy of science in a broad general philosophical context. The above-mentioned compromise between absolutism and conventionalism was one element of this environment. Rationalism was another. In an extensive study on rationalism and irrationalism [2.1], Dąmbska distinguished four kinds of irrationalism: logical (acceptance of contradictions), epistemological (acceptance of beliefs which are not intersubjectively testable and not intersubjectively communicable), metaphysical (recognizing the reality as irrational), and psychological irrationalism (a positive attitude toward irrational beliefs). *A contrario*, a rationalist is a person who accepts beliefs which are (a) coherent, and (b) intersubjective. According to Dąmbska, no kind of irrationalism is acceptable, while, of course, people sometimes accept irrational beliefs by mistake. In other words, a conscious irrationalism is not compatible with the scientific attitude. Thus, for Dąmbska, science is a perfect pattern of rationality. As an interesting historical corollary of her analysis of irrationalism, she pointed out that radical apriorism is usually a kind of irrationalism. Thus, one should not confuse epistemological rationalism (Plato, Descartes, for example) with rationalism proposed by Dąmbska, which may be termed "methodological rationalism". Another general Dąmbska's view consisted in a sharp distinction between the context of justification and the context of discovery. She did not claim that the philosophy of science should ignore the latter, yet both contexts are, according to her view, logically independent. Dąmbska treated the relation between the history and philosophy of science as a special instance of this view. She regarded the history of science as particularly important for the philosophy of science, yet argued for a logical independence of the philosophy of science with respect to the history of science. However, she did not work in formal logical methodology, even though she had a great respect for logic. Dąmbska's analyses were more conceptual than formal-logical. She represented a typical moderate reconstructionism in philosophy, not only in the philosophy of science but also in semiotics.

III. Let us now pass on to more special problems discussed by Dąmbska in her works in the philosophy of science. The concept of scientific law was the first such topic. In 1933, she published a long essay *O prawach w nauce* [Laws in Science], in which a variety of question concerning scientific laws was investigated, including the most fundamental one: laws in *a priori* sciences, laws in empirical sciences, laws in the humanities, laws and causality, and the possibility of defining the concept of scientific law. According to Dąmbska, every science (including the humanities) intends to formulate laws. On the other hand, she did not claim that all scientific laws fall under a general pattern. In other words, Dąmbska did not propose reducing all scientific laws to one laws of one kind, e.g. physical laws. Dąmbska, contrary to the novelties coming from the founders of quantum mechanics, defended causality against Heisenberg's criticism. She pointed out that Heisenberg misinterpreted the principle of causality giving its epistemological form: if we know the present state of the world, we can uniquely calculate its future. For Dąmbska, Heisenberg's criticism only limits the application of the classical concept of causality, and if we extend the concept of predictability, we can make causality coherent with new physics. The problem of truth of scientific laws was another problem considered by Dąmbska. She did not agree with the views denying that scientific laws could be evaluated as true or false in the classical sense. In order to argue for her view, Dąmbska insisted on a distinction between truth and the knowledge of truth. According to her position, instrumentalism is a result of a confusion between truth and its criteria. This view can be regarded as a special case of a compromise between absolutism and conventionalism.

Dąmbska was aware of difficulties connected with the definition of the concept of scientific law. She saw no possibility of defining this concept by selecting a single, simple property. Instead, she tried to define a set of properties as attributes of scientific laws. Finally, she proposed the following description. A sentence S of a language L is a scientific law if and only if: (a) S is a general conditional sentence about the succession of events; (b) S does not contain temporal coordinates; (c) the extension of S is an open class; and (d) S is an element of a scientific theory. The last condition implies that scientific laws inherit their justification from theories.

IV. The classification of reasonings was a favourite topic of Polish logicians and philosophers. It was an effect of the view that science is rational and proceeds by inferences. Jan Łukasiewicz and Tadeusz Czeżowski proposed a classification of reasonings which became very popular in Poland. Roughly speaking, reasonings are divided according to a logical relation between their elements, the premises and conclusion: a reasoning is deductive if its conclusion is logically entailed by premises, and it is reductive if its premises are entailed by conclusion. Unfortunately, this classification is incomplete, because it omits analogy as a kind of reasoning. Dąmbska tried to fill this gap by a closer logical analysis of analogy in [1.3, pp. 7-62].

According to Dąmbska, analogy is reducible neither to deduction nor to reduction (induction, in particular). She regards the essence of analogy in homomorphism between structures. More specifically, analogy between a set X and a set Y holds if and only if there is a homomorphic relation between the structure of the set X and the structure of the set Y. Assume that a set X is given, and there is a homomorphic relation between the structure of the set X and the structure of the set Y. Then, we can made comparative statements which establish properties of objects of the type Y on the base of given properties of the elements of the set X. This definition is sufficiently comprehensive to investigate the role of analogy in thinking, posing problems, constructing models (concrete as well abstract), creating concepts, or forming hypotheses. Basically, analogy has no important use in testing scientific statements. In this sense, analogy is an auxiliary method in science. However, it is a very important tool, because the cases of analogy very frequently occur in our thinking.

In 1967, Dąmbska published a book *O narzędziach i przedmiotach poznania* [On Tools and Objects of Cognition] [1.4]. Basically, it is a treatise in epistemology, but it has obvious metascientific aspects. This books develops an idea of epistemic (or cognitive) operators. The category of epistemic operators comprises various entities, of which models and language are the most important. In general, the knowing subject (scientific or not) always performs cognitive activities with the aid of operators. Language plays a special role in this idea. It is objectual (semantic) operator. According to Dąmbska, such a treatment of language is the only way out, if one wants to preserve realism. Clearly, this approach to epistemology is pragmatic in its spirit: it considers the knowing subject as acting with various operators.

V. Most Dąmbska's methodological views propose typical solutions for the philosophy of science of the first half of the 20th century. Realism, the context of discovery – context of justification opposition, rationalism or moderate conventionalism belonged to the standard (or received) theses of the philosophy of science, close to the Vienna Circle. However, Dąmbska, just as other Polish philosophers of science, did not propose radical solutions, perhaps except rationalism, which was a dogma in Poland. Some of Dąmbska's views were precursory and deserve attention still today, in particular her analysis of scientific laws, and her analysis of arbitrary and justified conventions.

SELECTED BIBLIOGRAPHY

1. **Books:**
    1. (1933). *O prawach w nauce* [On Laws in Science]. Lwów: Gubrynowicz i Syn.
    2. (1958). *Sceptycyzm francuski XVI i XVII w.* [French Scepticism in the 16th and 17th Century]. Toruń: Towarzystwo Naukowe w Toruniu.
    3. (1962). *Dwa studia z teorii naukowego poznania* [Two Studies on Scientific Knowledge]. Toruń: Towarzystwo Naukowe w Toruniu.

4. (1967). *O narzędziach i przedmiotach poznania* [On Tools and Objects of Cognition]. Warszawa: Państwowe Wydawnictwo Naukowe.
5. (1972). *Dwa studia o Platonie* [Two Studies about Plato]. Wrocław: Ossolineum.
6. (1975). *O konwencjach i konwencjonalizmie* [On Conventions and Conventionalism]. Wrocław: Ossolineum.
7. (1984). *Wprowadzenie do starożytnej semiotyki greckiej* [Introduction to Ancient Greek Semiotics]. Wrocław: Ossolineum.

2. Papers:

1. (1937). Irracjonalizm a poznanie naukowe [Irrationalism and Scientific Knowledge]. *Kwartalnik Filozoficzny* **25**, 83-118, 185-212.

Poznań Studies in the Philosophy
of the Sciences and the Humanities
2001, vol. 74, pp. 113-119

Tadeusz Batóg

SEWERYNA ŁUSZCZEWSKA-ROMAHNOWA – LOGIC AND METHODOLOGY OF SCIENCE

In the period of 1922-1928 Seweryna Romahnowa (née Łuszczewska, 10.08.1904, Mszana near Zborów – 27.06.1978, Poznań) studied philosophy at Lvov University, under the guidance of K. Twardowski, K. Ajdukiewicz and R. Ingarden, and mathematics under the supervision of H. Steinhaus and S. Banach. After graduation, she worked as a teacher in Lvov secondary schools. In December 1932, after completing an unpublished dissertation "O wyrazach okazjonalnych" [On Occasional Terms], she was awarded the degree of doctor of philosophy by Lvov University. The dissertation was supervised by Twardowski. The following year she was appointed assistant reader in the Chair of Philosophy headed by Ajdukiewicz.

In the autumn of 1939, Romahnowa was expelled from the university by the authorities of the Soviet Ukraine. She then started teaching in an Ukrainian secondary school in Lvov, but after the city was seized by Germans in 1941 she lost that job as well. In May 1943 she and her husband (dr E. Romahn, a philosopher) were arrested by the Gestapo. The next two years of her life were spent first in the Gestapo prison in Lvov, then in the concentration camp in Majdanek, then in the Ravensbrück concentration camp, and finally in a small Leipzig division of the Buchenwald camp. After the war, she worked as a clerk and a translator in UNRRA.

In late 1946, Romahnowa returned to Poland and settled in Poznań, where she was offered the position of reader in the Chair of the Theory and Methodology of Sciences (converted into the Chair of Logic), headed by Ajdukiewicz, in the Faculty of Mathematics, Physics and Chemistry of the University of Poznań. In 1954, she was awarded the title of associate professor, and in 1962 was appointed extraordinary professor. After Ajdukiewicz had moved to Warsaw, she took over the Chair of Logic (which was renamed the Chair of Mathematical Logic in the Institute of Mathematics in 1969). Three doctorates were successfully completed under her guidance: T. Batóg's, J. Czajsner's and M. Jarosz's. She retired in 1974.

Romahnowa's authored relatively few scholarly publications (which she always signed as Łuszczewska-Romahnowa). Most of them are works on the philosophy of science and its history, and only a few fall under mathematical logic.

Her most important writings on mathematical logic include the extensive article "Analiza i uogólnienie metody sprawdzania formuł logicznych przy pomocy diagramów Venna" [Analysis and Generalization of the Method of Verifying Logical Formulae by Means of Venn's Diagrams] [2]. In the article, she analyzes a graphic method of verifying the traditional syllogistic formulae of categorial propositions, and formulates a certain generalization of this method. Next, she establishes the discursive equivalent of this generalization and consequently arrives at a general method of deciding on the tautological nature of any formulae of a narrower one-argument predicate calculus.

Romahnowa's important idea was to link the concept of multi-level classification with the concept of distance, and in this way to search for the explication of the notion of natural classification, which is an important notion in natural sciences.

An n-level classification (where n is a positive integer) of a non-empty set X (termed the classification space) is constituted by every n-element sequence of non-identical families of sets F_1, F_2, ..., F_n, such that $F_1 = \{X\}$, all the sets which are elements of any family F_i ($i \leq n$) are non-empty and disjoint in pairs, the union of every family F_i is identical with the set X, and finally that every set Y belonging to a family F_i ($1 < i \leq n$) is a subset of some set Z belonging to the family F_{i-1}.

In other words, an n-level classification of a set X is a certain hierarchy of divisions (in the usual sense) of that set. The trivial division $\{X\}$ constitutes its peak, and each subsequent division is created from its predecessor through its fragmentation, i.e. through splitting at least one of its elements into at least two subsets. The subsequent divisions F_i are termed levels of the classification, and individual sets belonging to these levels are their members.

Every n-level classification sets a certain quasi-distance (or, in some cases, simply a distance) between any two elements of the classification space. In order to demonstrate this, the author first defines an ancillary concept of the index of the objects' similarity in terms of a specific classification.

Thus, if the sequence F_1, F_2, ..., F_n is an n-level classification of the set X, and x and y are any elements in the space of this classification, the number of the last level at which objects x, y still remain elements of the same member is termed the index of the similarity of those objects with respect to that classification. The index is designated with the symbol

$I(x, y)$. If the need arises, it may be indicated which classification is being referred to. The difference $n-I(x, y)$ is the quasi-distance between objects x and y set by the classification. If the difference is designated as $D(x, y)$, then it can be demonstrated that such quasi-distance has the following formal properties (for any x, y belonging to X):

$D(x, y) \geq 0$
$D(x, x) = 0$
$D(x, y) = D(y, x)$
$D(x, y) + D(y, z) \geq D(x, z)$

In a special case whereby each member of the last classification level is a one-element set, the following equivalence holds:

$D(x, y) = 0 \leftrightarrow x = y$

In such a case, the function $D(x, y)$ is simply distance in the ordinary mathematical sense, and the set X is a metric space, whereby the function is its metric.

On the other hand, objects that are studied and classified in empirical sciences are diversified in natural ways. They may differ from one another in some respect to a lesser or greater extent. Thus, for example, pairs of objects may be different with respect to the extent of difference of some type, or the extent of genealogical kinship, and so forth. In other words, nature itself sets certain distances between elements of various sets. The classification of such elements is a natural one if it reflects their real diversification. This goes to show that the concept of natural classification needs to be relativized. The concept may thus eventually be defined more precisely as follows:

An n-level classification of a set X is natural with respect to a distance $d(x, y)$ defined on the set X if and only if for any elements x, y, z, u of the set it is so that if $D(x, y) < D(z, u)$, then also $d(x, y) < d(z, u)$.

Some distance functions (the so-called classification functions) set relevant multi-level classifications for relevant sets unequivocally; other may merely be "reflected" by natural classifications.

This conception was elaborated upon in detail in Romahnowa's article "Classification as a Kind of Distance Function" [4]. The far-reaching generalization of the theory of multi-level classifications (taking into account, amongst others, classifications with a transfinite number of levels) was presented in an extensive bipartite contribution, "A Generalized Theory of Classifications" [8], co-authored by T. Batóg.

In one of her earliest articles, "Wieloznaczność a język nauki" [Polysemy and the Language of Science] [1], Romahnowa took up a certain issue in the philosophy of the language of science. She asserts that science by no means attempts to eradicate all the ambiguities that can be traced in its language, and – what is more – that eradicating them would in fact be impossible. Still, the language of science functions fairly well as an effective tool of communication. Romahnowa's explanation is that this is because certain properties of sentences or of their sequences – such as *a priori* truthfulness, empiricalness, logical consequence, mutual exclusion of sentences, and so forth – obtain for certain sentences or sequences of sentences of the natural language, irrespective of certain fluctuations of the meanings of terms found in that language which are permissible in that language. In other words, a reasoning using polysemous words may – within the domain of the language that it belongs to – allow only one decision regarding the truthfulness of the theses that are being put forward as well as the binding value of the vindications that are being used. Hence, polysemy which is embedded in a certain context does not have to be a potential cause of argument, in which words referring to that context would be challenged. Scientists may also deliberately take pains to formulate their reasoning carefully, so that polysemous words are only used in those contexts in which some possible fluctuations of the meaning of individual terms do not result in differences in deciding upon the truthfulness of the theses that are being propounded and the correctness of the vindications that are being developed.

In her article "Indukcja a prawdopodobieństwo" [Induction and Probability] [3], Romahnowa was concerned with the analysis and critique of the probabilistic approach to the problem of induction. In her view, the approach consists in adopting the following set of theorems: (1) The goal which is served by induction is that of obtaining true judgements on the world: true judgements in the absolute sense. (2) The conclusions of inductive inferences do not follow from their premises. Thus, while starting out from true judgements, one may arrive – in the course of inductive inferencing – at false conclusions. In this sense, inductive inferencing may fail. (3) In good inductive reasoning, the probability of the conclusion in view of the premises is sufficiently high; it normally approaches 1, and is at any rate higher than 1/2; this is a guarantee that by using such induction when inferring conclusions from true premises, one will sufficiently often arrive at judgements that are also true in the absolute sense.

Romahnowa asserts that this approach to induction is simply erroneous. In her view, even the best methods of inductive reasoning almost never lead from true individual knowledge to true general knowledge, and thus in every induction the probability of the conclusion in view of the premise approaches zero. The history of natural sciences bears testimony to this: these sciences must for ever correct their theories, as they do not match the facts. The basic source of the error in the probabilistic approach to induction lies in a false belief regarding the goals and results of human cognitive activity. The goal of this activity is not to arrive at absolute truth, but to obtain knowledge which will allow one to orientate oneself in the environment, in a broad sense, and to adjust one's actions to the conditions of that environment. False convictions (theories) would very often be valuable components of our knowledge; hence the induction which had led to them did not in fact fail. Consequently, the goal of induction must be defined differently from theorem (1). Also, the failure of induction must be understood differently than in (2). Finally, the theorem (3) must be simply rejected.

The solution to the problem of induction which emerges from Romahnowa's reasoning may be termed a pragmatic one.

Two Romahnowa's articles may be said to belong to rhetoric in a broad sense. These are "Pewne pojęcie poprawnej inferencji i pragmatyczne pojęcie wynikania" [A Certain Concept of Valid Inference and a Pragmatic Concept of Consequence] [5] and "Z teorii racjonalnej dyskusji" [From the Theory of Rational Discussion] [7]. They are included under rhetoric because they undertake to reconstruct in modern terms the classical theory of argumentation errors. Argumentation is understood as any finite sequence of qualified sentences, i.e. sentences which are simply asserted or which had been inferred from some specific earlier sentences. A stricter definition of this concept is as follows:

A finite sequence of qualified sentences

$$S_1(q_1), S_2(q_2), ..., S_n(q_n)$$

is an argumentation in favour of a thesis T if and only if the final sentence in it is T (i.e. $S_n = T$), if a qualification q_k of any sentence S_k in this sequence of sentences is either a qualification of assertion or it qualifies the sentence S_k as inferred from certain earlier sentences $S_{i_1}, S_{i_2}, ..., S_{i_j}$, and moreover subsequence of that sequence meets either of the two conditions listed above.

Romahnowa stresses that this formulation is a rather narrow one, as it does not allow formulae containing free variables as steps in the argumentation, and it does not account for apagogic argumentation. Next, she develops a theory of the so-called rejoinders to an argumentation (comprising the acceptance of an argumentation and various types of charges against it), where she draws on the traditional science of argumentation errors, such as *petitio principii, non sequitur* or material error.

Some attention should also be paid to the paper entitled "Czy filozofia obumiera?" [Is Philosophy Withering?] [6]. In the article, she defends the value of philosophy as a separate discipline and speaks against restricting its scope to the so-called little philosophy. She argues that little philosophy cannot be separated from grand philosophy in any tangible way, as the results that one arrives at in the former domain frequently lead to new solutions in the latter. Furthermore, Romahnowa challenges to much publicised view that philosophy – even when supported by contemporary science – has not yet matured enough to be able to resolve any "great" issues of world view (*Weltanschauung*). According to Romahnowa, this view is simply not adequately justified. The fact that age-old disputes on all most momentous philosophical propositions persist results not so much from any flaws in philosophical methods but from various social factors which remain outside science. In conclusion, Romahnowa points out that certain problems of grand philosophy may over time be reformulated with the assistance of new conceptual apparatuses of science (e.g. logic) and subsequently solved by means of scientific methods. In this context, she invokes the famous analyses of the issue of transcendental idealism that were conducted by K. Ajdukiewicz. At the same time, however, she agrees with those critics of philosophy who repudiate the systems of such classic thinkers as Kant, Hegel or Schopenhauer and dismiss them as muddy speculation.

All the remaining Romahnowa's publications are devoted to the history of logic and the history of the philosophy of science. In one of her contributions, she succinctly outlines the developments in the study of logic in Poznań, and two of her other writings deal with the Cartesian ideal of knowledge and with A. Arnauld and P. Nicole's logic (whose work she translated into Polish). Finally, in two of her publications she presents the programme of scientific philosophy proposed by K. Twardowski, the founder of the Lvov-Warsaw School, as well as his theory of knowledge.

SELECTED BIBLIOGRAPHY

1. (1948). Wieloznaczność a język nauki [Polysemy and the Language of Science]. *Kwartalnik Filozoficzny* **27**, 47-58.
2. (1953). Analiza i uogólnienie metody sprawdzania formuł logicznych przy pomocy diagramów Venna [Analysis and Generalization of the Method of Verifying Logical Formulae by Means of Venn's Diagrams]. *Studia Logica* **1**, 185-213.
3. (1957). Indukcja a prawdopodobieństwo [Induction and Probability]. *Studia Logica* **5**, 71-96.
4. (1961). Classification as a Kind of Distance Function. *Studia Logica* **12**, 41-81.
5. (1962). Pewne pojęcie poprawnej inferencji i pragmatyczne pojęcie wynikania [A Certain Concept of Valid Inference and a Pragmatic Concept of Consequence]. *Studia Logica* **13**, 203-208.
6. (1962). Czy filozofia obumiera? [Is Philosophy Whithering?]. *Studia Filozoficzne* 1(28), 187-196.

7. (1964). Z teorii racjonalnej dyskusji [From the Theory of Rational Discussion]. In: *Rozprawy logiczne: Księga pamiątkowa ku czci K. Ajdukiewicza* [Logical Treatises: A Festschrift in Honour of K. Ajdukiewicz]. Warszawa: PWN, pp. 103-112.
8. (1964) (with T. Batóg). A Generalized Theory of Classifications. *Studia Logica* **16**, 53-74, and **17**, 7-30.

Translated by Piotr Kwieciński

Poznań Studies in the Philosophy of the Sciences and the Humanities
2001, vol. 74, pp. 121-127

Artur Koterski

HENRYK MEHLBERG – THE REACH OF SCIENCE

I. In the inter-war period Henryk Mehlberg (7.10.1904, Kopyczyńce near Lvov – 10.12.1979, Gainesville) attended the famous seminars of the *Schlick-Kreis*. The first years after the Second World War he worked in Wrocław, then he left Poland for the USA. In the 1950s, Mehlberg succeeded Carnap by taking his chair at the University of Chicago. Later, he moved to Toronto. He died in Gainesville, Florida[*].

II. Mehlberg was a logical positivist. His conception of the reach of science was to clarify some concepts of logical empiricism, especially of verification and of empirical character of scientific theory. His research on this subject starts in 1946 [2.4] and assumes its final form in his *opus vitae, The Reach of Science* [1.1]. In Mehlberg's case, the principle of verification does not specify what kinds of sentences are meaningful, but it is to determine the range of science. That is why once can talk about a criterion of demarcation here.

Instead of asking about the explicit demarcation tool, Mehlberg raised the question about the range of science and its limits. He seeks to determine the class of problems that are solvable (and, on the other hand, those that are undecidable) by means of the scientific method: science consists of all solutions gained with the aid of such a method. Because of the requirement of the adequacy of any demarcation criterion, we have to take into consideration problems that have already been solved, those that may be solved with the aid of already known scientific methods, and even the problems the solution of which requires new methods that are currently unknown.

The term "the reach of science" is not unambiguous. Firstly, we may interpret it within the cognitive context. Then we speak of the *epistemological* reach of science – it is constituted by all problems that may be solved when particular scientific methods are applied. Next, this should be clearly separated from the *logical* reach of science, determined by available theories, logical tools etc., and the *objective* reach of science,

---

[*] We owe this information to Piotr Labenz and Anna Witeska [Editor's note].

which is the class of objects knowable through a scientific method. In his theory, Mehlberg looks for the epistemological reach of science.

If the reach of science is determined by solutions which one affords by means of a scientific method, we have to specify what kind of method it is. Unlike Bacon, Mehlberg didn't search for *the* one method, as it is obvious that scientists use many different methods. Though it is not possible to designate any scientific method as the "proper" one, we may look for characteristics that are common to all (or, at least, to the most important) of them. This is the only way to find *the scientific method*.

Introducing a new theory to science, we have to test it. Every new discovery ought to be verified by a competent scholar. And this is the essence of a scientific method: proof and verification. Solution to scientifically posed question consists in logically possible answer and description of the method of validation, i.e. how to obtain the evidence for the answer. Empirical evidence for P is a set of statements, Z, that secures the truth of information P; a scientist is forced to accept sentence S, the one that express P, because he believes in the truth of Z [2.9, p. 279]. Though "empirical evidence" cannot be defined univocally, there is a way to specify the method of validation independently of that concept, and indeed Mehlberg prefers the following solution: the method of validation for a problem of a kind K is a sufficiently reliable method that, given the problem P of the type K and a logically possible answer A to a question posed in P, allows to state a new statement B: "A is correct" (or "A is not correct"). Now we may say that one validates a sentence through the application of a reliable method. Thus, the definition of a solution of a scientific problem runs as follows: "A problem is scientifically solved if some logically admissible answer to it has been validated by a sufficiently reliable method" [1.1, p. 71].

If the reach of science is to be determined by the range of problems that may be solved by the scientific method, we can say that the reach of science is the reach of the scientific method.

III. Each method is a kind of prescription that, if fulfilled, brings about (always or at least in an appropriate number of cases) an event that is called the objective of the method. The class of objects that a method applies to constitutes the range of that method. Now Mehlberg understands the *method of science* as essential feature of all particular methods we encounter in science. Thus, the scientific method applies to all possible objects, and it is the widest possible method. And this amounts to the thesis of the universality of the scientific method: the solution to all genuine problems consists in a relevant and logically consistent answer, which is supported by a relevant method of validation. And this is exactly the gist of the scientific method; therefore, all solvable problems are simultaneously scientifically solvable.

Speaking about validation, we have to consider all types of this method. Mehlberg classifies them in the following way.

1. Firstly, the reach of science is determined by all *actually available* methods. The method is available if it was fruitfully applied. But considering the reach of science, we have to take into account also those problems that may be solved in the (unspecified) future, e.g. with the aid of unknown methods. Thus, we admit *merely possible* methods at this point. The method, understood as a sequence of operations, is possible if it is describable by precise statement.

2. Secondly, there are direct and indirect validating procedures. The method is *indirect* if its application requires inferential (e.g. deductive or inductive) reasoning. Otherwise, when gathering the evidence on the basis of sense-data observation or introspection, we talk about *direct* methods.

3. The solution of some problems may be obtained on the basis of singular statement, some require a law, and still other need a cluster of laws, i.e. a theory (or a set of theories). So, we may divide validating methods according to their objectives, which will be described by an existential statement, a singular law or a whole theory.

4. Next, Mehlberg distinguishes *special* and *basic* methods of science. The application of special methods is restricted to particular branches of science – for example, we make use of IQ test in the field of psychology and not in that of chemistry. On the other hand, there are methods that are applicable in every domain of science: observation, measurement and deductive or inductive reasoning.

5. The fifth division takes into account the relationship of a method to the trial-and-error procedure. Thus, when a method supplies admissible answer and evidence in favour of it, we call it a *discovery method.* Giving an answer to a question "What is the colour of this patch?" we look at it, discover what colour it is and at the same time we validate our answer (after Schlick we may call this kind of statements *Konstatierungen*). In more complex cases, when we cannot use this kind of methods, we have to appeal to *verification methods.* Then the answer is already given (a prediction) and our task is to verify it. In still more complicated cases only *heuristic methods* are applicable – they suggest the answer, though they cannot supply the evidence.

IV. In science, we encounter three groups of scientific statements, each with its own validation method, which are, however, closely connected.

Fact-like statements describe things and events. They specify a quality that may be said about an object or a relation that holds between two or more of them. When a fact-like statement is validated, it becomes a fact-statement (i.e. true fact-like statement).

A statement about facts may be verified directly or indirectly. It is verifiable directly if it is possible to ascertain its truth value without any inference – without the need to establish the truth or falsity of any other expression. Otherwise, it is indirectly verifiable. That is the case of even the simplest measurement.

If the set of statements S, which is needed to verify indirectly verifiable assertion A is finite, and S, supplied by analytic judgements, implies A, then we call A "finitely verifiable". This kind of verifiability specifies the reach of measurement methods and of methods of indirect observation. Finite verifiability, in opposition to direct verificational methods does not determine the truth value of A; a conjunction of two finitely verifiable statements may turn out to be unverifiable. Here, Mehlberg introduces a distinction: "Two finitely verifiable statements will be said to be jointly (or separately) verifiable if their conjunction is also (or is not) verifiable" [1.1, p. 295]. This requires additional restriction, because, as noticed above, two separately verifiable statements may turn out to be jointly unverifiable (e.g. one finds such pairs of statements in quantum mechanics). To fix this inconvenient possibility Mehlberg introduced the notion of joint verifiability *in the theory*. Accordingly, theory T is verifiable if any two consequences of T, p and q, are verifiable in T – that is, p and q have to be jointly verifiable, *provided* there is a set of observational statements that are compatible with T of which the conjunction p and q is a consequence.

Validation of a law-like statement reduces to validation of fact-like statements. This is possible because of methods of approximation. This is where we meet finitely unverifiable statements. Strict verification (or falsification) of any law is not possible. Accordingly, scientists only confirm their laws. At any rate, we are set against the "problem of induction", which, generally speaking, always involved difficulties in doctrines of logical empiricists. Mehlberg's proposals are quite peculiar; for instance, he considers quantifying space and time in regions limited to some thousands of years and light-years – that would reduce the scope of scientific laws from "strict" to "numerical" universality.

Universal statements are not finitely verifiable, but it does not mean that we cannot test them. Laws are inductively verifiable if any of their instances may be directly or finitely verified. Thus, we obtain two classes of verifiable statements. More generally we can say that A is empirically verifiable if A or its negation follows (in a strict or probable manner) from a consistent set (that may be infinite) of directly verifiable statements.

To validate a theory we have to validate its consequences. And Mehlberg identifies an empirical theory with all verifiable consequences that may be derived from a subset of statements which he termed an "axiomatic basis". In this way theories are at least inductively verifiable and they may be considered as true or false. Thus, we say that:

[Theory is validated if] a suitably selected sample of the statements which make it up is validated by the usual methods applicable to single statements. If so, we can formulate the general rule of validation. If all mutually independent consequences of the theory which have been examined so far by applying conventional [...] methods have turned out to be correct, then the theory itself may be considered to have been confirmed or rendered probable by this sample of its consequences to a degree depending upon the number and the mutual independence of the consequences included in the sample [1.1, p. 221].

The validation of the theory can be replaced by validation of a law. Thus, we formulate in the metalanguage a law correlated to the theory T: "All verifiable consequence of T are true." And if we substitute such a law for the theory, the validation of theory will be reduced to inductive validation of that law. Thus, we can understand empirical validation of a theory as a special case of inductive validation of a specific meta-law.

V. Any empirical theory can be axiomatized. If it is, then there is an axiomatic basis for it. An axiomatic basis for a given verifiable theory is a finite and consistent set of synthetic statements, consequences of which constitute that theory. A verifiable theory (which is not, strictly speaking, an axiomatic system) allows different axiomatic bases; it does not contain all consequences of its axiomatic basis, but only those that are empirically verifiable. This means that verifiable theory does not need to contain its own axiomatic basis. If it is the case, the theory has an *external basis*. Otherwise, when all axioms are verifiable, the basis is *internal* [2.4, pp. 257]. For example, physical geometry that we obtain from Euclid's axioms is a verifiable theory with external basis.

Mehlberg's principle of verification is not a criterion of meaning. It does not eliminate unverifiable sentences from science. It only states that unverifiable assertions have no logical value. The unverifiable exists in all sciences. However, according to Mehlberg's solution, it is still possible to keep metaphysics off science, as scientists – unlike metaphysicians – do not attribute the truth value to the unverifiable statements [2.6, pp. 92-93]. In opposition to the main currents in the Vienna Circle positivism, Mehlberg maintained that a theory with unverifiable assumption might be acceptable.

VI. Even though both sciences and metaphysics contain sets of unverifiable statements, the role they play is quite different. In science they are auxiliary devices, while in metaphysics they constitute the core of a system: But this is not the only way to distinguish an acceptable scientific theory. A good scientific theory should be characterized in the following way.

1. An acceptable theory should satisfy the economic principle (so it assures the richness of empirical laws, while it remains simple).
2. Of course, every good theory also has a predictive function. Mehlberg reminds us of the Comtean maxim *Savoir pour prévoir*, and, in the tradition of the Vienna Circle, especially of Neurath, views the possibility of predicting as the most important feature of science.
3. A genuine theory gives us explanation of the world (or any part of it).
4. A scientific theory allows testing of its predictions and explanations. In this way it specifies kinds of action, which let us control our environment.
5. Finally, a valuable scientific theory has the informational (cognitive) function. It "satisfies one of man's strongest needs, i.e. the need to know about his observable environment".

But we cannot forget that any empirical theory would not satisfy all those conditions unless accompanied by additional, empirically untestable elements. These are:

a) The mathematical and logical formalism of the theory.
b) Metaphysical statements, that is untestable assumptions of science. They are not decidable by empirical or logic-mathematical methods, but – accordingly with reformulated principle of verification – they are not senseless either: these are statements of an external basis.

VII. Verifiable theories determine the actual scope of science – the range of problems that may be solved scientifically. Verifiable statements are true or false, and if they are verified, they provide reliable information, genuine scientific knowledge. Thus verifiability is coextensive with epistemological reach of science. Unverifiable theories and statements have no definite truth value, though they perform auxiliary function in science and are indispensable there.

Mehlberg rejects the theory evaluation stated in Vienna in the early 1930s by Waismann (conclusive or strong verifiability) and Popper (the criterion of absolute falsification). These criteria are too restrictive and historically inadequate.

Being a positivist, he accepts the thesis of the unity of science. The unification is granted by the scientific method of solving empirical problems: arriving at a logically admissible answer through collecting relevant evidence. All branches of science, regardless of any suggested divisions, share this method, may it be *Natur- und Kulturwissenschaften*, pure and applied sciences, or object sciences and their metatheories. Mehlberg does not speak of the unity of language, but repeats Karl Pearson's thesis: *"The unity of all science consists alone in its method"*.

SELECTED BIBLIOGRAPHY

1. **Books:**
   1. (1958). *The Reach of Science*. Toronto: Toronto University Press.
   2. (1980). *Time, Causality, and the Quantum Theory*, ed. by R.S. Cohen, Robert, Boston: Reidel Publishing Company.

2. **Papers:**
   1. (1935/1937). Essai sur la théorie causale du temps. *Studia Philosophica* **1**, 1-141 (part I) and **2**, 111-231 (part II – sometimes cited as *Durée et causalité*, which is subtitle of this part).
   2. (1936). O przyczynowej koncepcji przestrzeni i czasu [On the Causal Conception of Space and Time]. *Przegląd Filozoficzny* **39**, 409 (with a short discussion on pp. 409-410).
   3. (1937). Sur quelques aspects nouveaux du parallélisme psychophysiologique. In: *Proceedings of the International Congress*, Paris: Hermann & Cie, pp. 77-84.
   4. (1946). Positivisme et science. *Studia Philosophica* **3**, 211-294.
   5. (1948). O niesprawdzalnych założeniach nauki [On Untestable Assumptions of Science]. *Przegląd Filozoficzny* **44**, 319-335; reprinted in: T. Pawłowski (Ed.), *Logiczna teoria nauki* [Logical Theory of Science]. Warszawa: PWN, 1966, 341-353.
   6. (1948). Idealizm i realizm na tle współczesnej fizyki [Idealism and Realism Against Contemporary Physics]. *Kwartalnik Filozoficzny* **1-2**, 87-116 (part I), and **3-4**, 202-239.
   7. (1954). The Range and Limits of the Scientific Method. *The Journal of Philosophy* **51**, 285-293.
   8. (1961). Physical Laws and Time's Arrow. In: H. Feigl and G. Maxwell (Eds.), *Current Issues in the Philosophy of Science*, New York: Holt Rinehart and Winston, pp. 105-138. (The book contains also Mehlberg's comments on other's papers – pp. 102*f*, 360*ff*.)
   9. (1962). Theoretical and Empirical Aspects of Science. In: E. Nagel, P. Suppes, and A. Tarski (Eds.), *Logic, Methodology and Philosophy of Science. Proceedings of the 1060 International Congress*. Stanford (CA): Stanford University Press, pp. 275-278.
   10. (1962). The Present Situation in the Philosophy of Mathematics. In: B.H. Kazemier and D. Vuysje (Eds.), *Logic and Language. Studies Dedicated to Professor Rudolf Carnap on the Occasion of His Seventieth Birthday*. Dordrecht: D. Reidel Publishing Company, pp. 69-103.
   11. (1966). The Relativity and the Atom. In: P.K. Feyerabend and G. Maxwell (Eds.), *Mind, Matter and Method: Essays in Philosophy of Science in Honor of Herbert Feigl*. Minneapolis: Minneapolis University Press, pp. 449-491.
   12. (1969). Philosophical Aspects of Physical Time. *The Monist* **53**, 340-384.
   13. (1977). Individuality, Reality and Space-Time. *Methodology and Science* **10**, 34-63.

Poznań Studies in the Philosophy
of the Sciences and the Humanities
2001, vol. 74, pp. 129-133

Mieszko Tałasiewicz

MARIA KOKOSZYŃSKA-LUTMANOWA – METHODOLOGY, SEMANTICS, TRUTH

I. Maria Kokoszyńska-Lutmanowa (6.12.1905, Bóbrka near Lvov – 30.06.1981, Wrocław) studied in Lvov under Kazimierz Ajdukiewicz's supervision, and in Cambridge, under Ludwig Wittgenstein's. Before the Second World War, she lived and worked in Lvov. After the war, as Lvov ceased to be a Polish city, she moved to Wrocław.

Along with such philosophers as Alfred Tarski, Izydora Dąmbska or Henryk Mehlberg, she belonged to the third (and some would say: the last) generation of the philosophical Lvov-Warsaw School, established by Kazimierz Twardowski in 1895. At the starting point of her scientific career in the late 1920s, the question of scientific language was of foremost importance in analytical philosophy and everybody's eyes were turned to the newly born science: logical semantics. Hence, this is what she began with.

II. In [1], which is one of her quite early writings, Kokoszyńska took on a somewhat preliminary problem: the question of meaning of propositional functions and of variables incorporated in these functions. She opposed the view that such variables are meaningless, and suggested that the meaning of a variable is a proper part of meanings of all the substitutions – a common part, at which meanings of all the substitutions overlap. In [2] and [3] she already engaged fully in the analysis of central semantic problems: the definiability (or rather non-definiability) of truth and other semantic notions in syntax and the question of the proper nature of metascience. She compared the views of Rudolf Carnap (from *Logische Syntax der Sprache*) and some other neopositivists (such as Otto Neurath and Carl Hempel) with those of Alfred Tarski and Kurt Gödel, and developed a serious critique of the neopositivist standpoint. No interesting philosophical investigation of science – she says – can be made in pure syntax. Carnap's programme of limiting metascientific considerations just to the syntactic aspect of scientific language is misleading. If the logic of knowledge is to be the only way of reasonable philosophical research, it has to be interpreted much more widely than Carnap and his colleagues are apt to. Namely, it has to encompass semantics. And it was Tarski who showed that semantics was indeed a scientific domain.

III. The Lvov-Warsaw School was often compared with the Vienna Circle – sharing with it a rationalistic, logical way of making philosophy, an interest in science and the scientific method, and a forceful critique of speculative philosophy. But there were some significant differences between the two movements, and Kokoszyńska became one of the leading polemists of the Lvov-Warsaw School engaged in a debate with logical positivism. She reviewed books and articles of Moritz Schlick, Carnap, Neurath, Hempel etc., kept track of the controversies among them and with the evolution of their standpoint, following them with sympathy and appreciation. Yet, she pointed out sharply and precisely many weaknesses and inconsistencies. In [5] for instance, apart from many efforts to highlight the strengths of the Vienna Circle programme, she encloses a warning: the programme is too radical. Some things are false, some unjustified, some promissory rather than promising. She repeats her doubts against pure syntax, she points out that the neopositivist drive for a unified scientific language is either sterile (for the empirical language), as no one has ever shown that all empirical scientific statements can really be expressed in one universal language, or wrong (for the full scientific language), as it has been shown that logical knowledge cannot be expressed in one language, because of the fact that for semantic issues we always need a metalanguage different from the language in question. (In spite of this warning, Kokoszyńska was quite positive about the integration of sciences as such, especially methodological integration, and she was continuously struggling for that even much later, e.g. in [17].) Finally, in [6] and [8], she took into consideration the neopositivist abomination for metaphysics. She distinguished two kinds of what is called "metaphysics": a methodological approach, according to which one asserts an undetermined (synthetic) proposition in a dogmatic, unjustified way – and a set of certain propositions. In the latter meaning, there are three cases: (a) metaphysical statements are determined (analytic) propositions (e.g. axioms of language); (b) they are undetermined and testable; (c) they are undetermined and non-testable. Kokoszyńska pointed out that what is beyond doubt unacceptable in science is a metaphysically dogmatic method. As she wrote, in contrast to what the members of the Vienna Circle maintained, metaphysics as a set of propositions should not be excluded from science automatically. Metaphysical statements of the types (a) and (b) were, in fact, presènt in sound science – and rightly so. She stressed that many philosophers (like Jan Łukasiewicz) were apt to regard metaphysics as a set of very general, but still theoretically testable propositions. Kokoszyńska would agree only that the third kind of metaphysical statements are useless in science and that no scientific interest should be given to them. But still she would disagree with neopositivists as regards to the question of meaningfulness or meaninglessness of these statements. Namely, she could see no reason for denying that they had some meaning, and regarded neopositivist views as merely a project of some future, ideal language, a project which could have many advantages, but quite a large number of drawbacks as well.

IV. With respect to semantic inspirations, Kokoszyńska took another problem, which she elaborated in several papers for many years [e.g. 4, 9, 10]: the problem of the relativity of truth. She performed an extensive analysis in order to establish and clarify the relativistic thesis (the moderated version of which she finally reproduced as: "1. The term "true" is an incomplete predicate; 2. $\exists X \{\exists Y,Z [(X$ is true with respect to $Y)$ $\wedge (Non$-X is true with respect to $Z)]\}$" [10, p. 94], and to consider the most common arguments for it. She criticized those arguments and the result of this criticism may be formulated as follows:

> The opinion that the standpoint of relativism with respect to truth is proved by the[se] arguments [...] is due to one of the following facts: (1) the expression "relativism of truth" is given a non-genuine meaning; (2) the term "true" is understood in an unusual way; (3) certain features of the language used are misinterpreted. [...] [As to the first part – the arguments appeal to the fact that not all theorems are necessary; that is true, they are not – says Kokoszyńska – but this does not support any kind of genuine relativism.] The second part views pseudo-proofs which are based on a psychological, sociological or pragmatic theory, on the consistency theory, or on a certain normative theory of truth [none of these theories of truth is adequate – she maintains] [...]. The third part is concerned with the misinterpretations arising from: the deictic character of certain expressions belonging to the everyday language, the changed definition of sentence in the language of epistemological relativists, or the need of introducing an infinite number of truth concepts (if Tarski's method of defining truth is adopted) [in all of these cases there is a "right" or "proper" interpretation, according to which relativism does not hold] [10, p. 139].

She pointed out the necessary assumptions that are to be made to defend the absolute concept of truth: the definition of truth must allow one to distinguish between the concept itself and the criteria of truth and fulfil the requirement of having the *T*-convention among its consequences (Kokoszyńska regarded the *T*-convention as a necessary condition of adequacy for any theory of truth) – and it must be possible to correct all occasional expressions (which possibility is denied by some philosophers, e.g. those who concentrate on the properties of everyday language).

V. Subsequently, Kokoszyńska turned to more strictly methodological questions. She conducted thorough research in the field of induction and different kinds of deductive procedures. In [11] she formulated a definition of induction as any non-deductive inference (where "deduction" is given a distinctive meaning, inspired by the works of Ajdukiewicz, as "reasoning that follows the relation of logical entailment"; thus many inferences that follow any non-logical entailment should be classified as inductions, even if they are infallible) and she proposed some criteria of correctness of inductive inferences, absolute and relative, based upon the theory of probability. In [12]-[15], she tried to clarify the discussion about deductive and non-deductive sciences and stressed that the *fundamentum divisionis* should be not the mere method of sciences in question. She pointed out that the so-called non-deductive sciences might and should use deductive procedures in some respects. It is valuable – she says – to make even deductive systems out of these non-deductive sciences in order to clarify concepts, establish logical relations, perform tests and engage in some heuristics. Nevertheless, they will not cease to be non-deductive ones, as induction is still immanent to them. On the other hand – she suggests – it is not absurd to assume that induction sometimes interferes with deductive sciences, such as mathematics, at least in their heuristics (that is why she maintains that methodological differences shall not prevent us from attempting to unify sciences, even deductive with non-deductive ones – as in [17]).

While analyzing the differences between deductive and non-deductive sciences, Kokoszyńska distinguishes two kinds of deductive justification: absolute and relative. The former is applicable only to analytic statements, whereas the latter can be very useful with regard to synthetic ones, especially those of non-deductive sciences. Thus, she finds herself in need of formulating a theory of analytic statements. In the writings mentioned above, and in [16], she answers this need and satisfies the requirement of providing such a theory. Openly engaging in a polemic with Willard Van Orman Quine, Kokoszyńska gives a clear notion of the analyticness of statements in semantic terms. She points out that rules of denoting, which define analytic statements, belong not to the language in question, but to its metalanguage, so that the definition is not circular (as Quine suggests). She also maintains that – contrary to logical tautologies – analytic statements are not completely independent of "reality". They are "independently" true always and only if respective rules of denoting (in metalanguage) stick to really existing objects, which may be the *questio facti*. On the other hand, for synthetic statements it is not enough to make sure about the existence of respective objects, but some other kind of justification is needed (although sometimes vainly) to establish their logical value.

VI. Although some of Kokoszyńska's methodological ideas may seem quite ordinary nowadays (and some, perhaps, outdated), it must be stressed that she was among the first philosophers who appreciated Tarski's semantic theory of truth and that her criticism about some ideas of the Vienna Circle became classic. She developed a very interesting theory of analytic statements, which is much less popular than it deserves to be, and engaged in a controversy that is now topical again: the question of relativism. Her investigations in this field are in urgent need of recalling.

SELECTED BIBLIOGRAPHY

1. (1932). Z semantyki funkcyj zdaniowych [On Semantics the of Propositional Functions]. *Księga Pamiątkowa Polskiego Towarzystwa Filozoficznego we Lwowie 12.02.1904–12.02.1929.* Lwów, 312-313.
2. (1936). Syntax, Semantik und Wissenschaftslogik. *Actes du Congrès International de Philosophie Scientifique.* F.3. Paris, 9-14.
3. (1936). Logiczna składnia języka, semantyka i logika wiedzy [Logical Syntax of Language, Semantics and Logic of Knowledge]. *Przegląd Filozoficzny* **39**, 1, 38-49.
4. (1936). Über den absoluten Wahrheitsbegriff und einige andere semantische Begriffe. *Erkenntnis* **6**, 143-165.
5. (1937). Filozofia nauki w Kole Wiedeńskim [Philosophy of Science in the Vienna Circle]. *Kwartalnik Filozoficzny* **13**, 1, 151-165; 2, 181-194.
6. (1937). Sur les éléments métaphysiques et empiriques dans la science. *Travaux du IXe Congrès International de Philosophie, Congrès* Descartes. F.4. Paris, pp. 108-117.
7. (1938). Bemerkungen über der Einheitswissenschaft. *Erkenntnis* **7**, 325-335.
8. (1938). W sprawie walki z metafizyką [On Fighting Metaphysics]. *Przegląd Filozoficzny* **41**, 9-24.
9. (1948). What Means Relativity of Truth?. *Studia Philosophica* **3**, 167-176.
10. (1951). A Refutation of the Relativism of Truth. *Studia Philosophica* **4**, 93-149.
11. (1957). O dobrej i złej indukcji [On Good and Bad Induction]. *Studia Logica* **5**, 43-70 (incl. summary in English).
12. (1960). O stosowalności metody dedukcyjnej w naukach niededukcyjnych [On the Applicability of the Deductive Metod in Non-deductive Sciences]. *Sprawozdania Wrocławskiego Towarzystwa Naukowego.* 15 A, 45-49.
13. (1962). O dwojakim rozumieniu uzasadniania dedukcyjnego [Two Concepts of Deductive Justification]. *Studia Logica* **13**, 177-196 (incl. summary in English).
14. (1963). O dedukcji [On Deduction]. *Studia Filozoficzne* 1, 77-86, (incl. summary in English).
15. (1967). W sprawie koncepcji nauk dedukcyjnych [About the Conception of Deductive Sciences]. *Studia Filozoficzne* 1, 57-63.
16. (1970). Złudzenia aprioryzmu [Illusions of Apriorism]. *Ruch Filozoficzny* **28**, 1-2, 72-78.
17. (1973). Metodologiczne problemy integracji nauk [Methodological Problems of Integration of Sciences]. *Studia Filozoficzne* 4(89), 47-56.

Poznań Studies in the Philosophy
of the Sciences and the Humanities
2001, vol. 74, pp. 135-139

Zygmunt Hajduk

STANISŁAW MAZIERSKI – A THEORIST OF NATURAL
LAWFULNESS

I. Stanisław Mazierski (9.10.1915, Koral near Włocławek – 23.06.1993, Lublin) graduated from the Faculty of Christian Philosophy of Warsaw University in 1950, after submitting the thesis "Koncepcje metafizyki hipotetycznej u M. Wartenberga" [Conceptions of M. Wartenberg's Hypothetical Metaphysics]. He obtained his PhD degree in 1951 [1.1] and his postdoctoral degree in 1961 [1.2]. He was appointed assistant professor in 1966, associate professor in 1971 and full professor in 1981. In the years 1962-1963, he was granted a scholarship from Louvain University, Belgium (Institute Supérieur de Philosophie). This scholarship (discussions with professors L. Raeymaeker, G. van Riet, J. Ladriér) resulted in a monograph [1.3.].

II. *Prolegomena do filozofii przyrody inspiracji arystotelowsko-tomistycznej* (Prolegomena to the Philosophy of Nature Inspired by Aristotle and St Thomas) is a volume that is the most representative for his metaphilosophical research – one of the fields of his scientific output. The monograph contains a proposition for the philosophy of nature as a section of the classical philosophy. The methodological status of this discipline has been constituted by four components: a starting point, the subject, the method of the philosophy of nature and its attitude towards philosophical and extra-philosophical sciences. The solutions of all those constituents are controversial. The starting point in the philosophy of nature is constituted by results of common and scientific experience interpreted in some way. The subject of research in this philosophical discipline is constituted by the most general properties of bodies in the *corpora ad sensum* meaning. The data in the sensual cognition of a body are stretched, spatial, temporal and variable. A research method specific for the philosophy of nature is a physical abstraction.

The constituents of the status of the philosophy of nature specified above are being established in a discussion with the representatives of a classical trend in the Thomist metaphilosophy. The names that are encountered most often there are: N. Luyten, A.G. van Melsen, J.J. Maritain, K. Kłósak, S. Kamiński, M.A. Krąpiec.

The expression of the connections of peripatetic philosophy with natural sciences and these of other sciences, and in particular of the related philosophical systems based on K. Ajdukiewicz's theory of open and closed languages seem to present fewer controversies and enjoy a broader recognition. The author declares his openness towards the philosophy of nature, its autonomy, and, at the same time, its independence – particularly of metaphysics.

III. The second plane of Mazierski's research contains the basic items of the philosophy of nature, also called "philosophical cosmology". For their most part, they result from the implementation of the metaphilosophical programme outlined above, and are presented in a textbook [1.4]. It contains the issues of the traditional philosophy of nature built on four cases contained in the scheme of categories of Aristotle's and St Thomas' philosophical system. They include: extent (quantity), space, time, change and, in a special case, motion. In the philosophy of nature, they constitute the attributes of the bodies existing in nature. Cosmological problems implied by these philosophical categories are presented according to the contemporary results of formal and natural sciences. It also concerns the problems of natural cosmology, together with its selected philosophical conclusions. S. Mazierski, presenting the theses of the traditional philosophy of nature in the context of science, appears to be a true continuator of the thought of Louvain Thomism. The confrontation of Thomist cosmology with the achievements of natural sciences forced philosophers to correct some of their theses. It concerns, for example, acknowledging the discontinuity of matter, and revising the opinions concerning the structure of the Universe.

The whole of philosophical cosmology includes, beside inanimate nature, the philosophy of animate nature. S. Mazierski is the author of several papers, in which he undertakes some of the issues of this part of the philosophy of nature: problems of mechanism, finalism, functionalism and laws in biological sciences. He is the editor of a collective work [1.5]. Beside the key issues of contemporary biology and the philosophy of biology, the book presents selected issues concerning the characteristics of life, biogenesis, evolution of biosystems, as well as the selected issues of worldview (monogenism, polygenism).

IV. The third field of S. Mazierski's achievements is located on the level of the philosophy of natural sciences, particularly physical sciences. When characterizing those achievements, it is worth noting that S. Mazierski was the type of a scholar who, to a great extent, took into account the results of contemporary natural sciences. He also appreciated the application of the research apparatus used in those sciences, and in particular that used in mathematical sciences, in the exploration of micro-, macro-, and mega-world (the cosmos). He also postulated the improvement of the knowledge of the methods that are genuinely used within the particular fields of scientific cognition. He voiced the need for making the implications of those fields more clear. He was conscious of the usefulness of explications and possible completions and corrections; they would be introduced into the traditional philosophical notions and theses in which they occur. They appeared to be partly questioned and partly controversial, particularly in the field of more recently developed branches of physics, e.g. physics of a microworld. S. Mazierski analyzed, among others, the categories of reality, substance, determinism, indeterminism, causality, truth, space, time, and eternity.

The problems that S. Mazierski considered the most extensively, i.e. in a separate monograph, *Determinizm i indeterminizm w aspekcie fizykalnym i filozoficznym* [Physical and Philosophical Aspects of Determinism and Indeterminism], and several articles were those of determinism, indeterminism, and causality. Thus, at the same time, he joined the discussion that had been under way since the 1930s in the community of philosophizing physicists, philosophers of science and philosophers of nature. A boundary is made between the epistemological plane (the possibility of projection) and the ontological plane, which considers an overlap of physical influences in the systems. The possibility of projecting/predicting is most often based on scientific laws and theories. We are dealing here either with the deterministic or with indeterministic laws or theories. Causality also occurs in various types: physical, cosmological and metaphysical. The questioning of the first of them on the grounds of a microworld (among others Heisenberg's principle of uncertainty) created an opportunity to put forward a conception of generalized causality, also called an unequivocal conception. The other two kinds of causality are being relativized to the relative fields of classical philosophy, where they are defined.

S. Mazierski analyzed the concept of law of nature. His last book, *Prawa przyrody: studium metodologiczne* [Laws of Nature: A Methodo-logical Study], is a collection of all his earlier papers concerning the subject (they appeared starting from 1963). Among other issues, it contains the genesis and development of the very notion of natural law (both deterministic and statistical). The legal formula is being reconstructed as a general conditional sentence, being an inductive generalisation. Induction also plays a role in their legalization. The prognostic role of laws is elaborated most satisfactorily. Besides the standard divisions, S. Mazierski considered a relatively developed classification of laws proposed by

M. Bunge. S. Mazierski treated induction not only as a tool of proving but also in a heuristic way. His position is, in principle, on the side of inductionism, with certain elements of hypothetism. Scientific theories or systems are of a hypothetical-deductive type. At the same time, he turns the attention towards the positive (confirmation) rather than towards the negative (falsification) results of the tests. Emphasis is placed on confirmation, i.e. on the positive justification of laws or hypotheses with a determined degree of probability.

V. S. Mazierski also analyzes the categories of time and space. They are relativized towards different branches of physics, including relativist physics with its epistemological implications. He stresses the difference between epistemological relativism and the relativism of the relativity theory. The turn in geometry caused by the discovery of non-Euclidean geometry appeared to be cosmologically essential. Time occurs in many interpretative versions in different branches of philosophy, including the classical one, e.g. as constituted by a philosophical analysis of eternity (Boethius).

The semantic aspect of the philosophy of physics is present in Mazierski's works in the analysis of the truthfulness of the sentences of classical and modern physics. The consciousness of the need to differentiate between truth and notorious or known truth provides an opportunity for discussing the issues of the theory of physical cognition, including the problem of interference of the subject of cognition with the object of cognition, the problem of the complexity of description of the condition of a microsystem, and the effort to retain the objectivity of such description. The author's standpoint is that of epistemological realism. He develops his standpoint more extensively in another context, when he analyzes Einstein's concept of the role of philosophy in physical sciences. Besides the theses of other philosophical systems it conditions, at least implicitly, the theories of natural sciences, particularly physical sciences. The analysis of the relativity theory reveals certain types of assumptions, which are also regarded as relevant ideas for those sciences. Among others, a thesis assuming the existence of the transcendental, in relation to the subject, real world, in which objective regularities occur. In a special case, those regularities assume the form of a causal nexus. Its elements are constituted by the objects of this world. Next, there is a group of epistemological assumptions. In the theory of physical cognition, a conviction is obligatory that the transcendental world is cognizable. Cognizability of the world is founded by the rationality of its structure. The compatibility of theoretical schemes with the objects of a real world is determined by the results of experience. The revolutions taking place in science occur at different levels of scientific cognition. Revolutions whose range also covers the changes in the assumptions of science belong to the most universal and epoch-making ones.

VI. The results of S. Mazierski's research which have been schematically outlined above, with emphasis placed on those which are included in the philosophy of natural sciences, allow one to sustain K. Kłósak's opinion, according to which he is above all a theoretician of physical cognition. The philosophy of nature constitutes his second field. In this aspect, he resembles F. Renoirte, another representative of classical philosophy.

SELECTED BIBLIOGRAPHY

1. Books:
1. (1958). *Pojęcie konieczności w filozofii św. Tomasza z Akwinu* [The Concept of Necessity in St. Thomas' Philosophy]. Lublin: Towarzystwo Naukowe KUL.
2. (1961). *Determinizm i indeterminizm w aspekcie fizykalnym i filozoficznym* [Physical and Philosophical Aspects of Determinism and Indeterminism]. Lublin: Towarzystwo Naukowe KUL.
3. (1969). *Prolegomena do filozofii przyrody inspiracji arystotelowsko-tomistycznej* [Prolegomena to the Philosophy o Nature Inspired by Aristotle and St Thomas]. Lublin: Towarzystwo Naukowe KUL.
4. (1972). *Elementy kosmologii filozoficznej i przyrodniczej* [The Elements of the Philosophical and Physical Cosmology]. Lublin: Towarzystwo Naukowe KUL.
5. (1980) (Ed.). *Zarys filozofii przyrody ożywionej* [An Outline of the Philosophy of Animate Nature]. Lublin: Redakcja Wydawnictw KUL.
6. (1993). *Prawa przyrody: studium metodologiczne* [Laws of Nature: A Methodological Study]. Lublin: Towarzystwo Naukowe KUL.

2. Papers:
1. (1988). The Newtonian Concept of Space and Time. In: *Isaac Newton's Philosophiae Naturalis Principia Mathematica*. Singapore: World Scientific. An International Publisher, 15-27.

*Poznań Studies in the Philosophy
of the Sciences and the Humanities
2001, vol. 74, pp. 141-151*

Andrzej Bronk

STANISŁAW KAMIŃSKI – A PHILOSOPHER AND HISTORIAN
OF SCIENCE

I. Stanisław Kamiński (24.09.1919, Raszyn Podlaski – 23.03.1986, Lublin) was a philosopher, philosopher of science and historian of science. He studied philosophy and theology at the Catholic seminaries in Janów Podlaski and in Siedlce in 1938-1946, and then at the Catholic University in Lublin (KUL), where, in 1949, he defended his doctoral dissertation on *Fregego dwuwartościowy system aksjomatyczny zmiennych zdaniowych w świetle współczesnej metodologii nauk dedukcyjnych* [Frege's Axiomatic System of the Sentential Variables in the Light of the Contemporary Methodology of Deductive Sciences]. Since 1957, he was head of the Chair of Methodology (the first one in Poland, founded in 1952 by J. Iwanicki) at KUL, associate professor at KUL since 1965, and full professor there since 1970. He was the dean of the Faculty of Philosophy at KUL for many years and laid the methodological foundations for the establishment of the "Lublin School of Philosophy".

II. Kamiński's main fields of interest were history of science and logic, general and special methodology, methodology of philosophy and (medieval) semiotics. He saw his main achievements in the domain of the theory of science and the methodology of classical philosophy, especially in the studies of the method and language of metaphysics. He gave a methodological description of general metaphysics, philosophical anthropology, ethics, philosophy of religion, philosophy of history and religion studies (religiology). He investigated the beginnings of the mathematical induction in the Middle Ages and in modern times, modern history of the theory of definition (Th. Hobbes, B. Pascal, J. Locke, E.B. de Condillac, J.D. Gergonne), the theory of argumentation (reasoning), the structure and the evolution of scientific theory, the deductive method (B. Pascal, G.W. Leibniz, G. Frege), and the achievements of logic and philosophy in Poland.

A characteristic feature of his philosophical and methodological approach was a special type of historicism, consisting of referring to the heritage of the past and at the same time to the latest achievements in logic and the philosophy of science. He looked to history for inspiration, for how to solve his own problems and for a partial confirmation of the legitimacy of his answers. He also used history to understand better the context of the problems under discussion. Kamiński was a kind of polyhistorian, an encyclopaedic mentality, with a tendency to notice every fact significant for the solution of some problems, interested in classifying and systematizing. His belief that a comprehensive study of a phenomenon has to take into account its genesis, structure and functions was of great didactic value.

III. Kamiński claimed to belong to two main philosophical traditions: to classical philosophy and to analytical philosophy, both in its scholastic model and in the model of the Lvov-Warsaw School. To the first tradition – the realistic theory of being and knowledge – he owed his philosophical and historical interests; to the second – his logical and methodological interests in science. In his philosophical and methodological evolution he went through three characteristic stages: from an anti-metaphysical attitude, where he was enchanted with the possibilities of formal logic, to a methodological and pro-metaphysical attitude, and in the end to a clearly philosophical and sapiential attitude. In the first period, he accepted and faithfully imitated the philosophy of science of the logical empiricism, which he later modified with Popperism, thereafter supported – in order to avoid scepticism – with intellectualism: a view that the person possesses an intellect (*intellectus primorum principiorum*) as a special cognitive faculty as well as discursive reason and senses.

Kamiński's research extended wide and deep also because of his philosophical provenance. Since he was interested in the problems of contemporary philosophy, he looked for inspiration and philosophical foundations for his studies of the conceptions of science in the ideal of classical philosophy inherited from Aristotle, with which he increasingly identified. In postulating practising philosophy within a methodological culture, he did it for the benefit of classical philosophy. He understood philosophy as a fundamental discipline whose aim was to detect the foundations of science and at the same time to lay foundations for scientific knowledge. He defended the autonomy of philosophy against empirical sciences and at the same time proclaimed the openness of philosophy to sciences, which form the heuristic basis of philosophy. It is clear that fallible results of science cannot be a foundation for a philosophy which attempts to supply the necessary knowledge as the classical philosophy wants to do.

Kamiński's methodological interests are characterized by a philosophical and historical approach. He had a broad concept of knowledge and was maximalist both in raising questions and in giving answers. He cultivated the ideal of rational knowledge. In accordance with classical philosophy, he saw the substance of the person as being *ens rationale,* a being realizing himself in a disinterested search for a theoretical truth, whose highest expression is philosophy. He stressed the epistemological and methodological plurality of knowledge and distinguished (with Kant) a material and formal part of knowledge, assuming that the formal element manifests itself as the logical form in the formal procedures of the (scientific) cognitive processes, that is in the (scientific) method and the (scientific) language. Making many attempts at a methodological characterization of different types of cognition and knowledge, he distinguished – besides common-sense knowledge that lies at the bottom of any other type of knowledge – scientific, philosophical and theological knowledge. None of them can be reduced to other types of knowledge, because each has its own problems, goals and methods. At the top is a kind of sapiential knowledge, which is much more than a simple generalization of all particular kinds of knowledge.

Believing that the task of methodology is to be helpful for all types of cognition, Kamiński was interested not only in the methodological (formal) dimension of science (in the scientific method, language and the structure of science) but also in the epistemological and cultural dimensions. In the spirit of the Lvov-Warsaw School, he emphasized the need for methodological reflection in every type of scientific work because every time questions are asked, definitions, distinctions, and divisions made, there is a need for arguments and reasoning. A rational, critical and responsible study of each type of knowledge is decisive for intersubjective meaningfulness and verifiability of scientific theories.

Kamiński was practising a form of the philosophy of scientific method. He claimed that the main task of methodological reflection was to study the (mainly non-evident) methodological and philosophical presuppositions of the scientific method, as well as to criticize any attempts of the ideological exploitation of science for purposes other than scientific. He emphasized the mutual dependence of philosophy and science but he did not explain that dependence in details. Not only does the concept of philosophy depend on the concept of science prevailing at a given time, but also the way that science is done is dependent on philosophy. For instance, the metaphysics ultimately explains the world (saying it is either monistic or pluralistic) and at the same time provides the ultimate foundations for scientific knowledge. Philosophy is present in science itself and in views (theories) about it. A scientific study free from assumptions is a methodological myth because behind every great scientific theory there is some philosophy largely unrecognized by the scientist. It is therefore impossible to study the philosophy of science without some philosophical preconception of science itself; it is impossible to examine the subject matter of science aphilosophically.

In the tradition of the Lvov-Warsaw School, Kamiński initially used the term *general methodology of science* to specify his own interests in science; later, under the influence of the Anglo-Saxon analytic philosophy, he abandoned it in favour of the term the *philosophy of science* or even the *theory of science*. Not always consistently, he distinguished the *methodology of science* (sometimes he also used the term *general methodology of science*) which has to study what is common to all sciences, from *methodologies of sciences* (from a *particular methodology of science*) which study what is specific to each of the particular sciences. He distinguished two principal concepts of the term *methodology of science*: 1) In a narrow sense (a somehow classical and somehow neopositivist concept of the sense of the *logical theory of science*), its methods consist mainly in the logical analysis as well as in the reconstruction of research activities and their codification. It aims at constructing general models of optimal (rational and efficient) scientific procedures and at their justification with respect to the goals of science. Thus, it is a descriptive, explanatory and practical discipline, because it justifies the reconstructed or projected scientific methods. 2) In a broad sense, it is the philosophy (theory) of science as the edifice of all philosophical, logical and humanistic interests in the phenomenon of science.

Kaminski also had difficulty in ordering the meanings of the term *philosophy of science*. He distinguished at least three (two metaobjective and one objective) meanings of this term: 1) in the broadest (scientistic) sense, the philosophy of science is located on the borderline between philosophy and science. It belongs simultaneously to philosophy, logic and metascience and can be either of strictly philosophical (epistemological) nature, of more humanistic-practical (as a *science of science*), or mainly of formal nature (as a *logical theory of science*). This kind of philosophy of science includes general methodology of science and logical semiotics, methodology of natural sciences and philosophy of nature, and the methodology of humanities and methodology of philosophy. 2) In a narrower and more proper (classical) sense, the term *philosophy of science* refers to the theory of science considered as an entity, to the theory of scientific knowledge (its origin, limits and value), and to a theory that ultimately explains science as a domain of culture (especially trying to determine the position of science in culture and its role in the contemporary world). 3) In the objective sense, the philosophy of science seeks to answer questions about what matter, time, space, structure, aim (*finis*), determinism, functionalism, etc., are.

For many reasons, Kamiński was convinced of the necessity and legitimacy of having many types of sciences of science. He claimed that this broad approach helped in understanding the meaning and place of science in culture (in understanding the scientistic-technical mentality and the cult of experts), in a theoretical explanation of the nature and foundations of scientific knowledge, in demonstrating the integrative function of methodological reflection for the unification of specialist sciences and for the cooperation of scientists, as well as in emphasizing the practical importance of the methodological awareness for various sciences. In the case of philosophy, methodological reflection is usually the only way to control philosophical theories.

Grouping various metasciences (sciences of science) according to the ontological point of view and to methods with which the sciences are studied, Kamiński arrived at a typology which was typical of him. He distinguished three main types of sciences on science: the humanistic (history, sociology, psychology, ethics, economy and politics of science), the philosophical (ontology, epistemology, the narrowly conceived philosophy of science, and the philosophy of culture), and the formal sciences (logic of scientific language, formal logic, theory of scientific argumentation and reasoning, and methodology of science). Admitting that the choice of terms: *methodology of science, philosophy of science, theory of science,* seems to be a matter of national custom or even convention, Kamiński stressed again and again that the study of the nature of science should take into account all aspects of science: the logical, humanistic and philosophical; however, the philosophical (epistemological) aspect of study is most important.

Kamiński's methodological interests, to which his historical interests are subordinated, were oriented mainly at the philosophical – and more precisely – the epistemological dimensions of science. He therefore postulated a philosophy of science as an autonomous discipline dealing with the ontological and epistemological foundations of science (sciences). He himself practised the philosophy (methodology) of science for the purpose of stating the external and internal presuppositions of science and understanding the various concepts and conceptions of science, "X-raying" them from the point of view of the classical philosophy, which defined the constant framework of all his methodological studies. Seeing the ambiguity (analogy) of main philosophical and methodological notions, S. Kamiński claimed that the task of a methodologist is, among other things, to detect this ambiguity and to evaluate in a possibly neutral, disinterested way the epistemic (formal) value of theses and theories put forward by various sciences and types of philosophy. Even if the methodologist *qua* methodologist does not give direct significant answers to questions put by scientists and philosophers, he can nevertheless, equipped with formal instruments delivered by the development of logic and methodology, study and disclose the presuppositions that lie at the bottom of the solutions of some scientific or philosophical problems, and he can evaluate the logical and methodological value of proposed scientific or philosophical theses and theories as well as the formal correctness of arguments given for their justification.

IV. Kamiński approached science from many angles and was interested in it at many levels: descriptive, explanatory, evaluative and normative. He emphasized the meaning of science as an important cultural and social fact. As an epistemic optimist, he believed in the cognitive possibilities of man and treated science as a natural prolongation of common-sense knowledge and as an eminent achievement of the human beings. Kamiński's methodological aims were partially normative, as he sought to develop evaluative standards of scientific knowledge. By examining the nature of science he hoped to establish what made the hard core of science. Against the background of classical philosophy as a background and its broad concept of knowledge as *epistéme* (so he knew *a priori* what a "good science" was) he sought to determine the methodological standards of scientific knowledge. He was prescriptive in his intentions as he recommended standards by which scientific knowledge should be evaluated. At the same time, Kamiński tried to uncover the methodological standards and procedures that actually form scientific practice. This is independent of the fact that the scientists in question may have been explicitly aware or unaware of them.

According to the classical approach, Kamiński was mainly interested in what was constant in the stormy and changing vicissitudes of the evolution of science and of the concept of science, and in what he termed, following classical philosophy, the nature of science. He stressed what was common to various sciences and not what set them apart. Rejecting T. Kuhn's thesis of paradigm incommensurability, he believed that even if there are many essentially different concepts of science, there nonetheless remaines between them a genetic identity, a bond between succeeding concepts, and a functional identity, that there is a permanence in function that they had in culture. The aim of science is always to conceptualize methodologically and to explain rationally what was acquired by experience, and to unite in a systematic and profound vision of reality the different acts of cognition, so that they could somehow serve people. Perceiving various links between philosophy and science, Kamiński warned against isolationism in science. As a philosopher, he wanted to find a remedy for the ongoing specialization in science, for the epistemic disintegration of sciences or rather for the disintegration of the scientific view of the world. Therefore, he stressed the unity of scientific knowledge while at the same time emphasizing the plurality and the methodological autonomy of each of the sciences.

His understanding of science indebted Kamiński to contemporary philosophy and to classical philosophy. He distinguished (at least) three concepts of science: 1) the broadest concept, in the classical, Platonic-Aristotelian sense of *epistéme*, i.e. well justified, empirical-rational knowledge, distinct from doxal knowledge; (empirical) sciences, philosophy and theology belong in this category as well; 2) a more narrow concept, referring only to real (empirical) sciences: the natural and human (social) sciences; 3) the narrowest concept, which he rejects because of its scientistic narrowness, where the concept of science is referred only to natural science or exclusively to physics.

The nature of science was determined by Kamiński from the point of view of its subject matter, aims, methods, logical structure and genesis. The question of what science is concerned with, i.e. what its subject matter is, is a philosophical question; it presupposes an appropriate understanding of the nature of the world. Kamiński opted for a pluralistic approach to the world: the principal object of science is the objective world but so are subjective states of man and products of his mind and his language. The best diagnostic test of the scientific character of science is the scientific method. Also here – rejecting the scientism (i.e. the view that the scientific method and knowledge is the model and the measure of each knowledge) – Kamiński takes a pluralistic attitude. Accepting that the scientific method is not simple, that there is not one uncomplicated ideal way of doing science, and that because of the multiplicity of questions and scientific aims it would be difficult to construct one universal scientific method as a uniform set of rules for every kind of science, Kamiński assumes that different subject matter and different goals of scientific cognition require different research strategies and types of cognitive procedures. One manifestation of Kamiński's methodological pluralism is his antinaturalistic position in the theory of the humanities where he supported the thesis of their methodological autonomy – with regard to natural sciences. Closer to the philosophical cognition and knowledge, the humanities do not meet and cannot meet the conditions imposed upon the natural sciences, as they differ from them in the subject matter (the world of culture) and hence in method (understanding, interpretation) used.

In the spirit of explanatory realism, Kamiński treats scientific knowledge realistically. The main task of science is to explain observed data. Scientific cognition is objective even if reality can be grasped not only by direct observation but also indirectly. Kamiński claimed that both the concepts and the statements specific to physics were neither simple empirical nor simple *a priori*. Fulfilling its cognitive aims, science connects the objective world and the subjective factors of cognition by its observational-theoretical language and therefore the scientific system (scientific theory) comes into being as if by a dialogue between the mind and the facts. Scientific cognition never consists in copying the "ready-made" world grasped by the senses but rather in interpretation in the form of a scientific theory.

V. The methodological aspect of science manifests itself in respecting methodological rules; the logical nature of science is secured by appropriate methods of formulation and justification of scientific statements and theories. Generally speaking, scientific statements and theories have to be obtained by explicitly given methods that are in accordance with the methodological rules that can be applied in a certain domain of knowledge, justified in an intersubjectively verified way, i.e. independently of any emotional-volitive states of the subject, and expressed in an intersubjectively meaningful language. Kamiński was a methodological purist and demarcationist; he demanded that the methodological autonomy of each science must be respected because this helps to control its results. Yet he was not an extreme demarcationist, for he believed that even the border between opposite types of knowledge was not obvious. Similarly, ascribing different methodological standards to philosophical and scientific knowledge, he conceded that in concrete cases it was difficult to indicate the borderline between philosophy and empirical science.

Kamiński's views on the place and function of the concept of truth in science underwent some evolution. He was persuaded that the search for truth cannot cease to be one of the main aims of science as truth was a fundamental value satisfying the person's natural intellectual desire to truly know the world. Kamiński always pleaded for the classical (correspondence) concept of truth according to which truth belongs directly only to statements (propositions). Scientific theories are true only indirectly. Initially, Kamiński accepted that the statements of empirical and theoretical physics are objectively true as well. In the course of time, he realized the controversial nature of the concept of truth in science in which a moderate active role of mind and the influence of a conceptual apparatus on scientific results were present. Moreover, the truth of scientific theses is neither the object nor the goal of physics, the truth is postulated but not guaranteed.

Kamiński paid a lot of attention to the category of rationality and to the role of rational factors in scientific knowledge; he detected them mainly in the theoretical dimension of science. After studying the criteria of rationality in the philosophy of science during the first three decades of our century, he discovered that rationality as a philosophical problem appeared only at the beginning of the twentieth century. He saw three traditional controversies in the history of philosophy: between rationalism and empiricism (apriorism and aposteriorism) as a controversy about the source of knowledge; between rationalism and intellectualism as a controversy about the cognitive faculties; and between rationalism and irrationalism as a controversy about valuable knowledge, i.e. its ultimate foundations.

Distinguishing metaphysical rationalism (a teleological order of the world) and epistemological rationalism (knowledge based on senses and reason but independent of emotional and volitional states of the subject), Kamiński connected the rationality of science with epistemological intellectualism: the view of the intellect as a third, along with senses and reason, intuitive, self-reflexive and by the same token self-controlling cognitive faculty, mediating, among other things, the dialogue between empirical and theoretical factors in science. He believed that intellectualism overcame the difficulties connected with the search for the principle of induction and its justification. If empiricism is not supported by intellectualism, science remains a set of disconnected data ("facts"). Rationalism alone, when it minimizes the cognitive role of senses and experience, leads to idealism. For this reason, Kamiński rejected both narrow inductivism of the positivists, for it did not allow general knowledge and Popperian fallibilism and probabilism, for they ascribed a provisional character to scientific knowledge.

Science, regardless of its manifold limitations, remained for him an ideal of rational knowledge and he always believed at heart that the aim of science is the true knowledge and that of philosophy is the true necessary knowledge.

As the leading function of truth in science was weakening, Kamiński opted for a kind of restrained cumulativism which does not presuppose a strict formal correspondence between the succeeding conceptions of science or between scientific theories but accepts a continuity of the scientific tradition, which manifests itself mainly in the objective goal of science: in searching for the most general and fundamental aspects of the reality, the goal which unifies science and allows it to explain and to understand that reality.

VI. Kamiński proposed an innovative understanding of the history of the concept of science in its historical modifications. His main work *Nauka i metoda. Pojęcie nauki i klasyfikacja nauk* [Science and Method. Concept of Science and Classification of Sciences], can be seen as an original attempt of a paradigmatic, more philosophical than historical and sociological, presentation of different conceptions of science.

He distinguished four fundamental transformations in the conception of science during its history, determined by Aristotle, Galileo Galilei, A. Comte and K. R. Popper; four great conceptions (paradigms) of science are connected with their names: the classic, modern, positivistic and Popperian. The classical conception of unified science and uniform knowledge has its origin in the antiquity and was mainly the work of Aristotle. It is the conception of science as an intuitive-deductive cognition (philosophic-scientific) with the aim to grasp the essence of things (genetic empiricism and methodological intellectualism and rationalism). In the 17th century (Galileo and Newton), as a result of gradual emancipation of empirical sciences and development of mathematics, the conception of knowledge as a deductive theory which was empirically tested, was created. The third conception of scientific knowledge was born mainly thanks to the positivists (A. Comte). Science should refer exclusively to the phenomena grasped by experience. The aim of science is to seize inductively the regularities of events and to test them empirically. Science is not interested in the question "why?" but limits its interest to the question "how?". Valuable knowledge is only that which makes life easier and satisfies the basic needs of man. The fourth conception of science is elaborated by K. R. Popper. The starting point of science are not mechanically assembled and dogmatically treated observations but creatively understood problems that are born on the basis of all existing knowledge (there is no dichotomy between the empirical and theoretical part of knowledge). What is characteristic of scientific activity are bold general hypotheses tested and falsified by experience. The task of science is to find satisfactory explanations and to approach truth although not giving true theories about the world. The actual science is one of the possible sciences as a means to explain the phenomena better and better.

Even if Kamiński did not creatively practise contemporary formal logic, he had some ideas on the modernization of the traditional (Aristotelian) logic of syllogism, on generalization of the laws of the logical square (figure of oppositions) in the language of the propositional calculus to all laws of the direct conclusion, on rules of syllogism for sentences with the negation of their subject, on showing the discrepancy between the traditional and modern quantification, on the role (in logic) of the so-called first principles of thinking. He also studied some philosophical implications and consequences of Gödel's theorem. He believed that the whole formal logic is a philosophical discipline in the broad sense of the term *philosophy* because of ' its general and speculative as well as apodeictic nature and an universal use of its results. In his works on logical semantics, S. Kamiński was concerned with a definition of formal logic, a function of operators in logic and common-sense language, linguistic fallacies, a conception of logical error and the systematization of typical logical errors and fallacies.

SELECTED BIBLIOGRAPHY

1. Books:
1. (1958). *Georgonne'a teoria definicji* [Georgonne's Theory of Definition]. Lublin: TNKUL.
2. (1961). *Pojęcie nauki i klasyfikacja nauk* [The Concept of Science and the Classification of Sciences]. Lublin: TNKUL.
3. (1962). [with M. A. Krąpiec]. *Z teorii i metodologii metafizyki* [From the Theory and Methodology of Metaphysics]. Lublin: TNKUL.
4. (1989). *Dzieła zebrane.* T. 1: *Jak filozofować? Studia z metodologii filozofii klasycznej* [Collected Papers. Vol. 1: How to Philosophize? Studies in Methodology of Classical Philosophy], ed. by T. Szubka. Lublin: TNKUL.
5. (1993). *Dzieła zebrane.* T. 2: *Filozofia i metoda. Studia z dziejów metod filozofowania* [Collected Papers. Vol. 2: Philosophy and Method. Studies from the History of the Method of Philosophizing], ed. by J. Herbut. Lublin: TNKUL.
6. (1994). *Dzieła zebrane.* T. 3: *Metoda i język. Studia z semiotyki i metodologii nauk* [Collected Papers. Vol. 3: Method and Language. Studies from Semiotics and Philosophy of Science], ed. by U. Żegleń. Lublin: TNKUL.
7. (1992). *Dzieła zebrane.* T. 4: *Nauka i metoda. Pojęcie nauki i klasyfikacja nauk* [Collected Papers. Vol. 4: Science and Method. Concept of Science and Classification of Sciences], ed. by A. Bronk. Lublin: TNKUL.
8. (1998). *Dzieła zebrane.* T. 5: *Światopogląd – Religia – Teologia* [Collected Papers. Vol. 5: Worldview – Religion – Theology], ed. by M. Walczak and A. Bronk. Lublin: TNKUL.

Poznań Studies in the Philosophy
of the Sciences and the Humanities
2001, vol. 74, pp. 153-161

Mieczysław Omyła

ROMAN SUSZKO – FROM DIACHRONIC LOGIC
TO NON-FREGEAN LOGIC

I. Roman Suszko (9.11.1919, Podobora – 3.06.1979, Warsaw) was one of the most fascinating personalities in Polish academic community after the Second World War and one of the most outstanding logicians of the time. He was above all a scientist but he also participated in academic life. He was Dean of the Faculty of Philosophy at Warsaw University for two terms of office. He studied abstract problems of logic, but also played a part in the satirical film *Rejs* [The Cruise] directed by M. Piwowski.

Suszko was involved in various scientific problems, for example: logical syntax of natural language, liar antynomy, logical probability; but two of his achievements have the greatest value for philosophy of science, i.e., diachronic logic and non-Fregean logic.

Like Ferdinand de Saussure, who, in his monograph *Cours de Linguistique Generale* (Lausanne 1916), made a distinction between synchronic and diachronic linguistics, Suszko draws a distinction between synchronic and diachronic formal logic. Diachronic formal logic was for Suszko the application of the models theory to formalized languages in order to describe the abstract structure of the development of knowledge. Non-Fregean logic, on the other hand, was a term used by Suszko to refer to classical logic enriched by identity connective and quantifiers binding sentential variables. The identity connective joins two sentences into a true sentence when sentences describe the same situation. It turns out that non-Fregean logic is such a general logical calculus that the classical predicate calculus, classical sentential calculus, Łukasiewicz's finitely-many-valued logics and some modal systems are particular cases of non-Fregean logic. Because of some interpretational difficulties concerning the notion of situation, the logic has not gained among logicians the regard which it deserves.

II. Suszko began his studies in 1937 at the Poznań University. During the Nazi occupation, he lived in Cracow where he worked in the municipal power station and simultaneously attended the clandestine study classes organized by the professors of the Jagiellonian University. He studied logic under the supervision of Professor Zygmunt Zawirski. Between 1946 and 1953 he worked at Poznań University in the Department of Theory and Methodology of Science headed by Kazimierz Ajdukiewicz.

In 1948, he was awarded a doctoral degree for his dissertation *O systemach normalnych i pewnych zagadnieniach logiki elementarnej* [On Normal Systems and Some Problems of Elementary Logic] written under the supervision of Kazimierz Ajdukiewicz. In 1951, he qualified as assistant professor for his postdoctoral dissertation *Canonic Axiomatic Systems.*

In 1953, he moved to the Faculty of Philosophy at Warsaw University, to the Department of Logic. From the beginning of his work in Warsaw he had close contacts with the Polish Academy of Arts and Sciences, first with the Group of Algebra headed by Jerzy Łoś, and then with the Group of Logic of the Institute of Philosophy and Sociology. In 1957, based on the law of the time, he was awarded the degree of candidate of philosophical sciences at the Department of Philosophy of Warsaw University for his thesis "Logika formalna a niektóre zagadnienia teorii poznania. Diachroniczna logika formalna" [Formal Logic and Some Problems of Theory of Knowledge. Diachronic Formal Logic] [2].

In 1958, he obtained Ford's grant to conduct his research in the USA. In 1959, he qualified as professor. In 1966 he quit the post at Warsaw University and moved to the Institute of Philosophy and Sociology of the Polish Academy of Arts and Sciences, where he headed the Group of Logic from 1964 to 1967. From 1967 to 1969 and from 1970 to 1973 he gave lectures in the Stevens Institute of Technology in Hoboken, New Jersey. On returning to Poland, he worked in the Group of Logic mentioned above. During his entire academic career he was a member of editorial boards of several scientific periodicals, including *Studia Logica.*

III. In his research, Suszko concentrated on the study of the logical calculi, the set theory and the model theory. His logical work was closely connected with the classical problems of philosophy as Suszko attempted to express some philosophical ideas such as some assumptions concerning the development of science and some concerning Wittgenstein's ontology, using a precise, formalized language.

There were two most creative periods in Suszko's research. The first was the period directly after his arrival in Warsaw, when he conducted research on the model theory together with Łoś. The other was the so-called *non-Fregean period.*

In the 1950s, Łoś and Suszko obtained a number of valuable results concerning the extension of models for first-order predicate language and their common work "Remarks on Sentential Logic" contributed greatly to the development of Polish research on sentential logic. Ten years after publishing it, the authors introduced many important logical notions in the above-mentioned dissertation, such as the notion of the structural operator of consequence and the notion of matrix strongly relevant to a given sentential logic. That work is still quoted in almost all publications on sentential calculi. The problems brought forward in the work were solved a relatively short time ago.

IV. In the late 1950s and early 1960s, Suszko wrote many articles concerning the applications of logic, and, first of all, the model theory to formulate and describe strictly the problems concerning the development of knowledge. His 1950 contribution mentioned above was the first work in the world in which the model theory of formalized languages was applied to the analysis of the development of knowledge, and one of the first works in which the methods of the model theory were used to examine the problems outside mathematics; in this case it was a description of philosophical problems. This work applied the methods of the model theory to methodological research, which has been productively developed later by other logicians.

For a description of development of knowledge, Suszko considers a sequence of epistemic oppositions: E, E^*, E^{**}...; each of these oppositions is a pair $E = \langle S, O \rangle$, where S is the subject or the mind and O is the object or the world. The subject S is understood by Suszko as pure mind, i.e. Suszko put aside all extracognitive, for example pragmatic, relations between the subject and the world. The main component of the subject S is the formalized language L. Moreover, the subject is also equipped with logical and extralogical principles of thinking. The logical principles of thinking are marked, first of all, by the operator of consequence Cn defined on the language L, while the extralogical principles of thinking are identifiedy, in the first place, with the set of analytic axioms A. The subject also accepts a certain set of sentences T, as theorems about the world. It is obvious that $Cn(A) \subseteq T$.

The subject in Suszko's system is, therefore, represented by the scheme: $S = \langle (L, Cn), A, T \rangle$ while the object of knowledge O, i.e., a certain fragment of the world, is represented in epistemic opposition E by a certain model M of the language L, which is a system of the type $(U, R_1, R_2, ..., R_n)$ where $R_1, R_2, ..., R_n$ are denotations of all extralogical constants of the language L; they are, therefore, some singled out objects in set U, some features of objects from set U and some relations among these objects. With $Ver(M)$, it is marked the set of true sentences of the language L in model M. The development of knowledge, according to Suszko's accurate remark, consists in acquiring more and more truths in a process of learning about the world.

Suszko distinguishes two essential types of the development of knowledge: evolutionary, when the object of knowledge does not change, and revolutionary, when the object expands. Evolutionary development of knowledge is based, first of all, on the fact that the sequence of sets of sentences accepted by the subject: T, T^*, T^{**}, includes more and more true sentences about the same object. In the course of evolutionary development of knowledge, it also happens that some extralogical principles of thinking change, i.e., there is a change of axioms from A to A^*. Suszko considers the following two cases of this kind:

(1) $A \neq A^*$ and $Cn(A) = Cn(A^*)$ (axioms systematization)
(2) $A \subset A^*$ and $Cn(A) \neq Cn(A^*)$ (axioms reinforcement)

The case (2) includes, likewise, the following case

$$(L, T, A) \ / \ (L, T^*, A^*) \ / \ (L, T^{**}, A^{**})$$

A certain sentence α of the language L, originally is not considered true, i.e., $\alpha \notin T$, and the development of knowledge continues so that α becomes first a non-analytic statement, i.e. $\alpha \in (T^* - Cn(A^*))$, and in the next stage of knowledge it is included in the set of analytic sentences of the language, i.e. $\alpha \in Cn(A^{**})$. This development of knowledge was, according to Suszko, noticed by conventionalists. Evolutionary development of knowledge is what T. S. Kuhn calls the stage of *normal science* in his book *The Structure of Scientific Revolutions* (Chicago 1962). Revolutionary development of knowledge, on the other hand, is a change in a object of knowledge represented here by the model of language. The change is: M/M^*. Suszko describes two types of such changes:

(1) The universe of objects U does not change, but new features and relations of objects from this universe are discovered and consequently a new model M^* is the extension of model M, i.e.:

$$M = (U, R_1, R_2,..., R_n) \ / \ M^* = \langle U, R_1, R_2,..., R_n, Q_1, Q_2,..., Q_n \rangle$$

Such an extension of knowledge results in the extension of language L to language L^* corresponding to model M^*.

(2) The universe of the model $M = (U, R_1, R_2,..., R_n)$ extends and $M^* = (U^*, R_1^*, R_2^*,..., R_n^*)$ where $U \subset U^*$, and $U \neq U^*$ and relations R_i^* for $i = 1, 2,..., n$ are extensions of corresponding relations R_i. Model M is then a submodel of model M^*.

An epoch in development of knowledge is, according to Suszko, a period of time when there occurs only evolutionary development of knowledge, i.e. the period in which the object of knowledge does not change and so the language used by the subject seeking knowledge does not change either. If the change of epistemic oppositions

$$\langle (L_t,\ Cn),\ A_t,\ T_t,\ M_t \rangle\ /\ \langle (L_t',\ Cn),\ A_t',\ T_t',\ M_t' \rangle$$

is that $M_t \neq M_t'$ and M_t is a submodel of M_t' or M_t' is an extension of M_t, then t, t' are different epochs in development of knowledge.

While discussing the problems of the development of knowledge, Suszko uses a suggestive terminology introduced by Ajdukiewicz in his work: "Das Weltbild und die Begriffsapparatur" (*Erkenntnis* **4**, 1934), changing its sense only a little.

Language L is a notional apparatus built upon model M_t and model M_t is called by Suszko the layer of the world which was reached by science in the epoch t. The set T_t, in turn, is *a picture of the world in epoch t*.

Thus, *the world-perspective in epoch t* is a family of all theories:

$$T_t = \{T^i_t : T_t \subset T^i_t\}$$

Set $T_t \cap Ver(M_t)$ is called by Suszko the real knowledge of the world in epoch t; similarly, he defines potential knowledge of the world in epoch t as a family of sets of sentences: $Ver(M_t) \cap T^i_t$, where the sets T^i_t are all theories belonging to the world-perspective in epoch t.

According to Suszko, for a given subject, only these fragments of the real world which are included into the model of language L, which the subject is equipped with, can be the object of the knowledge.

Originally, Suszko identified the object of knowledge with the intended model M of language L, and later with the semantic closure of the model M which arises by "enlarging the model M by adjoining to it all extensions (generalized denotations) which are definable in language L with respect to considering the model M".

Intuitively speaking, the model M of language L contains, apart from the universe of the model, denotations of all simple extralogical symbols. The model is, therefore, a set-theoretical structure type: $(U,\ d(C_1),\ d(C_2),...)$, where U is the universe of language and $d(C_k)$ is a denotation of an extralogical constant C_k occurring explicitly in the vocabulary of the language L. The semantic closure of the model M includes all set-theoretical formations which can be defined on the set U with the use of any expressions of language L. In the semantic closure of the model M, nominal expressions are assigned to corresponding elements of the universe U of the model M whereas sentential formulas with one nominal variable are assigned to the sets of objects which satisfy a given formula. If there are no free nominal variables in a sentential formula, then this formula is either satisfied by each element from the set U and then it is true, or it is satisfied by no element and then the sentence is false. Therefore, in the semantic closure of the model for the predicate first-order language there are two generalized denotations for sentences and all true sentences have the same generalized denotation, and all false sentences also have one common denotation. Generalized denotation of a true sentence is the universe U of model M, while generalized denotation of a false sentence is the empty set $\varnothing$. The assumption that all true sentences have one common denotation and that all false sentences have one common denotation, likewise, was termed by Suszko the Fregean axiom. The model theory for predicate languages is termed by Suszko Fregean model theory.

V. In 1966, via Bogusław Wolniewicz's monograph *Rzeczy i fakty. Wstęp do pierwszej filozofii Wittgensteina* [Things and Facts, Introduction to the First Philosophy of Wittgenstein], Suszko became acquainted with Wittgenstein's metaphysical views included in *Tractatus Logico-Philosophicus*. Since that moment, a new, so-called non-Fregean period in Suszko's work began. In his *Tractatus*, Wittgenstein expressed his belief that names designate objects and sentences describe situations. According to Suszko, in order to formulate adequate statements concerning objects as well as situations in a formalized language, there must be two types of variables in that language: nominal variables running through the universe of objects and sentential variables running through the universe of situations.

Then, according to the famous principle *No entity without identity*, Suszko introduced into the language of classical logic the identity connective which is used to mark that two sentences describe the same situation. Such logical calculus was termed by Suszko non-Fregean logic. The name "non-Fregean logic" originates from the fact that it is not an assumption of this logic that the universe of sentential variables, i.e. the universe of situations is two-elements set. Non-Fregean logic is the most general extensional and logically two-valued logical calculus. Its logical two-valuation consists in that each sentence of the language for which this logic is valid, for all interpretations allowed by this logic, is either true or false. Particularly, the assumptions of this logic are classical laws: the law of the excluded middle and the law of non-contradiction. Extensionality, on the other hand, consists in that what any complex expression relates to is marked by the particular constituent expressions of this complex expression. The generality of this logic consists among others in that the principles of that logic do not put any limits on universes, and that the universe of nominal variables is not empty and the universe of sentential variables has at least two elements. The principles of non-Fregean logic require only consequent application of symbols not leading to non-contradiction, while they do not require us to assume any equality which does not result from previously assumed identities.

This logic has interpretations by which it is entirely characterized from the formal point of view and which may be applied to constructing the models of the theory of situations.

In my opinion, the formalization of all ontologies assigning sentences semantic correlates that are different from their logical values and not changing, in any essential aspect, the intuition that they are the correlates of sentences and not of names, requires non-Fregean logic or, at least, a theory based on it. However, non-Fregean logic is a logical calculus and – just as any logical calculus – it can be developed independently of its origins and philosophical motivations and without making any assumptions concerning reality. Non-Fregean logic does not establish univocal meanings of such terms as "situation" and "object", just as geometry does not establish univocal meanings of such terms as "point" or "straight" line and despite it geometry is applied to the description of the world. What is the universe of nominal and sentential values depends on the application. Similarly, in geometry or the standard models theory, it is the particular application that determines what the point or the object is.

The non-Fregean period was the most creative time in Suszko's life: during that period he wrote 36 scientific papers, all of which concerned non-Fregean logic (compared to the total of 85 publications which he wrote in his lifetime), he supervised 7 doctoral dissertations during that time, 5 of which concerned non-Fregean logic.

VI. In Suszko's logical papers there are many remarks concerning the philosophy of language and science, particularly many general views on logic. We will introduce those which, it seems, inspired his work on formal logic.

According to Suszko, the subject of logical studies is constituted by all the notional structures arising from the knowledge of the world. The whole of the structures he terms *logical material*. The material is of linguistic nature and is disseminated as scientific papers, philosophical treatises, and more casually as discussion and lectures. The state of logical studies at a given time is determined to a large extent by the logical material known at the time and by accessible research equipment. It is so because logical structures must be based on the structures originating as a result of direct knowledge of the world. For Suszko, logic is a science which is closely connected with the theory of knowledge, and in some of his papers he terms it a part of the epistemology or even formal epistemology. According to Suszko, the source of the intersubjective sense of expressions of any language is the reference of the language to corresponding fragments of the reality. And so, understanding the expressions is secondary in comparison to semantic relations occurring between the language and the objective domain R. In order to study semantic relations of the language, one must formalize it, and put the reality into certain set-theoretical framework. In logic, theory of sets and relations is usually assumed as formal theory of reality. Suszko even termed the set theory formal ontology. Formalized languages examined in logic are, with regard to their syntax and semantics, more or less exact reflections of fragments of natural languages or the languages of particular sciences, or they are hypothetical assumptions concerning these languages. Logical research shows that all ethnic and scientific languages have some common syntactic structure. Formal logic extracts from the variety of natural languages and particular scientific branches their common syntactic structure as well as a framework of semantic reference common for all the languages.

As a hypothesis which could explain the existence of these structures Suszko assumes in [2] that there is a structural syntactic framework, by means of which consciousness can grasp reality. This framework has been determined by the *surface of the world*, as Suszko calls it. The surface of the world is anything that has been an objective correlate of discursive human consciousness, emerging from the remotest past.

> It consists of a universe whose elements are things, not too big and not too small, persisting in the spatio-temporal environment of the primitive man, as well as of a characterization including simple external features of those things and the relation between them. Upon this model the primitive conceptual apparatus has been built, whose syntactic structure mirrors the ontological structure of the surface of the world [2, p.84].

In science, it happens that an accurate description of certain problems goes beyond the potentiality of natural language. For example, the notion of continuity could not be precisely expressed until the quantifiers were introduced. According to the standards of contemporary logic, in a given formalized language we consider as many categories of beings as many types of variables there are in a given language. In a language in which there appear only various non-sentential variables, we cannot, in turn, formulate philosophical statements concerning the world as a whole. According to Suszko, the invention in the history of human thought of such conceptions as Fregean sentential logic, Leśniewski's protothetics and Wittgenstein's *Tractatus* are important, among others, as their formalized versions require languages with sentential variables contrary to theories known from mathematics and other sciences, which require only various nominal variables. Suszko, while inventing non-Fregean logic, had apparent semantic and philosophical reasons, two of which are, in my opinion, most important:

(1) the conviction that reality should be regarded not only as a universe of objects possessing certain features and connected with certain relations, but, for more complete description of the world, reality should be also regarded as a universe of situations some of which at least are describable by the sentences of a certain language;

(2) ontological propositions can be divided into two types: a) those which are manifested in logical syntax and semantics of a language, particularly in relationships of logical consequence holding in a given language, b) those which are explicitly expressed in a object-language and which are ontological propositions that are not of metatheoretical nature. According to Suszko, in the language of non-Fregean logic we can precisely formulate theorems concerning the world perceived after Wittgenstein as the whole of facts.

SELECTED BIBLIOGRAPHY

1. (1955). [with J. Łoś]. On the Extending of Models (II). Common Extensions. *Fundamenta Mathematicae* **42**, 343-347.
2. (1957). Logika formalna a niektóre zagadnienia teorii poznania. Diachroniczna logika formalna [Formal Logic and Some Problems of the Theory of Knowledge. Diachronic Formal Logic]. *Myśl Filozoficzna* 3(29), 34-67.
3. (1958). [with J. Łoś]. Remarks on Sententional Logics. *Indigationes Mathematicae* **20**, 177-183.
4. (1967). An Abstract Scheme of the Development of Knowledge. *Actes du XI Congrès Internnational d'Histoire des Sciences, Varsovie-Cracovie*, 24-31 août 1965, Wrocław: Ossolineum, pp. 52-55.
5. (1967). An essay in the Formal Theory of Extension and of Intension. *Studia Logica* **20**, 7-36.
6. (1968). Formal Logic and the Development of Knowledge. In: I. Lakatos and A. Musgrave (Eds.). *Problems in Philosophy of Science. Proceedings of the International Colloquium in the Philosophy of Science, London 1965.* Vol. **1.** Amsterdam: North-Holland, 210-222.
7. (1968). Ontology in the *Tractatus* of L. Wittgenstein. *Notre Dame Journal of Formal Logic* **9**, 7-35.
8. (1975). Abolition of the Fregean Axiom. *Logic Colloquium*, ed. by R. Parikh, *Lecture Notesi in Mathematics* **453**. Springer Verlag 1975, 169-239.
9. (1977). The Fregean Axiom and Polish Mathematical Logic in the 1920s. *Studia Logica* **36**, 376-380.
10. (1994). The Reification of Situations. In: J. Woleński (Eds.). *Philosophical Logic in Poland.* Dordrecht: Kluwer Academic Publishers, 247-270.

Poznań Studies in the Philosophy
of the Sciences and the Humanities
2001, vol. 74, pp. 163-168

Anna Jedynak

HALINA MORTIMER – THE LOGIC OF INDUCTION

I. Halina Mortimer (11.04.1921, Warsaw – 8.05.1984, Warsaw) attained her BA degree just before the Second World War and then, during the war and for several years afterwards, worked as a clerk at various institutions. She studied law and philosophy at Warsaw University, attending Kazimierz Ajdukiewicz's lectures. In 1953, she completed her studies of law and in 1955 attained a Master's degree in philosophy. She started working in the Polish Academy of Arts and Sciences and since 1958 – at Warsaw University, Institute of Philosophy, Department of Logic. In 1963 she attained her doctoral degree based on the dissertation "Definicja probabilistyczna na przykładzie genotypu" [Probabilistic Definition Exemplified by the Definition of the Genotype]. She frequently participated in various conferences and symposia. She was periodically Vice-Dean and Deputy Head of the Institute.

II. Mortimer was interested in the logic of induction. Her scientific work developed in two directions. On the one hand, she was an expert in the logic of induction, systematizing and popularizing it. Her main work *Logika indukcji* (*The Logic of Induction* in the English translation) shows the main problems, controversies and directions of development in that domain. Her task was not only to present different views, but also to appraise them. On the other hand, her profound knowledge of the subject in question, including shortcomings of the views held by many methodologists, inspired her to propose her own resolutions of some problems. *Logika indukcji* is not only a historical work; it is also a critical appraisal of inductive logic and a proposal of some improvements. The ideas of those improvements were developed in some other her papers.

III. Let us see how Mortimer presented the logic of induction in her main study. The aim of inductive logic is to reconstruct such rules of inference, which are non-deductive, but rational, i.e. in a sense methodologically correct.

1. The very possibility of building a logic of induction is problematic. Some philosophers question this possibility.

a. D. Hume initiated a pessimistic view that no such rules could be reconstructed; we can justify induction on a psychological level, but not on a logical one.

b. Nowadays the views of P. Strawson and S. Toulmin converge with those of Hume: they find induction an informal reasoning and claim that we can describe inductive procedures, but we cannot submit them to any norms.

c. K. Popper claimed that it is not induction which is used in science, but a hypothetical-deductive method, and thus no logic of induction is needed at all. Mortimer showed that in fact Popper's views were not as far from inductionism as he was convinced they were, because both for Popper and for the inductionists induction based on probability was not a tool of scientific discovery, and procedures of testing hypotheses are similar in both conceptions.

2. The attempts at reconstructing inductive reasonings are based on the notion of probability, because for such reasoning, although experimental evidence does not guarantee the truth of the conclusion, it makes it more probable. Various attempts of building a logic of induction are based on different notions of probability, for there are different views on what probability is, which features of objects it is applied to and how to measure it.

a. According to statistical interpretation, the probability of a given sentence is a relative frequency of true sentences in a set of sentences of the same type as the given one (H. Reichenbach). What can be used to estimate the methodological correctness of a scheme of inductive inference is a relative frequency of true conclusions in a set of inferences represented by the scheme in question (K. Ajdukiewicz). Mortimer claimed that induction based on thus interpreted probability fell into a vicious circle, for calculation of a relative frequency is itself based on an inductive procedure.

b. The logic of induction based on logical probability refers to the degree of confirmation of some sentences by other ones, especially the confirmation of hypotheses by experimental evidence. The degree of confirmation can be expressed only comparatively (J. M. Keynes) or as a number, according to an accepted method of calculating the conditional probability of some sentences with respect to other ones (R. Carnap, J. Hintikka). The exponents of this line aim to keep the measure of confirmation in accordance with an intuitive estimation of inductive inferences, but they find it difficult to determinate any natural criterion of calculating such a measure; such criteria are always in a sense arbitrary. R. Carnap was the first philosopher who devised a method of calculating the measure of confirmation. However, his method is not free of shortcomings: it can only be applied to very simple languages; in the case of an unlimited universum, the degree of confirmation of any general sentence by experimental evidence always equals zero; and the very degree of confirmation of non-general sentences depends on the choice of a parameter: the higher the latter, the higher our tendency to generalize the results of experiments. J. Hintikka modified Carnap's theory, so that general sentences were confirmed in a non-zero degree. The convictions based on such an inductive logic are submitted to graduation. According to Carnap, the only task of the logic in question is to determine the rules of confirmation, and not to decide which sentences should be accepted.

c. Induction based on subjective probability (with respect to a person) is interested in determining the degree of a justified belief that given sentences are true (F. Ramsey, R. Jeffrey). On the grounds of this notion of probability unconditional or *a priori* probability of sentences makes sense. Thus, Bayes's theorem, which presupposes that the knowledge of an *a priori* probability of hypotheses is given, can be used to calculate the degree of confirmation of hypotheses by experimental evidence. This concept of induction is used in the theory of making decisions. The subjectivity of probability is limited only by the requirement that a function of beliefs of any person should fulfil the formal characteristics of probability and that new experience should modify those beliefs. Mortimer claimed that these foundations were too weak for inductive logic to be based on them. Another shortcoming of this concept is the requirement that everybody should be able to express the degree of own beliefs as numbers.

IV. The theory of subjective probability uses the quantitative notion of the acceptance of sentences: sentences can be more or less accepted. However, the logic of induction uses also the qualitative notion of acceptance of sentences. The latter notion is troublesome: is it at all possible to speak about the acceptance of sentences, the truth of which we are not definitely convinced of? Representatives of inductive logic answer this question in different ways, which however always involve some paradoxes.

1. One of the directions of developing the logic of induction is to formulate rules of acceptance (*resp.* rejection, *resp.* no acceptance and no rejection) of hypotheses, depending on relations between them and experimental evidence. Mortimer shows that some quite intuitive rules of this type entail paradoxes. According to the probabilistic rule of acceptance a hypothesis should be accepted always and only if its probability with respect to the premises is higher than an established degree. If we assume C. Hempel's postulates concerning the mutual consistence within a set of beliefs and the acceptance of the consequences of a formerly accepted set of sentences, this rule entails the so-called paradox of lottery (we are ready to admit that one of the lots will gain, but

being asked about each of them separately, we never express our belief that this particular lot will gain, so, consequently, we reject the conjunction of sentences predicting about each particular lot that it will gain). Mortimer presents various ways of avoiding this paradox:

a. One way is to weaken Hempel's postulates as suitable for deductive systems, but not for empirical science (H. Kyburg).

b. The second way is to strengthen the probabilistic rule of acceptance (I. Lévi, R. Hilpinen, J. Hintikka). If the rule is strengthened, fewer hypotheses should be accepted, and especially those which lead to paradox should not be accepted. Mortimer analyzed such modified rules and pointed out their shortcomings: for instance, given Lévi's rule, the acceptance of a hypothesis depends on whether concurring hypotheses are expressed separately, as many possibilities, or rather conjointly, as one alternative possibility.

2. The representatives of the theory of confirmation (e.g. R. Carnap) did not take up the problem of acceptance of sentences on the ground of inductive logic, for they assumed that acceptance of inductive conclusions is never rational. A high degree of confirmation of a given hypothesis justifies only a high degree of our belief in it, but not our acceptance of it. According to the behaviourist interpretation of induction (J. Neyman, R. Carnap) inductive procedures do not involve inferences at all; they do involve inductive behaviour, that is decisions of acting in a particular way with no certain knowledge about the situation. Controversies on the ground of the theory of confirmation involve mainly the measure of confirmation. The theory leads to many paradoxes. Mortimer described them and presented various ways of avoiding them. She analyzed these views, estimated and supplemented them, if needed. The paradoxes are as follows:

a. Sentences which are logically equivalent should be confirmed by the same observational sentences. The sentences "Every raven is black" and "Every object which is non-black is non-raven" are logically equivalent. The first one is confirmed by cases of black ravens and the second by cases of objects, which are non-black and non-ravens. But, since the sentences are logically equivalent, the latter cases should confirm the first sentence as well. However, this is paradoxical.

b. It is not intuitive that some observational data confirm mutually inconsistent hypotheses, especially if one of them contains unpredictable predicates, e.g., "blue in the 22nd century". Cases of green emeralds confirm both the sentence „All emeralds are green" and the sentence "All emeralds are green till the end of the 21st century and blue in the 22nd century".

c. It is assumed that every sentence confirms to a certain degree its reasons as well as its outcomes. Thus H confirms its reason H and G, and the latter confirms its outcome G. So every sentence confirms every other sentence.

d. Let us assume that we know the probability of A on the condition that B, the probability of A on the condition that C (which is not equal to the former probability), and we know that both B and C hold. When the condition is fulfilled, the absolute probability takes the measure of the conditional one. Thus, the absolute probability of A should take two different measures of conditional probabilities (with respect to B and C).

V. This is the way in which Mortimer systematized and presented the large area of the logic of induction. Let us now present some ideas of her own.

1. She modified the rule of the acceptance of inductive conclusions, stated by Hintikka and Hilpinen. She proved that the rule fulfilled Hempel's postulates at the cost of some limitations: namely, the rule could be applied only to such functions of confirmations, in case of which: a. an unlimited growth of the number of observations entailed approximation of the confirmation of the hypothesis to 1; b. the number of observations really grew unlimitedly. Mortimer modified the rule in order to expand its scope. Her rule is as follows: A hypothesis should be accepted always and only if its confirmation on the ground of the given experiment is higher than an established degree and the hypothesis is consistent with all world descriptions (expressed in terms of emptiness or non-emptiness of each predicate), consistent with the experiment.

2. Mortimer dealt with the problem of the analytic character of some theorems in natural sciences. She distinguished main variants of the problem in question. Two of them (namely: "Are some physical theorems terminological postulates?" and "Is it suitable to treat them as postulates?") were, in her opinion, insufficiently discussed and no answer was justified. The third one: "If they are interpreted as postulates, can they ever be falsified by experimental evidence?", she resolved in the affirmative. She developed Ajdukiewicz's view, supporting it by examples drawn from scientific practice.

3. Mortimer examined the role of probabilistic definitions in science, that is, definitions taking the form: "An object x has the defined theoretical property T always and only if the probability of the property Q in the set of all objects being in the relation R to x equals the number p". She presented a probabilistic definition of genotype on the ground of a formally reconstructed fragment of Mendel's genetics. She proved that an adequate empirical explanation of the concept of genotype requires a probabilistic definition, which follows from this very genetic theory. She also proved that other definitions of genotype led to contradiction in genetics.

4. She examined the conditions for the acceptance of probabilistic postulates – first for two of them, then for more, in a more general form. She agreed with D. Kaplan and H. Mehlberg that theoretical and observational predicates are not bound by any dependence strong enough to guarantee the truth of partial definitions of the former in terms of the latter; according to our knowledge, the definitions in question are only probable, not true. Their form is as follows: the probability of $T(x)$ on the condition that $Q(x)$ is equal to p. If we admit more than one probabilistic postulate for the term T, the problem of the existence of objects being T arises. What dependence should hold between $Q_1...Q_n$ to guarantee non-emptiness of the denotation of T? Mortimer formulated a proper condition concerning the relations holding between observational predicates in terms of which T was characterized. The condition was proved to be a necessary one; whether it was sufficient as well was only highly probable. Mortimer demonstrated how to separate analytic and synthetic components within probabilistic postulates.

SELECTED BIBLIOGRAPHY

Books:
1. (1982). *Logika indukcji*. Warszawa: PWN. English translation: *The Logic of Induction*. Chichester: Ellis Horwood Lim. 1988.

Papers:
1. (1960). Uwagi na temat sporu o analityczny charakter nauk przyrodniczych [Remarks on Controversy over the Analytic Character of Natural Sciences]. *Studia Filozoficzne* 1, 71-106.
2. (1964). Definicja probabilistyczna na przykładzie definicji genotypu (Probabilistic Definition Exemplified by the Definition of the Genotype). *Studia Logica* 15, 103-161.
3. (1964). O pewnej definicji genotypu [On a Definition of Genotype]. In: *Rozprawy logiczne: Księga pamiątkowa ku czci Profesora Kazimierza Ajdukiewicza* [Logical Treatises: A Festschrift in Honour of Kazimierz Ajdukiewicz]. Warszawa: PWN, 113-122.
4. (1966). O warunkach przyjmowania postulatów probabilistycznych dla terminów teoretycznych [On Conditions of Acceptance of Probabilistic Postulates for Theoretical Terms]. In: H. Eilstein and M. Przełęcki (Eds.) *Teoria i doświadczenie* [Theory and Experience]. Warszawa: PWN, 75-87.
5. (1968). Logiczne podstawy 'Analizy ukrytej struktury' [Logical Foundations of the 'Latent Structure Analysis']. *Studia Filozoficzne* 2, 95-124.
6. (1970). Conditions of Acceptance of Probabilistic Postulates. *Studia Logica* 26, 87-97.
7. (1973). A Rule of Acceptance Based on Logical Probability. *Synthese* 26, 259-263.
8. (1973). Logika indukcji Carnapa i Hintikki [The Logic of Induction in Carnap and Hintikka]. *Studia Filozoficzne* 1, 125-150.
9. (1978). Logika odkrycia naukowego Karla Poppera [The Logic of Karl Popper's Scientific Discovery]. *Studia Filozoficzne* 4, 171-176.

Poznań Studies in the Philosophy
of the Sciences and the Humanities
2001, vol. 74, pp. 169-172

Jan Woleński

KLEMENS SZANIAWSKI – RATIONALITY AND STATISTICAL METHODS

I. Klemens Szaniawski (3.03.1925, Warsaw – 5.03.1925, Warsaw) began his study of philosophy at the clandestine Warsaw University during the Second World War. He was taught mainly by Tadeusz Kotarbiński, Jan Łukasiewicz, Maria and Stanisław Ossowskis, Władysław Tatarkiewicz, and Henryk Hiż. After the war, he continued his studies at Łódź University; his master thesis was devoted to Mandeville's moral thought. In 1950, he obtained his PhD, based on the dissertation on the concept of honour in the Middle Ages; Maria Ossowska was the supervisor. In the 1950s, he became a member of the Department of Logic at Warsaw University. He obtained his postdoctoral degree in 1961. In 1969, Szaniawski was promoted to the rank of professor. He died in Warsaw on 5th March, 1990. A collection of Szaniawski's most important philosophical papers is published in Polish [1] and in English [2].

II. Szaniawski's first philosophical interests concerned ethics. However, the period of the early 1950s in Poland was not propitious for independent research in the field. Hence, Szaniawski decided to move to formal logic and the philosophy of science. When Kazimierz Ajdukiewicz came to Warsaw in 1955, Szaniawski began to work with him on the theory of fallible reasonings, i.e. reasonings in which non-deductive inferences are essentially involved. Both Ajdukiewicz and Szaniawski intended to develop a general theory of inductive procedures. Ajdukiewicz's programme of the so-called pragmatic methodology, which offers a picture of science as an activity with an explicit epistemic dimension constituted by acts of assertion of sentences, became a general background for that research. Essentially, the theory of induction focuses on five main problems: (a) the definition of induction, (b) types of inductive inferences, (c) the criteria of acceptance for inductive conclusions, (d) the justification of induction, (e) a general description of science as based on induction. Szaniawski as a person educated in the tradition of the Lvov-Warsaw School considered inferences as central in rational procedures. On the other hand, he did not believe in inductive logic, because he saw no possibility of any attribution of quantitative confirmation to general hypotheses on the basis of empirical data. Hence, he was not attracted by heroic attempts to solve the problem of induction undertaken by Hans Reichenbach and Rudolf Carnap. However, he also rejected Karl Popper's scepticism towards induction. What remained?

Szaniawski, following Ajdukiewicz, strongly believed that decision theory and modern statistics could provide proper tools for the analysis of induction. In order to employ those devices, he performed a profound analysis of the principal foundational controversies in statistics. There are two central problems in the foundations of statistics. The first concerns the criteria of accepting statistical hypotheses. Let us say that a statistical rule R is a function from a set of empirical data E to another set, namely that of statistical hypotheses. Some statisticians, such as Jerzy Spława-Neyman, maintain that inductive rules are accepted as maximizing utility (the theory of decision function). On the other hand, other statisticians, such as Ronald Fisher, argue that statistical rules maximize probability (the theory of estimation). The second debate concerns the issue of whether inductive rules are inferences or not. Now, the theory of decision function is usually connected with the view that statistical rules are not genuine inferences, but the estimation theory is interpreted as inferential as far as the matter concerns statistical procedures. In fact, both cases of the great protagonists, that is Spława-Neyman and Fisher, very clearly confirm these ideological correlations because the former argued for a non-inferential theory (statistical rules determine inductive behaviour), and the latter attempted to justify the view that rules applied in statistics are inferences. In Szaniawski's view, those coincidences, perhaps not accidental in everything, still admit a sound compromise. He defended an intermediate way, namely the theory of decision function but with inferential treatment of statistical rules. He also tried to reconcile Bayesian and non-Bayesian ideas in statistics and probability theory. According to Szaniawski, this is a chance of overcoming the dilemmas of objectivism and subjectivism in the related fields. He made a very interesting observation, namely that the problem of rationality constitutes a common denominator of conflicting theories concerning the foundations of inductive inference. Taking this idea as crucial, Szaniawski attempted to demonstrate that the opposition of inferential and behaviourist account of induction was fairly apparent. Szaniawski also worked on more specific and formal problems in theoretical statistics. In particular, he introduced an original β-criterion of rational decision making and generalized the method of sequential tests.

Szaniawski also continued Ajdukiewicz's way of thinking about the criteria of accepting sentences. It consisted in connecting rationality with an estimation of profits and losses, relatively to accepted or rejected statements. However, Szaniawski did not agree with Ajdukiewicz on one very important point. According to Ajdukiewicz, it is possible to define a scale of degrees of infallibility of inductive inferences completely independently of individual subjective factors. Szaniawski, following modern statistics, rejected this view. This means that he attributed greater importance to pragmatic elements in inferences than Ajdukiewicz himself did. It is an interesting historical fact which shows how developments in a particular field (statistics in this case) force changes in philosophy. Although it can hardly be said that Szaniawski successfully solved the old problem of induction in the context of statistics and decision theory (perhaps it is an unsolvable question altogether), he certainly identified its critical point, namely found that it was located at an interplay of inference and acceptance.

III. In the 1960s, Szaniawski began to work on a general approach to science in which scientific knowledge was conceived as an information-searching process. He believed that this approach helped in illuminating many controversies in the philosophy of science. Just as in the case of the foundations of statistics, he intended a reasonable compromise between various alternatives to science that seemed radically and ultimately incoherent. Discovery or justification? Falsification, verification or both? Dynamic or static approach to science? Progress or not? Szaniawski's general approach to science was as follows. Science is an information process directed by epistemic decisions of scientists. Any epistemic decision is undertaken under uncertainty. It is the key fact which cannot be neutralized by idealizing assumptions proposed, for example, by Carnap or Popper. Hence, gaps created by uncertainty cannot be filled by estimations of confirmation or corroboration abstracted from the real cognitive situations. And consequently, pragmatic parameters are unavoidable in the analysis of science. This general picture was then applied in investigating more concrete problems. For example, the correctness or rationality of fallible inferences becomes a special case of decisions undertaken under uncertainty. Szaniawski planned an extensive book about this programme, but his death interrupted the ambitious task.

IV. Szaniawski always considered applications of formal methods in social sciences, mainly economics and sociology, psychology and ethics. He believed that formal models, for example models of distributive justice or distributing goods, could considerably help in implementing the basic principles of rationality in various social activities. In fact, rationalism, understood by Szaniawski as anti-irrationalism, was the very guiding idea of his philosophy. This principle, derived from Ajdukiewicz, demands that only those statements and opinions deserve to be accepted as rational that are intersubjectively expressible, communicable and testable. Szaniawski believed that social life could be essentially improved by the propagation of rational standards, although, on the other hand, he perfectly understood how many obstacles there were on this way. Anyway, the maxim *Plus ratio quam vis* was especially important for him, especially in social practice. He devoted much of his life to showing that it is more than utopia.

SELECTED BIBLIOGRAPHY

1. (1994). *O nauce, rozumowaniu i wartościach* [On Science, Reasoning and Values], ed. by J. Woleński, Warszawa: PWN.
2. (1998). On Science, Inference, Information and Decision Making. In: *Selected Essays in the Philosophy of Science*, ed. by A. Chmielewski and J. Woleński, Dordrecht: Kluwer Academic Publishers.

Poznań Studies in the Philosophy
of the Sciences and the Humanities
2001, vol. 74, pp. 173-181

Krystyna Zamiara

JERZY GIEDYMIN – FROM THE LOGIC OF SCIENCE TO THE THEORETICAL HISTORY OF SCIENCE

I. Jerzy Giedymin (18th July 1925, Kleck near Nieśwież – 24th June 1993, Piła) studied philosophy, specialising in the field of English philology, first at the Jagiellonian University in Cracow and then at Poznań University. At the same time, he studied at the School of Commerce (presently the University School of Economics) in Poznań, from which he graduated in 1950. He worked at the School from 1948 to 1954. From 1958 to 1967, he worked at the Chair of Logic of the Philosophical-Historical Department at the Adam Mickiewicz University in Poznań. He received his doctoral degree in 1951, and was awarded the title of full professor in 1966. In the years 1957-1958 and 1959-1960 he stayed at London University as a scholar of the Ford Foundation, where he conducted research on hypothetism under the supervision of Karl R. Popper. He lived in Great Britain since 1966, lecturing in logic and methodology of science and later also in the history and philosophy of physics at the universities in Durham, London and Brighton. He was professor of Sussex University in Brighton since 1967, and in 1986-1990 he was head of the Department of Logic and Methodology of Science at the School of Mathematical and Physical Sciences of this University. Giedymin was a member of many learned societies both in Poland and abroad. In 1983-1986 he held the position of the chairman of The British Society for the Philosophy of Science. He was the member of the Editorial Board of *The British Journal for the Philosophy of Science*. He was one of the initiators of Archives-Centre d'Études et de Recherche Henri Poincaré in Nancy, France.

II. Giedymin worked in the following fields:
(1) general methodology of sciences,
(2) methodology of liberal arts (more precisely: social and historical sciences),
(3) history and philosophy of physics.

It is possible to distinguish two periods in his research activities, which more or less coincide with the time when he was employed at the Adam Mickiewicz University in Poznań (the first or "Polish" period) and with his emigration to Great Britain (the second or "British" period), respectively. In the first period, he conducted research exclusively in the first two fields, applying himself to the "logic of science" based on the model adopted from Kazimierz Ajdukiewicz; the model was enriched by elements of Karl Popper's hypothetical methodology. His work from that period testifies best to his assertion that the scholars mentioned above were his true teachers. In the second period, he still took up general methodological issues, but gave up detailed methodological reflection, turning to the liberal arts instead. He took up studies of the history and philosophy of physics in the 19th and 20th century. This resulted in momentous findings on the subject of conventionalism.

III. The logic of science, practised by Giedymin in the Polish period, is the study of science, perceived as the set of characteristic research activities and their products (taking the form of some notions or statements). The method of rational reconstruction played a basic role in those studies. It made use of logical and mathematical tools for the formal expression (in the appropriate language) of the reconstruction of the researched aspects of the scientist' procedures or knowledge created by them. The employment of the conceptual apparatus of the decision theory and the information theory, apart from the habitual logical-mathematical tools, was a special feature of the reconstruction language applied by Giedymin. The results of the reconstructing actions took the shape of explications (the bands of explications) or the so-called theory of rational activity. The extrapolation of a given solution upon other fields, the generalization of the reconstruction result and the testing of its appropriateness constitute the characteristic features of Giedymin's research procedure. It is to be credited with the fact that the studies in the field of the methodology of various branches of science referring to historical sciences or social sciences (mainly economics), generally speaking humanities, resulted in solutions in the field of general methodology of science. Thus, general methodological studies cannot be separated from the special methodological ones in the Giedymin's works originating from the period under discussion. Sometimes, the starting point of his methodological examination is constituted by a certain problem referring to science in general, completed in its demonstration in the context of research proceedings in a certain field of the humanities. Sometimes, the starting interest in a certain human phenomenon leads to the result of general-methodological significance. This kind of practice is justified by the position of methodological naturalism, which was advocated by Giedymin.

This methodology generally corresponds with those which Ajdukiewicz assumed talking about "pragmatic methodology" or "traditional methodology", respectively. Consequently, Giedymin took over from his master the postulates that a given science should serve as the starting point in methodological studies, reconstructing the principles which drove the scientists (though the scientists may not always realize them), that the methodological norms were not dictated to scientists by a methodologist but suggested as a more accurate (theoretical, logical) interpretation of the ones which they themselves followed. Giedymin practised descriptive methodology. However, this does not mean that it was free from normative regulations. Within the area in which he applied and developed the postulates defining the research procedures of a methodologist, the nature of his logical studies of science deviates from the ones that we can encounter in R. Carnap or K. Popper. Although these disparities are not clearly visible, they are significant. Whereas neopositivist methodologists locate themselves in relation to science in the position of unshaken authority and perform the role of an exquisite arbiter on the issues of research design, descriptive methodologists as defined by Ajdukiewicz would locate themselves in the position of a benevolent and understanding researcher discovering only something that exists in the proceedings of scientists, even though they themselves may not always notice it.

All this allows one to comprehend the ease of Giedymin's transition from the logic of science to the theoretical history of science at the moment he shifted his cognitive interests. At first, they referred to strictly methodological topics. Then (in the British period), methodological problems got entangled in the considerations of the annals of science. In the latter studies, the knowledge in the field of the general methodology of science played the role of theory enabling both precise conceptualisation of the historical phenomena and the revelation of certain hidden interrelations among them. Here it should be noted that Giedymin did not establish a theory of science which would be regarded as a systematic basis for his factual considerations on the annals of physics. His studies were conducted on the theoretical level transcending by far what is suggested by the familiar view of the researched phenomena. He was always a critic of the idiographic model of studies in history, advocating the type which is close to the so-called idea history. Investigating an event of considerable significance for the annals of physics, he applied the "case study" approach. Perceiving a mentioned event, he was trying to grasp certain relations from the annals of physical, philosophical and methodological conceptions. "Hidden" theory, which he was definitely guided by in those studies, was an original combination of different trends of thought. These were taken over mainly from Ajdukiewicz, Popper and Poincaré. The whole tradition of the contemporary philosophy of science was its background. Giedymin was clearly familiar with it.

It is difficult to discuss in a concise way the results of the detailed logical analyses conducted by Giedymin in the first part of his scientific activity. They can be outlined only. Chronologically speaking, the following achievements are to be considered the most significant:

(1) the conception of the historical source and the characteristics of inferring on the basis of the sources [1.1],

(2) the theory of the reliable informant/observer [1.2; 2.4],

(3) the erothetic logic [2.2],

(4) the considerations on the empirical theory: its logical structure, philosophical presumptions at its basis, cognitive status, relation to experience, empirical testability, etc. [2.1; 1.2; 2,5; 2.6].

The consideration of the methodological and epistemological problems of the historical sciences "from the logical point of view" drew Giedymin not only to the revision of the traditional views on the historical source or the so-called witness, as well as the determination of based-on-the-sources deductions (the result in the field of methodology of history). It also led to the statement that the analogous procedures (deductions and types of interpretations) are applied in other branches of science and their role is similar (the result in the field of general methodology of sciences). Giedymin's considerations on the criticism of sources and witness interpretation constitute a particularly interesting exemplification of the statement that certain research activities (along with methodological rules defining them) traditionally tied with the historical study only, in fact exist in other sciences or are characteristic of science in general. The notion of the reliable informant (critic) plays a vital role in these considerations. Giedymin noticed in them the relation between the procedure referred to as the analysis as well as the criticism of sources and the explanation and application of interpretation hypotheses, in which the formal notion of a reliable informant plays a vital role. Taking up the attempt to explicate this notion in a few works, Giedymin drew on the conviction that "the notion of the reliability of the informant [...] is necessary in the rules relating to the activity of any observer, experimenter, scientist testing the hypotheses" [1.1, p. 4]. As a result, there developed a methodological framework which was applicable in any empirical discipline. The system of reliable informant notion explications has the form of the theory of rational action (in a sense Giedymin attributed to this term in his work: [2.3]). The theory can be regarded as the introductory outline of the (formal) theory of confirmation [1.2, pp. 106 ff]. In other considerations on the subject of the so-called observational terms and propositions, Giedymin builds the criticism of positivistic (in a broad sense) conceptions of experience, the idea of theoretical and observational language division (according to the popular Carnap's model) with the following statement:

The impressions received by the observer or even by many experimentally independent observers constitute neither sense criteria nor validity criteria of the observable expressions of the empirical sciences language. However, the statements on the reactions of the observers, on the reliability of the observers or on their systematic errors are the ingredients of testing procedure of hypotheses and theories expressed in the observational statements. The elements are constituted by the explicitly formulated premises or by the silently adopted assumptions [2.6, p. 94].

It may be noted that the contributions to the confirmation theory (Giedymin is to be credited for them) exceed his conception of informant observer reliability. However, it is impossible to discuss all of them here.

The specification of the relations between different matters identified thus far is Giedymin's ordinary practice. It should not seem strange, then, that building the "logic of interrogative propositions", he indicates that one of the possible applications of the proposed constructions is their employment in the characterisation of the hypothesis-testing and text analyses procedures. The accomplishment of Giedymin's interrogative proposition analyses seems to be based upon the focus applied to the methodological aspect of asking questions and searching for answers to them. Giedymin put forward the erothetic logic in the form of the system of explications. The erothetic logic is assumed to constitute the set of formal tools used for the cognitive situation analysis in scientific practice. Scientific practice is to be understood both synchronically and diachronically in the process of development.

The question arises of what the embracing factor of the variety of results recalled here is. The results which built the methodological characteristics of the basic procedures applied in any empirical discipline or typical of certain group of sciences (humanities) or of individual sciences from this group (i.e. history, economics). One of the factors is constituted by the applied method (discussed above) and the adopted philosophical assumptions. The latter include, to recall the most significant ones: methodological naturalism (mentioned above), antiphenomenalism and anti-inductionism, hypothetism (all theorems of the empirical sciences are only hypotheses: revocable propositions), realism with respect to the cognitive status of theoretical theorems (opposed to instrumentalism), methodological individualism. "The critical test" is the only valuable method of hypotheses (theories) testing. It is the main assumption of the falsificationism (in the weak version not accompanied by the declaration of the ultimate character of any theoretical system falsification). It is visible that the basis of Giedymin's philosophy of science was taken from Popper. It is partly due to the original combination of the method worked out in line with the hints of Ajdukiewicz's and Popper's philosophy of science that Giedymin's works from the first period are especially strong (in the second period the influence of Popperism is somewhat smaller; it is the conventionalist frame that prevails). The values of this combination are most clear in the analyses in the field of methodology of humanities. This branch, largely underdeveloped compared with the methodology of natural sciences, was undoubtedly provided by Giedymin with a new, more advanced dimension.

IV. The second, British period in Giedymin's scientific activity is marked by the abandonment of the logical studies on liberal arts, the shift from the logical to the philosophical-historical reflections and directing them to physics, i.e. science in the strict sense. Some prior cognitive interests, visible in the kinds of the problems that were taken up, preserved their significance. It is only their cognitive context that was broadened. The research problems solved by Giedymin usually refer to the role of convention and theoretical theorems in the interpretation of the empirical material. Then it was obvious to Giedymin that no version of the radical empiricism, the view according to which the experience univocally determines the shape of the scientific theories, could be advocated. Pursuing this track, he paid close attention to the arising (at the beginning of the century) conventionalist conceptions both in the variants constructed by Poincaré and Duhem and Ajdukiewicz's radical conventionalism. At the same time, he retained sympathy for Popper's hypothetism (particularly for the idea of empirical testing of hypotheses by means of the "critical test"), which he invariably adopted, developed (proposing the explications of certain notions of the "logic of scientific discovery") and popularised among Polish philosophers of science.

The questions on the substance of scientific theory, on the relation of theory to experience, and on the cognitive status of theory always belonged to Giedymin's main methodological interests. Investigating those problems in the British period, he leaned towards the view that the differential-equations configuration is the substance of theory. Here, he referred to the tradition started by Hamilton, Hertz and Poincaré. Subjecting the views of those scholars to the scrutiny of detailed examination, Giedymin tried to reconstruct the notion of theory as presupposed by them. In this way, he achieved the conception of "the theory family". Such family is constructed by the theories based on the same mathematical structure, observationally equivalent (providing the basis for the same empirical anticipations), but admitting, in the form of theoretical assumptions, various ontologies superstructured upon them. Giedymin considered Helmholtz's and Maxwell's theory of the electromagnetic field as an example of such theories family [2.10]. The particular theory of relativity in the alternative formulations of Poincaré and Einstein constituted another, possibly more philosophically interesting example of the pair theory in the considered sense of the equivalents [1.3, pp. 149-195). The conception of theory reconstructed on the basis of the works of the enumerated physicists was to be a starting point for Giedymin. It led to the development of the new general scientific theory conception. Giedymin called it a pluralist or poligenetic conception [2.11].

Researching the genesis of the particular theory of relativity, Giedymin remarked that Lorentz's discovery of the group of transformations specific to the theory constituted the essence of the theory. Starting with this point, he repeatedly stressed the critical significance of the earlier mathematical results for the development of the 20th-century physics. Here, he highlighted the Lie group theory, Lobachevsky's and Riemann's geometries as well as Klein's research programme. Summarizing his studies on the genesis of STR, Giedymin concluded:

> In 1905 [...], i.e. simultaneously with Einstein's (1905) – though the two papers were independent of each other – Poincaré produced most of the mathematics of special "relativity", *viz.* precise Lorentz transformations, the postulate of relativity in terms of the form invariance of laws with respect to Lorentz transformations, the group properties of Lorentz transformations (the invariants of the Lorentz group), the covariance of Maxwell's and Lorentz's equations with respect to the Lorentz group, the elements of the four-vector formalism (in the context of a Lorentz invariant gravitation theory), together with the idea of Lorentz transformations as rotations in four-dimensional space. So far as these components of special relativity are concerned, one can only speak of simultaneous discovery of Einstein and Poincaré [1.3, p. 189].

Nevertheless, Giedymin makes the reservation that the equivalence of Poincaré-Einstein results can be advocated to certain extent only. It depends in particular on the assumed conception of theory and interpretation. If it is assumed that one of the possible interpretations of a given theory is valid (Einstein's in this case), Poincaré cannot be included among the discoverers of STR. However, if we adopt the understanding of the theory according to the views of Hamilton-Hertz-Poincaré, the results obtained by Poincaré and Einstein constitute an example of a simultaneous discovery.

The studies on mathematics and physics led Giedymin to the general formulation of the conventionalism and geometrical empiricism opposition. Giedymin claimed that geometrical empiricism, at least in Poincaré's sense, is to be understood as the view according to which spatial geometry can be univocally stated on the basis of measurements. These measurements are generalised by the induction [2.10, p. 2]. In my opinion, Giedymin's interpretation of Reichenbach is misguided. Undoubtedly, Gauss supported the empiricism as defined above. According to Giedymin, the view that is antithetic to empiricism is expressed by the following main thesis of the geometrical conventionalism, which he himself supports:

(C4) Physical geometry is a family of observationally equivalent systems of geometry plus physics which differ among themselves with respect to experimentally indistinguishable ontologies [2.10, p. 8].

In parallel with the discussion on conventionalism and geometrical empiricism, Giedymin also considers the view which he himself terms physical conventionalism. As it turns out, it is exactly the same conception which I mentioned before, discussing the pluralist interpretation of the scientific theory:

(C5) A physical theory is a family of observationally equivalent theories which share the same mathematical structure (or: whose mathematical structures are equivalent) and which differ with respect to experimentally indistinguishable ontologies [2.10, p. 15, C-5].

Giedymin claims that geometrical conventionalism constitutes a particular case of physical conventionalism when the pair: "geometry–physics" will be treated as a physical theory [2.10, p. 16].

The empirical content of a theory identified with the set of empirical laws entailing by the theory is invariant. It remains unaltered in the case of a convention shift: ontology or geometry. For Giedymin, this statement, treated as an important element of Poincaré's conventionalist philosophy, constitutes the basis of numerous conclusions. One of them points to the disparities between phenomenalistic and conventionalist understanding of the theory. Another refers to the issue of realism, still another takes up the problem of comparability (commensurability and incommensurability) of historically subsequent physics theory (understood in the standard way).

In Giedymin's interpretation, conventionalism seems to lead beyond the realism–instrumentalism opposition, as understood by Popper or Feyerabend. Instead, he approaches this stand towards antirealism viewed as a pragmatic suspension of judgement on the issue of theoretical hypotheses. Giedymin defines the conventionalist view on the cognitive status of theory, operating with constructionist concept of truth, as "structural realism" or interchangeably "structural conventionalism". According to him, the empirical content of the theory that is pluralistically understood (polytheory) is co-defined by the observational consequences class and mathematical structure which ties all elements of the theory with the appropriate relations. The following statement belongs to his implications: "only those models of the observational part of the theory are admitted which are extendible to the full models of the (whole) theory" [2.10, p. 16].

In the debate on the problem of incommensurability of the historically successive physics theories (i.e. Newtonian and relativist mechanics), Giedymin locates himself somewhere between adherents and opponents of the incommensurability thesis. Appealing to the notion of polytheory, he demonstrates that the theories, incommensurable due to various ontologies can be comparable that provided, they are combined by the same mathematical structure and empirical content. Empirical content appears invariant with regard to the change of the conventional elements of the polytheory. This is a really interesting result, which compels one to give much thought to the whole history of argument on the incommensurability of theories.

SELECTED BIBLIOGRAPHY

1. Books:
1. (1961). *Z problemów logicznych analizy historycznej* [From Logical Problems of Historical Analysis]. Poznań: PTPN-PWN.
2. (1964). *Problemy – założenia – rozstrzygnięcia. Studia nad teoretycznymi podstawami nauk społecznych* [Problems – Assumptions – Resolutions. Studies in Logical Foundations of Social Sciences]. Poznań: PTE-PWN.
3. (1982). *Science and Convention. Essays on Henri Poincaré's Philosophy of Science and the Conventionalist Tradition.* Oxford. Pergamon Press.

2. Papers:
1. (1960). A Generalization of the Refutability Postulate. *Studia Logica* 10, 97-110.
2. (1960). Confirmation, Critical Region and Empirical Content of Hypotheses. *Studia Logica* 10, 122-125.
3. (1960). Koncepcja racjonalnego działania i charakterystyka metodologiczna opartych na niej teorii [The Conception of Rational Activity and Methodological Characteristics of the Theories Founded on This Conception]. *Roczniki PTE. Rocznik 1958/59,* 61-81.
4. (1963). Reliability of Informants. *The British Journal for the Philosophy of Science* **13**, 52, 287-302.
5. (1964). Strength, Confirmation, Compatibility. In: M. Bunge (Ed.). *The Critical Approach to Science and Philosophy.* New York: The Free Press of Glencoe, 52-60.
6. (1966). O teoretycznym sensie tzw. terminów i zdań obserwacyjnych [On the Theoretical Sense of the So-called Observational Terms and Propositions]. In: H. Eilstein and M. Przełęcki (Eds.). *Teoria i doświadczenie* [Theory and Experience]. Warszawa: PWN, 91-109.
7. (1968). Revolutionary Changes, Non-Translatability and Crucial Experiments. In: A. Musgrave, I. Lakatos (Eds.). *Problems of the Philosophy of Science.* Amsterdam: North-Holland, 223-227.
8. (1973). Logical Comparability and Conceptual Disparity Between Newtonian and Relativistic Mechanics. *British Journal for the Philosophy of Science* **24**, 270-276.
9. (1978). Radical Conventionalism, Its Background and Evolution: Poincaré, LeRoy and Ajdukiewicz. Editor's Introduction. In: J. Giedymin (Ed.): *Kazimierz Ajdukiewicz. The Scientific World-Perpective and Other Essays 1931-1963.* Dordrecht-Holland, Boston USA: D. Reidel Publishing Company, XIX-LIII.
10. (1991). Geometrical and Physical Conventionalism of Henri Poincaré in Epistemological Formulation. *Studies in History and Philosophy of Science* **22**, 1, 1-22.
11. (1992). Conventionalism, the Pluralist Conception of Theories and the Nature of Interpretation. *Studies in History and Philosophy of Science* **23**, 3, 423-443.

II

SCIENTISTS

Poznań Studies in the Philosophy
of the Sciences and the Humanities
2001, vol. 74, pp. 185-188

Władysław Krajewski

MARIAN SMOLUCHOWSKI – A FORERUNNER OF THE CHAOS
THEORY

I. Marian Smoluchowski (28.05.1872, Vorderbühl near Vienna –
5.09.1917, Cracow), an eminent Polish physicist, was born and educated in
Vienna (his father was the chief of the Polish section of Emperor's
Chancellery). He received his PhD from Vienna University in 1895. He
then worked in physics laboratories in Paris, Glasgow, Berlin and Vienna.
In 1899 he became professor of physics at the University of Lvov, in 1913
at the Jagiellonian University in Cracow. Both of them were Polish
Universities in Galicia, which was an autonomous part of the Austro-
Hungarian Empire. In 1917, he was elected Rector of the Jagiellonian
University, however, he died of dysentery soon – one year before Poland
regained independence.

Smoluchowski was a follower of Ludwig Boltzmann, although not
directly his student. The main scientific works of this Polish physicist,
both theoretical and experimental ones, were devoted to the kinetic theory
and especially to the theory of gas fluctuations, initiated by Boltzmann. In
1906, Smoluchowski disclosed the nature of the Brownian motion (simultaneously with Albert Einstein but in another way, which was very highly assessed
by Einstein). He investigated the phenomena of opalescence, coagulation, and
others. He explained, by means of the fluctuation theory, why the sky is blue.
Marian Smoluchowski discussed various philosophical problems of science.
He was engaged in the struggle for atomism when many scientists and
philosophers (E. Mach, W. Ostwald, P. Duhem) denied the existence of
atoms and molecules. He analyzed the experimental results of J. Perrin and
T. Svedberg which demonstrated the existence of atoms. Smoluchowski
showed that there are phenomena which could not be explained by
classical thermodynamics but only by kinetic theory, and wrote about the
limits of the former [1.1]. He criticized not only the anti-atomistic
phenomenalism but also, in general, the positivism which "neglects
speculations" and "clips the wings of reason". Furthermore, Smoluchowski
stressed that experiments must be directed by hypotheses, that deduction in
physics is "no less valid than induction", etc. [2.2].

II. Smoluchowski's most important contribution to the philosophy of science is his analysis of the concepts of chance and statistical laws in physics. I shall present them using two sources: an earlier Polish paper [2.1] and a more elaborated German paper published posthumously [2.3].

Smoluchowski was a strict determinist, like almost all scientists before the rise of quantum mechanics. However, he attached great importance to the concepts of chance, probability, and statistical laws. Their role in physics increased during that time but not many scientists took this into account. Smoluchowski dealt with these concepts more than any other physicists in his time (his colleagues noticed this after his death).

Smoluchowski strongly opposes a popular view held by many philosophers that the concepts of chance and probability are a result of our partial ignorance of the causes of events. He points out that when scientists do not know the cause of an observed phenomenon they do not state that it is due to chance. For example, when a deviation of Uranus' orbit was observed, astronomers did not hold it to be due to chance but searched for its causes. Usually, a physicist speaks about chance and about accidental events when s/he can estimate their probability. And this probability depends upon conditions in which the events occur but not upon our knowledge. If we had complete knowledge about phenomena and their causes we could deal without concepts of chance probability, the probabilities estimated by us and the discovered statistical laws still being valid. In that sense they are objective [2.3].

III. Smoluchowski starts to discuss the problems of introduction of the concepts of probability and statistical laws in physics, posing two basic questions:

1. How is it possible that the results of chance can be calculated, i.e. that *accidental causes have lawful effects*?

2. How does chance arise when all events in nature are determined, submitted to laws, i.e. how *lawful causes can have accidental effects*?

If we consider chance as an opposition to law, we will not answer these questions. If we hold chance to be a result of our ignorance, we will not answer them either. Some authors, says Smoluchowski, appeal to the law of great numbers, as the basis of the probability calculus, however, it is not a solution but a replacement of the question. It is necessary to search for another solution.

Smoluchowski takes into account only the "regulated chance" (*der geregelte Zufall*), i.e., the chance submitted to statistical laws. It is, he says, a special kind of causal link. The main idea can be formulated as: *small causes – great effects*. He quotes Henri Poincaré who also pointed out this feature of a kind of chance (Poincaré 1912). However, according to Smoluchowski, there is another essential feature of regulated chance that was not noticed by Poincaré. The same effect is brought about by different causes; they are "divided" by causes which bring about another effect. As a result, the causal link "oscillates".

The simplest example (mentioned both by Poincaré and Smoluchowski) is the phenomenon of the roulette. Very small differences in the initial speed of the wheel decide whether the result will be favourable or not. Many intervals of values of this speed will give a favourable result. They are divided by intervals giving the opposite result.

Smoluchowski formulates a more exact definition of regulated chance:

1. Very small differences of initial conditions, i.e. of the values of the magnitudes relevant for these conditions, lead to great differences in the effects, decide whether a favourable effect will occur or not (*small causes – great effects*).

2. Many intervals of values of these magnitudes lead to the same result, i.e., the favourable effect may be obtained by many different initial conditions (*different causes – same effects*).

3. The shape of the function connecting causes (initial conditions) and effects must be regular, not have many sharp extremes. This condition is, however, rarely mentioned elsewhere.

Smoluchowski adduces various examples of regular chance. In each of them he points out that the two definitions are fulfilled. We shall consider only one of them which is most interesting for us.

Imagine a hollow vessel of a deliberately irregular shape with ideally refracting internal walls and with only one small hole. A particle, e.g. a gas molecule, falls with great velocity into the vessel through this hole. It is refracted many times from the walls. Finally, after some interval t, it escapes by the same hole. When? It is a matter of chance. The interval t may be very short or very long. Both main parts of our definition are fulfilled: 1. small difference in the direction of motion can cause a big difference in the value of t. 2. many different directions lead to the same result, the same value of t.

For each value of t a definite probability W may be calculated. Smoluchowski shows that when we designate the velocity of the molecule by c, the volume of vessel by V, the area of the hole by ω, we may prove that the probability of t is:

$$W = \frac{\omega c t}{4V}$$

and the average time T after which the molecule leaves the vessel:

$$T = \frac{4V}{\omega c}$$

We see that W and T do not depend upon the initial conditions, i.e. upon the initial direction of the motion of the molecule.

Smoluchowski exploits this model to explain a possible mechanism of radioactivity. As is known, the time t after which an atom of a radioactive element will disintegrate is a matter of chance, but there is a stable probability of the decay in given t. This phenomenon seems to be indeterministic, subject only to statistical laws. However, the above described vessel-model shows that probabilistic laws of radioactivity can arise on the basis of strictly deterministic motion of particles. If we had many such vessels with holes, we could predict how many particles will go out in the time t. Of course, Smoluchowski did not suppose that atoms are hollow vessels. He simply showed that it was possible to obtain a stable probability, a statistical law on the basis of strict deterministic movement of particles.

We see that the chaos reigning in a system containing many events (in our example – many reflections of the particle) can arise on the basis of deterministic movement of particles. And this is exactly the main idea of the theory of deterministic chaos. Smoluchowski pointed out that a small cause may bring about a great effect. The same idea was expressed by E. Lorenz in his famous example of a butterfly causing a hurricane. However, neither Lorenz nor the other founders of the chaos theory mentioned Smoluchowski. They probably did not read his work.

REFERENCES

Poincaré, H. (1912). *Calcul des probabilités*. Paris: Gonthier-Villars.

SELECTED BIBLIOGRAPHY

1. Books:
    1. (1914) *Gültigkeitsgrenzen des zweiten Hauptsatzes der Wärmetheorie*. Leipzig und Berlin: Teubner.

2. Papers:
    1. (1916). Uwagi o pojęciu przypadku w zjawiskach fizycznych [Remarks about the Concept of Chance in the Physical Phenomena]. In: *Księga Pamiątkowa ku czci Bolesława Orzechowicza* [Festschrift in Honour of Bolesław Orzechowicz]. Lwów, 445-598.
    2. (1917). Przedmiot, zadanie, metoda oraz podział fizyki [The Object, the Task, the Method, and the Division of Physics]. In: *Poradnik dla samouków* [Self-Teaching Handbook]. Warszawa: Wydawnictwo A. Heflicha i S. Michalskiego, 3-62.
    3. (1918). Über den Begriff des Zufalls und den Ursprung der Wahrscheinlichkeitsgesetze in der Physik. In: *Pisma* [Writings], Vol. III, Kraków: PAU, 1928, 87-100.

Poznań Studies in the Philosophy
of the Sciences and the Humanities
2001, vol. 74, pp. 189-196

Alina Motycka

CZESŁAW BIAŁOBRZESKI'S CONCEPTION OF SCIENCE

I. Czesław Białobrzeski (31.08.1879, Povshekhon near Jaroslav in Russia – 12.10.1953, Warsaw) studied physics in Kiev. In 1879-1910, he worked in Paris under the guidance of P. Langevin. In 1913 he became professor of physics and geophysics at Kiev University. From 1921, he was a professor of theoretical physics at Warsaw University. His discoveries concerning the internal structure of stars and laws of the radiation pressure in them achieved international renown. In 1931, his monograph *La thermodynamique des etoiles* appeared. Later on, until his death, he dealt mainly with the interpretation of quantum mechanics, especially the wave-particle duality and the related philosophical questions.

II. Białobrzeski rejected the mechanistic and positivistic demand of the observability of phenomena investigated by science. His view can be formulated as the following thesis (1): in the modern physical sciences, entities that are inaccessible to visual representation have become the objects of investigation.

Absolute space and time have been rejected by the theory of relativity. According to this theory, the form of the world is a four-dimensional space-time with non-Euclidean geometrical properties (it is curved and, maybe, finite). Even though the mathematical apparatus of the non-Euclidean geometry allows for a detailed description of these properties, they are, nonetheless, inaccessible to a visual representation, which means that the four-dimensional space-time so characterized is entirely non-visual and inaccessible to representation.

Seeking an explanation of the wave-particle duality of radiation and matter has been one of the biggest problems faced by physical sciences. Quantum mechanics presents the duality as consistent but at the cost of substantial relinquishments: it is obviously impossible to realize objects bearing features of both a corpuscle and a wave. Objects of the microworld become unimaginable; science only provides a mathematical description of properties of an atom, and even in imagination it is inconceivable to reconstruct atomic phenomena with the use of concepts applied to describe empirical data. It shows that sense representation of an atom cannot be attained.

In modern science, only a small range of the investigated reality is accessible to senses. In this way, the fundamental philosophical requirement applying here has been irrevocably questioned, i.e. limited to the same extent to which modern science has limited the importance of classical physics.

III. Białobrzeski's works also lead to the enunciation which will be stated as thesis (2): in modern science, description no longer functions as an empirical report.

In classical science, a description of any physical system should fulfil the laws of Newtonian mechanics and provide precise knowledge of a material system in a given moment, which would allow full definition of its performance at a later moment. Such a description must therefore be expressed in the so-called observational language. The classical description of the state of a material system specified the dependence of every element of the system on time, position and speed counted in accordance with the motion equations.

Białobrzeski observes that modern physical sciences view an atomic nucleus, a chemical molecule and an atom as systems. From that it follows that in quantum mechanics the wave is not motion but serves to present an atomic system as being a "super-spatial reality". The observational language of classical physics does not apply to the description of such a reality; state as a super-spatial creation is not a measurable value; the state of a system is rendered by a specific function, and a quantum description of a state expressed by one function does not separate distinctively all the elements of the system, these being subordinated to the whole under description. Thus, the state of an atomic system consist of the physical content but cannot be ascribed physical values, which could be done in classical physics. A peculiarity of a quantum description resides in the fact that the function describing the state shows only possible outcomes of experiments, and the number of possible values whose probabilities are derived from the function of the state is infinite. A description of an atomic system does not fully define properties of the system. Therefore, it is a description which violates descriptive canons of classical physics in which a description concerned the motion of bodies and was expressed in measurable values (vectors, scalars, tensors). As a function, a description of the state of a system in quantum mechanics is only a mathematical account and does not belong to measurable values.

IV. Science has created a means (a method) which protects it against idle speculation and allows to respect the criterion of the binding truth (the scientific truth). It is a procedure of verification by experiment, which has always functioned in science. The requirement of scientific results being in accord with empirical data curtails excessive and superfluous freedom, imagination and empty speculation. This method, not only enabling science to maintain its identity over the centuries but also to make progress, is based on both sheer experience and mathematics. Apart from the concrete empirical data, the scientific method also has a mathematical side to it; the latter enables the laws and processes of nature to be expressed in schemes of mathematical patterns, which is due to the fact that mathematics supplies science with forms of thinking about quantitative relations of phenomena. Science requires that its results be presented in a precise form ensuring precision of thinking, which means that this form is mathematical. That is why in physical sciences the physical content is fused with a mathematical scheme. In the light of quantum physics, there emerge questions of whether the two aspects of the scientific method – empirical data and mathematics – influence each other and whether pure empirical data (i.e. unconditioned by theory) is possible at all. Classical physics and the whole of science took empiricism to refer to something accessible to senses, something visible, and something which could be perceived with senses or their technical extensions. However, modern science has entered the world of the invisible (of electrons, quanta, etc.) and this world has become an object of investigation in physical laboratories. That is why the concept of empirical data has not only expanded but has also undergone a certain qualitative change. Penetrating more and more into the world inaccessible to senses, the physics of the 20th century still cannot entirely abandon the requirement of an agreement between its enunciations and empirical data and cannot rely only on mathematical equations. This leads to the following question: what should this agreement stand for in modern times? As the language describing "pure" empirical data is impossible, the empirical evidence produced in modern physico-chemical laboratories is more of a theoretical structure than pure perception. That is why – as Białobrzeski observes – in modern science empirical data are much more theory-laden than in earlier periods. This allows one to formulate successive thesis (3): in modern science, empirical data have become more theory-laden.

Thesis (3) is also enhanced by Białobrzeski's argumentation for the role of empirical data in solving problems of the wave-particle duality. Agreeing to the complementarity of the wave aspect and the corpuscular aspect of matter and radiation, modern physics recognizes those aspects as mutually complementing each other, so that empirical data shows one or the other. The situation of this complementarity points both to the complexity of the construct known as experiment, and to the theoretical load of the experiment so conceived.

V. The originality of Białobrzeski's solution of the problem of causality derives from the fact that in the new model of science the scientist, instead of eliminating the concept, supports its expansion. Therefore, thesis (4) can be formulated in the following way: in the light of the achievements of new physics, the old principle of causality which is binding in science requires indispensable expansion.

Causality in classical physics is connected with the possibility of strict prediction with determinism about the absolute necessity of the running of events in the world. On the other hand, quantum mechanics supplies foundations for concluding that in the atomic world there are no strict causal relations in the sense in which they are understood by classical physics. Quantum mechanics is not able to state what will happen to an atom the next moment; its state may change in various ways and its future is not unequivocally determined. Such a situation should necessitate indeterminism and a need to revise scientific concepts of both causality and prediction.

Białobrzeski introduces a subtle differentiation between various forms of indeterminism (time indeterminism, measurement indeterminism, indeterminism indicated by Heisenberg's principle of indeterminacy) occurring in the atomic world and indicating an acausal character of occurring events. He proposes a modification of the philosophical category of causality, generalizing this concept to cover all phenomena of physical sciences. He distinguishes two kinds of causality: deterministic and indeterministic. The modification towards a more general understanding of the category of causality is connected with adjusting the content of the concept of causality to a new system of concepts in modern physics. In accordance with this modification, causality indicates – in general terms – accepting a particular state to be an indispensable condition for another specified state to occur a moment later. This condition, therefore, can be expressed either strictly (according to the requirements of classical physics) or in an undefined way (which occurs in quantum mechanics and means that a produced state belongs to a particular set of states, every one of which having a particular probability of its realization).

VI. Contemporary physics penetrating into the depths of the microworld is interested in the atom and its structure in an entirely different way than it could be expected from the perspective of classical science. This construct is no longer treated in a mechanistic fashion. In the eyes of modern physics, the whole is not a set of elements or some machinery. The whole is more than its elements, it determines their meaning and – as Białobrzeski says – becomes a unity of a higher order, a unity which constitutes laws governing its elements. It is a structuralist understanding of a whole, but as Białobrzeski does not use the concept of structure, he introduces his own terminology. In his terms, a structuralistically understood whole is called a "system" and the occurrence of particular elements in systems is called "systematicity" which refers to specific combinations of elements in the structure of the atomic world. Therefore, in Białobrzeski's view the overall conception of modern science is structuralist.

An important notion in this conception of science is constituted by the concept of potentiality, i.e. such a characteristic of every system in the microworld whereby the state of a system (e.g. of the atomic system) is a system of possibilities. As potentiality is a reality containing numerous possibilities, it is logically prior to a concrete realized order; it is a primary factor, it is active and shapes atomic structures. In an individual act only one of possibilities is realized. A reduction of a wave bundle can serve here as an example: while a measurement is taken, one of numerous possible states is realized. The act in which potentiality is realized is termed by Białobrzeski an "act of determination" and is conditioned by the properties of the whole system. Thus, thesis (5) can be stated as follows: modern science views the object of its investigation as a dynamic structure.

The conception of "systematicity – potentiality – determination" is an original idea and testifies to its author's deep intuition and professionalism – it is a structuralist conception expressed in the spirit of dynamic structuralism. It is the so-perceived dynamic structures that are – according to Białobrzeski's works – an essential feature of the atomic world.

VII. Classical physics, together with its mechanistic outlook on the world, established a specific way of explaining phenomena, with its procedures being almost entirely analytical. Nonetheless, the analytical method proved useless for examining atoms and the microworld and had to give way to such a method of explaining the world explanation which can be called – following Białobrzeski – holistic, organic, systemic and synthetic. In this sense, quantum mechanics manifests synthetic tendencies. Therefore, our next thesis (6) is the following: in modern science the method for explaining phenomena is structuralist.

VIII. For Białobrzeski, science is a complex and heterogeneous process. Some of its episodes concern phenomena which are frequently mutually exclusive. That is why an exclusive determination of any of them does not exhaust the complexity of the scientific process and additionally posed a threat of one-sidedness of the approach and of authoritarianism. When it is perceived that science is both revolutionary and evolutionary, its progressive and regressive tendencies can be demonstrated. A period called a scientific crisis is a period full of intensive discoveries and of various propositions for solving scientific problems. The crisis situation derives from the fact that the old theory is resourceless in some ways and it neither explains problematic issues nor provides foundations for any further explanations; the problems themselves appear to contradict the existing science. Therefore, a revolutionary change undermines not only the existing theory but also the respective philosophical conceptions and a new theory emerges as a carrier of new doctrinal conceptions with separate ideological motifs. This outlook on scientific revolutions allows to formulate thesis (7): scientific revolutions bring about changes of general philosophical nature.

In order to emphasize the alleged paradox of the scientific process, especially in revolutionary periods, it should be added that in Białobrzeski's assessment new physics does not negate the old one and does not abolish it. A revolutionary change in science is an event whose content is prepared beforehand; besides, revolutionary episodes yield material for a diachronical study of science.

IX. Białobrzeski claims that from the historical perspective it is seen that science preserves previous theories and adapts them (though in a modified form) to new ones. Therefore, next to the revolutions in science, his works refer also to a parallel evolutionary trend. This dual orientation in philosophizing about science does not lead to any contradiction. Both of the methods of representation are necessary, as in essence science is both revolutionary and evolutionary. In the evolutionary model, what is of the main importance are concepts, their formation and transformations. Thus, thesis (8) can be formulated as follows: the evolutionary aspect of science reveals itself especially in the process of concepts formation.

The historical material of science indicates the transience of its concepts. Studying the history of such notions as *body, matter, space, time, causality* and relying on actual data, Białobrzeski shows that forms used by the scientific thought become flexible and are able to undergo evolutionary adjustment to new domains of experience, that science frees itself from conceptual apparatus whose ability to acquire new facts has become exhausted and introduces new notions. From this perspective, Białobrzeski is inclined to view the crisis in which science started to plunge at the turn of the 19th century as a "mutation process", a process of deep transformations of nature concepts.

X. According to Białobrzeski, the revolutionary change in science of the 20th century impels one to revise the existing philosophical conception of science, especially positivism. True, it has never been an adequate philosophy for science, i.e. it has never corresponded to it at any level of development. Białobrzeski's assessment of positivism is conducted from the point of view of scientific needs (and even if he does not specify what positivism he refers to, it is quite evident that his criticism concerns mainly logical positivism). First of all, it is a mistake to transfer the requirements which apply to science into the philosophical field. On the other hand, the research duty of epistemology is to critically assess the foundations of a given science, which means those general philosophical doctrinal assumptions on which the science (its stage under study) is based. Such research – which has never been conducted by positivism – would show that (contrary to positivism) science includes metaphysics whose elements were inherent in, for example, common-sensical concepts of the realistic outlook. The new physics not only opposes naive realism but also replaces it with a system of concepts of the realistic outlook which is not common-sense realism. This criticism also serves as an argument against instrumentalism which often accompanies neopositivist views. The criticism concerns also the positivistic division of the language of science which distinguishes perceptual statements (the so-called observational language). As the confrontation with science shows, such a distinction is groundless and it must ultimately drive this doctrine to a conflict with science. Positivism is a normative programme and instead of studying what is done in science it proclaims how things should be done. The normativeness of the positivistic doctrine derives only from its programme and not from the description itself. That is why this programme comes into conflict with science, its fundamental contravention being just neglect of the presence of metaphysical element in science. As positivism has not grasped this aspect of science, it is not able to comprehend that, in a certain period, science relied on the foundations provided by naive realism. Positivism has been equally unable to realize that after the scientific revolution of the 20th century those foundations have changed in science and do not refer to naive realism any longer. Little has the realism of modern science to do with a common-sensical outlook. Białobrzeski's characterization of the realism of modern science derives from his multirange argumentation, springing from his truly profound studies.

SELECTED BIBLIOGRAPHY

1. Books:
  1. (1939). *New Theories in Physics*. Paris: International Institute of Theoretical Co-operation.
  2. (1956). *Podstawy poznawcze fizyki świata atomowego* [The Cognitive Foundations of Physics of the Atomic World]. Warszawa: PWN.
  3. (1964). *Wybór pism* [Selected Writings]. Warszawa: IW Pax.

2. Papers:
  1. (1934). Sur l'interprétation concrète de la mechnique quantique. *Revue de la Metaphysique et de Morale* **41**, 84-103.
  2. (1938). Le science de la culture. *Organon* **2**, 17-42.

3. (1927). O pojęciach ciała i materii według fizyki współczesnej [On the Concepts of Body and Matter According to the Modern Physics]. *Kwartalnik filozoficzny* **2**, 1, 79-106.
4. (1936). Nowe drogi współczesnego przyrodoznawstwa [New Paths of the Modern Natural Science]. *Nauka Polska* **16**, 1-14.
5. (1947). Synteza filozoficzna i metodologia nauk przyrodniczych [The Philosophical Synthesis and the Metodology of Science]. *Nauka Polska* **25**, 37-45.

Poznań Studies in the Philosophy
of the Sciences and the Humanities
2001, vol. 74, pp. 197-205

Wojciech Sady

LUDWIK FLECK – THOUGHT COLLECTIVES AND THOUGHT STYLES

I. Ludwik Fleck (11.07.1896, Lvov – 5.06.1961, Nes-Siyona) worked in Lvov and Przemyśl after receiving a medical degree from Lvov University. During his lifetime, Fleck published over 130 papers on various subjects in medicine. In 1935, he published his only book on the philosophy of science titled, *Entstehung und Entwicklung einer wissenschaftlichen Tatsache. Einführung in die Lehre vom Denkstil and Denkkollektiv*. At that time, his ideas passed unnoticed among philosophers of science. An important philosophical discussion between Fleck and Tadeusz Bilikiewicz about scientific realism was published in Polish, in August 1939, but it appeared too late to create any interest.

Fleck spent two years in concentration camps in Auschwitz and Buchenwald. Then, in Poland again, he was appointed professor of medicine at the Maria Curie-Skłodowska University in Lublin. In 1957, Fleck emigrated to Israel, where he died.

II. It was a common view at the beginning of the 20th century that the explanation why people created or accepted opinions of certain kinds was to be of psychological or sociological character: religious beliefs or philosophical systems are results of coercion exerted by social structures, economic interests, etc. on the thinking of individuals. But in the case of science, things were to be completely different: the *content* of scientific theories was to be independent of psycho-sociological factors; it was to be determined by (inductive and/or deductive) logic and experience.

The conviction of the decisive role of logic and experience was undermined by the conventionalist philosophy of Poincaré and Duhem at the end of the 19th century. Fleck accepted their claim that reality can be described in many incompatible ways, but rejected their view that we are free to choose between theoretical systems. He applied the above-mentioned view that our choices in the field of religion or philosophy are socially determined to the analysis of the growth of knowledge that is usually called "scientific". Historical studies on the genesis and development of the concept of syphilis and the discovery of the

Wassermann reaction convinced him that some elements of our knowledge – the most fundamental ones – are accepted, modified or rejected under the influence of social and cultural situations in which communities of scientists work. It is essential for scientific knowledge, that this knowledge is developed by communities – "thought collectives" – rather than by individual researchers.

A thought collective is defined as "a community of persons mutually exchanging ideas or maintaining intellectual interaction" [1.1, p. 39]. Thought collectives have their own structure that gives to our knowledge its particular character and determines the way of its evolution. There are relatively small esoteric circles of experts and much bigger exoteric circles of school teachers and people applying scientific knowledge in practice.

The system of beliefs common to members of a given thought collective – "a thought style" – is defined as "directed perception, with corresponding mental and objective assimilation of what has been so perceived".

It is characterized by common features in the problems of interest to a thought collective, by the judgement which the thought collective considers evident, and by the methods which it applies as a means of cognition [1.1, p. 99].

The training introducing one into a thought style is of a dogmatic character. Students attain competence in applying some principles, but their critical attitude to those principles is out of question. If they do not accept the set of beliefs common to all members of a given thought collective and if they do not master the same set of skills, then they will not be admitted to the community. Students are going through the process of initiation which introduces them into circles where everybody thinks in the same way. Disputes are possible about particular applications but not about basic principles and that is why what they believe seems obvious to them.

The individual within the collective in never, or hardly ever, conscious of the prevailing thought style, which almost always exerts an absolutely compulsive force upon his thinking and which it is not possible to be at variance [1.1, p. 41].

It is to be understood not as a set of *constraints* that society imposes on our cognitive practices, but as something that makes cognitive acts *possible* at all. The word "knowledge" is meaningful only in relation to a thought collective. And if for any reason someone formulates ideas that are beyond the limits of what is socially acceptable at a given time, they will remain unnoticed or misunderstood.

III. Within any system of knowledge *active* and *passive* elements can be distinguished. The active part consists of what Kant (wrongly interpreting classical mechanics as absolutely true) called synthetic *a priori* propositions, or of "principles" in the sense of Poincaré. It is the product of our collective imagination. But when we apply active elements that we invented ourselves to the description of phenomena, we discover associations that we experience as "objective reality". It is psychology and/or sociology that can explain the genesis and primary acceptance of active elements of our knowledge, but passive elements are, so to say, results of the interaction between a conceptual system and the world. Passive associations are ascertained by individual researchers according to the framework inculcated in them by the thought collective.

The evolution of a thought style usually starts from some pre-ideas originating in a wide cultural surrounding of emerging discipline. Not in every culture could science have arisen, but only within the one in which suitable germs of modern theories appeared. If the culture is not homogeneous, then at the beginning many competing schools of thought emerge that perceive phenomena from different points of view and discover different associations on the basis of (sometimes the same) experiments. Various conceptual systems single out within our sense data different things, properties and relations, so in some important sense, facts emerge together with our words and "changes in thinking manifest themselves in changed facts" [1.1, p. 50].

It is not possible to see anything definite simply by looking. We need specific mental readiness to notice new objects or processes, to separate them from attendant phenomena, to describe them, and to turn them into subjects of collective investigations. And this readiness is taken over from our social environment. Facts are not something that is *given*, that can simply be pictured in (observational) statements. When we follow the evolution of experimental research concerning some phenomenon, we find out that experimental reports formulated at various times can be very different and even contradictory. The whole process begins with an observational *chaos*, blindly looking for something that does not change according to our wishes. In the initial stage feelings, will and reason play a role. The results of experiments often cannot be repeated. The simplest reports are of a hypothetical character. As scientists feel that hypothesis are their own – "subjective" – contribution, they try to get rid of them and to reach "objective" reality in this way.

> The researcher [...] looks for that resistance and thought constraint in the face of which he could feel passive. [...] The work of the research scientist means that in the complex confusion and chaos which he faces, he must distinguish that which obeys his will from that which arises spontaneously and opposes it [1.1, pp. 94-95].

Something that from the point of view of a given system seems not to obey our will is called a *fact*. Any description of a fact consists of both active and passive elements, and as the active ones constantly evolve that implies transformations of passive ones, so "the results can be no more expressed in the language of the initial observations than, *vice versa*, the first observations in the language of the results" [1.1, p. 89]. The final results are products of long efforts, numerous trials and errors, acquired skills both manual and intellectual, and implicit as well as explicit knowledge.

A single experiment is of very little cognitive value; it is only the whole system of experiments that makes it possible to distinguish what is really important, to check the soundness of assumptions, to evaluate errors of measurements, etc. Thus not only active elements are results of collective efforts but the thought collective also elaborates "the solid ground of facts".

> Only through organized cooperative research, supported by popular knowledge and continuing over several generations, might a uniform picture emerge [1.1, p. 22].

In turn, the thought collective can work efficiently only when it gets suitable encouragement or stimuli from the exoteric circles of science.

People are talking to each other but mutual understanding is never perfect: the listener's associations are more or less different from those that the speaker intended. Circulating from individual to individual and changing their meaning in the process, ideas sometimes merge, sometimes displace each other, sometimes are introduced into other thought styles. That is why concepts are ultimately products not of any individual but of a collective. Finally, the body of knowledge emerges that nobody predicted or intended. Today, we cannot point a single scientist who invented the concepts of force or mass, as they function within classical mechanics and there was no one discoverer of the oxygen theory of combustion or the bacterial theory of disease.

An instructive illustration here is provided by the difference between a scientific paper published in a journal addressed to members of the esoteric circle of a given discipline and the knowledge as presented in textbooks or popular presentations directed towards people from exoteric circles of science. Journal papers are of a provisional and personal character. The results of individual research, selected and modified in the process of interpersonal exchanges and misunderstandings, become more and more impersonal – and obligatory. In this way, *textbook knowledge* emerges which must be mastered by everyone who tries to enter the esoteric circles of science:

> The preliminary signal of resistance has become thought constraint, that determines what cannot be thought in any other way, what is to be neglected or ignored, and where, inversely, redoubled effort of investigation is required. The readiness for directed perception becomes consolidated and assumes a definite form [1.1, p. 123].

When the first textbooks are written – and first thought constraints appear – a discipline passes from the *prescientific* to the *scientific* stage.

As the result of sustained collective research carried out under social pressures, a thought style is created. The dominance of a given thought style does not mean that some universal arguments in its favour were found. All arguments are formulated within the very system that they were supposed to justify. We cannot say that a theoretical system is verified or confirmed by facts, because facts arise and develop together with the said system.

Facts also cannot (in spite of what Popper claimed in *Logik der Forschung*, 1934) disprove a thought style. Fleck mentioned five ways in which a well-developed system resists anomalous facts. (1) For adherents of a thought style some facts are unthinkable, unimaginable, so they do not look for them. (2) When they accidentally come up against facts that could undermine their theories they simply do not notice them. (3) If they even notice them they sometimes keep anomalous facts in secret. (4) If the existence of perplexing facts is admitted, then great efforts are undertaken to explain those anomalies so as to reconcile them with the system. (5) Sometimes scientists see and describe things that do not exist but would have existed if the thought style was right. All arguments for a given system of beliefs are accepted as rational only when the system is already accepted and converted into a thought constraint and then all arguments are in fact superfluous.

But there are scientific revolutions (in Kuhn's sense) that change prevailing thought styles. How does it happen that in spite of all psychological and sociological pressures a group of scientists can create new active ideas and develop them into a new thought style, which takes hold of the minds of the scientific community? There are only a few scattered remarks about it in Fleck's book. For example, we find a comment that, as historical research shows:

> Every comprehensive theory passes first through a classical stage, when only those facts are recognized which conform to it exactly, and then through a stage with complications, when the exceptions begin to come forward. [...] In the end, there are often more exceptions than normal instances [1.1, pp. 28-29].

But what makes it possible to notice exceptions and even concentrate our attention on them? The basic stimulus has to originate from outside the results of experiments themselves.

There is an important suggestion in footnote 4 to §4:

> For the sociology of science it is important to state that great transformations in thought style, that is, important discoveries, often occur during periods of general social confusion. Such "periods of unrest" reveal the rivalry between opinions, differences between points of view, contradictions, lack of clarity, and the inability directly to perceive a form or meaning. A new thought style arises from such a situation [1.1, pp. 177-178].

Another important hint can possibly be provided by Fleck's remarks that the passive element can sometimes be transformed, within a different thought style, into an active one. Elements developed within older thought styles, becoming autonomous, could give rise to new systems. Possibly, an important role is played by misunderstandings during the intercollective exchange of ideas. Words change their meanings in many ways and this in turn creates new facts and opens new cognitive possibilities. In this way, an avalanche of transformations can begin, as within scientific systems there are internal connections so that every new fact changes all facts known before.

Probably no more hints about the mechanisms of scientific revolutions can be found in Fleck's book. His remarks on what happens in the result of such transformation are much more important. Fleck opposes the view that old, false statements are replaced by new ones, more true then their predecessors. Before and after the scientific revolution "the same" words have different meanings, so we do not talk more truly about the same facts, objects, etc. but rather we talk in a different way about different things.

There are numerous connections between consecutive thought styles, but they are of a historical rather then of a substantial character. We can trace the developmental paths of particular ideas, but the fact remains that meanings of words in old and new thought style are different, so succeeding thought styles are *incommensurable* – the Polish counterpart of the famous Kuhn's term can be found in Fleck's paper of 1939, "Nauka a środowisko" [Science and the Environment] [2.6].

Because a thought style is a directed perception, its change leads to a change of the ways of perceiving the world. Adherents of a new style are no longer interested in some phenomena and other, previously ignored facts become important. The way scientists think about the world changes as well. Sentences that identical in their external form will lead to different conclusions after the revolution. From the points of view of incommensurable thought styles the world is composed of elements of different kinds and possible and actual relations of those elements are different. Different truths will be obvious and various states of affairs will be considered as impossible. Different questions will be asked and different methods will be used to answer them.

As a result, a comparison of the cognitive advantages of incommensurable theoretical systems is not possible. All debates between adherents of different thought styles consist almost entirely of misunderstandings. Members of both parties are talking of different things (although they are usually under an illusion that they are talking about the same thing). They are applying different methods and criteria of correctness (although they are usually under an illusion that their arguments are universally valid and if their opponents do not want to accept them, then they are either stupid or malicious). In his "Nauka a środowisko" [2.6], Fleck claimed that even such a sentence as "A normal human hand has five fingers" cannot be understood in the same way by all people. In some tribal languages it could not even be formulated: some of them do not contain numbers bigger than three, others define "five" as "a hand" (using such a language, we can only formulate a tautology: "a hand has hand fingers"). There are numerous languages in which there is no general concept of "a finger". Concepts of "normal" are also different. Hence, the above sentence that for us represents an obvious fact, can be meaningless from another point of view.

IV. There are not only *gains* but also *losses* involved in the changing of a thought style. We become able to see new facts but at the same time we lose ability to perceive something that was perceived by our predecessors. For ancient thinkers things of which their world was composed had deep, symbolic meaning – those things were related to gods, good and evil, and destiny. Within some thought styles numbers were not only tools of description but were significant in themselves and formed meaningful connections. All those senses disappeared in our times. Contemporary thinkers read old books with the feeling of superiority – for they cannot understand that ancient people had more to say about what *was of superior value for them.*

The accepted knowledge always seems to be obvious, useful, justified; alien views appear doubtful, useless, unjustified. But contemporary concepts are as they are only due to an accidental historic development. If different active associations were developed in some distant past, then we would have a different, but also a harmonious system of knowledge today. We should remember about it when we evaluate old or alien thought styles. We should also be aware that probably our concepts sooner or later will be replaced by others. Our successors will find our ideas as artificial and unjustified as we find those of our predecessors.

The illusion that science constantly approaches its final goal, namely the true picture of reality, is strengthened by the fact that because of the collective nature of knowledge and mechanisms of its development individual scientists are not aware of the nature of processes in which they take part. They systematically misunderstand past thought styles. Even scientists own memories are not reliable. Fleck sums up the story of the discovery of the Wassermann reaction as follows:

> *From false assumptions and irreproducible initial experiments an important discovery has resulted after many errors and detours.* The principal actors of the drama cannot tell us how it happened, for they rationalize and idealize the development. Some among the eyewitnesses talk about a lucky accident, and the well-disposed about the intuition of a genius. It is quite clear that the claims of both parties are of no scientific value [1.1, p. 76].

"In science, just as in art and in life, only that which is true to culture is true to nature" [1.1, p. 35]. But such remarks refer to active associations only and not to passive associations discovered in the process of applying active ones that are objectively true or false within the system. As Fleck emphasizes, truth is neither "relative" nor "subjective",

It is always, or almost always, completely determined by a thought style. One can never say that the same thought is true for *A* and false for *B*. If *A* and *B* belong to the same thought collective, the thought will be either true or false for both. But if they belong to different thought collectives, it will just *not* be *the same* thought! It must either be unclear to, or be understood differently by, one of them [1.1, p. 100].

V. It is of course meaningless to ask whether a thought style as a whole is in agreement with reality. It would be better if we completely avoided using the word "reality", although the grammar of our language that requires the occurrence of a subject in a sentence forces us to do so, claimed Fleck in his paper "Nauka a środowisko". In "Problemy naukoznawstwa" [Problems of the Science of Science] [2.7] after criticizing the principle of the unity of science, he expressed an extreme attitude: no scientific discipline contains the objective picture of the world in the sense of one-to-one semantic mapping of it and it does not even contain any fragment of such a picture. If it contained it, there would exist a constant, unchanging part of science, and scientific knowledge would grow simply by the adding of new information, whereas historical research shows that it constantly changes as a whole.

In spite of the above, Fleck emphasized the uniqueness of both scientific thought styles and scientific thought collectives; although in his opinion differences between science and philosophy, art or religion are only of quantitative, and not of qualitative, nature.

Scientific thought styles are distinguished by the larger number of passive elements relative to the number of active ones. There are passive elements in every thought style, even in myths or fairy-tales. However, internal connections within mythical systems are more detached and that is why the world appears to the adherents of such thought styles as unstable and full of miracles. In contrast, scientific thought styles are characterized by a relatively large degree of internal connections, and this leads to the belief that there is objective reality that exists independently of our thoughts, feelings and wishes.

The peculiarity of *scientific thought collectives* consists mainly in their relatively *democratic* character. All scientists have equal rights to express their opinions, and arguments are never evaluated on the basis of the position of their author within the scientific community. Besides, scientists, contrary to peremptory and dogmatic priests, are in partnership with members of exoteric circles. Instead of instructing them, scientists try to gain laymen's recognition by presenting their achievements in a popular form in order to make them available to everyone.

We can also talk of emotions typical of scientific investigations (the concept of thinking free of emotion is meaningless for Fleck), of the specific intellectual mood of modern science. The mood is intimately connected with the structure of scientific thought collectives. "It is expressed as a common *reverence* for an ideal – the ideal of objective truth, clarity, and accuracy" [1.1, p. 142]. "All research workers [...] in the service of the common ideal, must equally withdraw their own individuality into the shadows, as it were" [1.1, p. 144].

SELECTED BIBLIOGRAPHY

1. Books:
1. (1935). *Entstehung und Entwicklung einer wissenschaftlichen Tatsache. Einführung in die Lehre vom Denkstil und Denkkollektiv.* Basel: Bruno Schwabe und Co.; English ed. by T.T. Tren, R.K. Merton, *Genesis and Development of a Scientific Fact.* Chicago: Chicago University Press, 1979.

2. Papers:
1. (1927). O niektórych swoistych cechach myślenia lekarskiego [On Some Specific Features of the Medical Way of Thinking]. *Archiwum Historii i Filozofii Medycyny oraz Historii Nauk Przyrodniczych* **6**, 55-64; English translation in: R.S. Cohen, T. Schnelle (Eds.). *Cognition and Fact – Materials on Ludwik Fleck.* Dordrecht: Reidel 1986, 39-46.
2. (1929). Zur Krise der 'Wirklichkeit'. *Die Naturwissenschaften* **17**, 425-430; English translation in: R.S. Cohen, T. Schnelle (Eds.), *op. cit.*, 47-58.
3. (1935). O obserwacji naukowej i postrzeganiu w ogóle [On Scientific Observation and Perception in General]. *Przegląd Filozoficzny* **38**, 57-76; English translation in: R.S. Cohen, T. Schnelle (Eds.), *op. cit.*, 59-78.
4. (1935). Zur Frage der Grundlagen der medizinischen Erkenntnis. *Klinische Wochenschrift* **14**, 1255-1259.
5. (1936). Zagadnienie teorii poznawania [The Problem of Epistemology]. *Przegląd Filozoficzny* **39**, 3-37; English translation in: R.S. Cohen, T. Schnelle (Eds.), *op. cit.*, 79-112.
6. (1939). Nauka a środowisko. Odpowiedź na uwagi Tadeusza Bilikiewicza [Science and the Environment. An Answer to Tadeusz Bilikiewicz's Remarks). *Przegląd Współczesny* **18**, 8, 149-156; 9, 168-174.
7. (1946). Problemy naukoznawstwa (Problems of the Science of Science). *Życie Nauki. Miesięcznik Naukoznawczy* **1**, 322-336; English translation in: R.S. Cohen, T. Schnelle (Eds.), *op. cit.*, 113-128.
8. (1947). Patrzeć, widzieć, wiedzieć [To Look, to See, to Know]. *Problemy* **2**, 74-84; English translation in: R.S. Cohen, T. Schnelle (Eds.), *op. cit.*, 129-152.
9. (1986). Crisis in Science. Towards a Free and More Human Science. Previously unpublished manuscript, Ness-Ziona 1960; English translation in: R.S. Cohen, T. Schnelle (Eds.), *op.cit.*, 153-158.

Poznań Studies in the Philosophy
of the Sciences and the Humanities
2001, vol. 74, pp. 207-212

Michał Tempczyk

LEOPOLD INFELD – THE PROBLEM OF MATTER AND FIELD

I. Leopold Infeld (20.08.1898, Cracow – 15.01.1968, Warsaw) was one of the most prominent Polish theoretical physicists of the 20th century.

He studied physics at the Jagiellonian University and his PhD thesis written in 1921 was one of the first in Poland after the country regained its independence. He proved that geometrical optics is the zero-order approximation to Maxwell equations for large frequencies in both relativity theories.

After a few years work as a schoolmaster, Infeld started his academic career as a research assistant at Lvov University. He worked mainly on two subjects: the uncertainty principle in quantum mechanics and the geometrical framework of General Relativity Theory (GRT). In his first well-known paper, he derived Dirac's equations in the gravitational field. Following that line of research, he subsequently formulated, jointly with van der Waerden, spinor calculus in GRT.

In 1933, Infeld went to Cambridge, where he worked on non-linear electrodynamics with Max Born. They published eight papers together and formulated a theory now known as Born-Infeld electrodynamics. It removes some of the shortcomings of ordinary electrodynamics, but its influence on contemporary physics has been rather small. Upon returning to Poland, Infeld failed to become an university professor and in 1936, thanks to Einstein's help, he got a grant by the Institute for Advanced Studies in Princeton, where he began his collaboration with Einstein. Using a new approximation method, they proved that in GRT the laws of motion follow from the field equations. Next, they solved the problem of the motion of double stars. Einstein and Infeld published three papers together.

In 1939, he was offered a professorship of theoretical physics at the University of Toronto, where he worked till mid-1950, when he decided to return to Poland; there he organized an Institute of Theoretical Physics at Warsaw University. His main fields of scientific research were GRT, electrodynamics, and cosmology. He was also interested in quantum physics, he devoted much of attention to it as a young scientists and in his popular books; however, later his scientific interests were concentrated mainly on GRT and related topics. He was the author of some books about the history of physics and mathematics, the role of GRT in modern physics, the new physical picture of the matter emerging from relativity theory and quantum mechanics and about the social and philosophical role of physics. His best known book of this kind was *The Evolution of Physics* written with Einstein in 1938. The book was in fact written by Infeld; Einstein only added some remarks to his manuscript.

II. When Infeld started his collaboration with Einstein, GRT was a widely accepted fundamental theory of modern physics and only few physicists wanted to develop further its conceptual and mathematical foundations. There were no empirical and theoretical reasons to try to change its general scheme; however, both Einstein and Infeld were aware that such a change would be very useful for our scientific picture of the world. Modern physics began with the rise of classical mechanics, which described the world composed of individual mechanical entities, such as atoms and molecules. In the 19th century, Faraday and Maxwell introduced a physical object of new kind: electromagnetic field that cannot be reduced to corpuscles. Contemporary physics uses corpuscular matter and fields as fundamental independent elements of the world and this duality was a big problem for Infeld. Corpuscular matter and fields have in relativistic physics opposite properties. The old mechanical world was consistent, being composed of one kind of physical objects – corpuscles. Infeld wrote in his autobiography [3, p. 255]:

> We now know for sure that the old mechanical concepts are insufficient for the description of physical phenomena. But are the field concepts sufficient? Perhaps there is a still more primitive question: I see an object; how can I understand its existence? From the point of view of a mechanical theory, the answer would be obvious: the object consists of small particles held together by forces. But we can look upon an object as upon a portion of space where the field is very intensive or, as we say, where the energy is especially dense. The mechanist says: here is the object localized at this point of space. The field physicist says: field is everywhere, but it diminishes outside this portion so rapidly that my senses are aware of it only in this particular portion of space.

This quotation expresses the main philosophical point of Infeld's scientific work: is it possible to elaborate a concise homogenous picture of the physical world? Is it possible to reduce matter and field to one kind of objects? From the logical point of view three views are possible:

1. The mechanistic view: everything can be reduced to particles and forces acting among them, depending only on distance.
2. The field view: everything can be reduced to field concepts concerning continuous changes in space and time.
3. The dualistic view: matter and field are independent of each other.

It is now very popular in the philosophy of science to speak about beauty and simplicity as essential features of all scientific fundamental theories. For Infeld, the mixture of matter and field was something temporary, accepted only out of necessity, because we have not yet succeeded in forming a consistent picture of the physical world based on field concepts only. He believed in the possibility of forming a pure field theory and to reduce that what appears as matter to the field. The search of simplicity in GRT was the main motivation of his work with Einstein. They focused their efforts on the problem of motion.

The GRT formulates equations of the gravitational field that describe changes of the field in space and time. From the point of view of the theory, matter can be considered as placed in regions where the Einstein's equations break down, where they are not valid any more. But the equations do not define the motion of the particles. It is assumed in GRT that this motion obeys a special law, the law of the "geodesic line". In other words, the equations of the gravitational field define the geometry of space and particle trajectories are the shortest curves in this geometry. Thus, the mathematical structure of GRT contains equations of two kinds: equations of the field and equations of motion. This combination of two kinds of equations was regarded by Infeld and Einstein as logically unsatisfactory. They tried to answer the following question: is it possible that equations of motion are already contained in the field equations but we are unable to deduce them? Field equations may impose some restrictions upon the motion of the particles making the additional law of motion unnecessary. Einstein and Infeld tried to interpret the law of motion as a first approximation of a deeper law which can be deduced from the field equations. Their efforts resulted in a new approximation method called Einstein-Infeld method.

Although Einstein and Infeld succeeded in deducing the law of motion from the field equations, their work has not been widely accepted by the community of physicists, who prefer to use the standard interpretation of GRT with two kinds of equations. Infeld was aware of the reasons for that approach. It is more convenient to work with both kinds of equations then to deduce the law of motion from the field equations. The methodological aspect of this situation is not new. The simpler our picture of phenomena and assumptions are from the logical point of view, the longer the chain of reasoning leading from the principal assumptions to results that can be compared to observations. Infeld wrote: "Paradoxically enough, modern physics seems difficult and complicated because it is so simple". The simplicity of the GRT mathematical structure is important for scientists who want to understand the structure of the matter and for philosophers, even though its application in scientific practice ran against difficulties.

III. From the philosophical point of view Infeld's second important contribution to modern physics, non-linear Born-Infeld electrodynamics, belongs to the same approach to the problem of matter and field. Classical electrodynamics was the main domain of the dualistic worldview. At the beginning of the 20th century, Lorentz formulated the field model of the electron. In that model, the electron is considered as a ball with an electric charge on its surface. The energy of the field generated by the charge is proportional to $1/r$, where r is the radius of the electron. Comparing that energy to the energy of the electron rest mass, given by the famous Einstein's formula $E = mc^2$, one can calculate the so-called classical radius of the electron. The model relates the mechanical property of the electron, its radius, with the electromagnetic field generated by the particle; however, it is still dualistic because the relation does not follow from the mathematical structure of electrodynamics. Another drawback of the model is that the energy of field generated by point particles is infinite.

Infeld and Born decided to formulate non-linear electrodynamics in which the radius of particles is finite and related to fundamental properties of the field. The field generated by an electron is non-linear in a very small region, whose radius is equal to the classical radius of Lorentz's model. Outside that region, the field behaves approximately in the way described by the standard Maxwell's theory. In the Born-Infeld model, the electron is actually considered as a special region of space in which the field is exceptionally strong. It is very similar to the description of particles in GRT. Both theories see the corpuscular matter in the same way. Particles are reduced to field concepts, which is consistent with Infeld's philosophical program. The reduction is possible because of the non-linearity of both classical fields.

In his popular books, Infeld used to emphasize the role of both scientific revolutions, the relativist and the quantum ones, in the first half of the 20th century. At the beginning of his scientific career, he worked in both domains. For example, in 1932-1933 he wrote papers devoted to the Dirac theory of the electron in the GRT. In the following years, having formulated the fundamental mathematical structure of non-linear electrodynamics, Born and Infeld proposed the method of its quantization, similar to the quantum version of classical electrodynamics.

IV. The collaboration with Einstein deeply influenced Infeld's scientific and philosophical views, but it does not mean that he shared all Einstein's opinions, especially those concerning the nature and role of quantum mechanics. Einstein's critical approach to quantum mechanics and his discussions with Bohr and Heisenberg in 1930 are widely known. He considered quantum mechanics as an incomplete theory and he believed that the quantum indeterminism would be removed from the theory with the so-called hidden parameters. All his attempts to prove the quantum incompleteness with ingenious mind experiments were unsuccessful; however, he did not change his critical opinion about quantum physics.

Infeld's approach to quantum mechanics was more modest and standard. He considered quantum physics as a step ahead from the old mechanical view, and a retreat to the former position seemed unlikely to him. He wrote that in its early states physics was sure of itself; it was convinced that by perfecting the knowledge of the laws of nature, it would in the long run lay bare the detailed future of systems. Physicists knew that in every measurement there is always some element of error; however, they supposed that measurement modifies the system only imperceptibly, and that one can imagine more precise measurements increasing the accuracy with which the initial state of the system can be determined.

The quantum physics broke with the deterministic method of classical physics because its principles do not permit us to determine the initial state of a microsystem to any desired degree of accuracy. The errors in measurement must satisfy Heisenberg's uncertainty principle. The difference between classical and quantum physics consists in radically different models of measurement. In classical mechanics one assumes that the intervention of measuring instruments modifies the system only negligibly, and that one can make this modification arbitrary small. This assumption is not valid in the microworld, where the process of measurement substantially modifies the system. At the end of his popular book *The World in Modern Science* [1, p. 273], Infeld wrote:

> Now let us see what the enthusiast of modern physics has to say. He knows that the sceptic's arguments are meaningless and pointless. What is the point of saying that the microphysical world is governed by some deterministic laws which are still unknown to us? Such a statement is merely an expression of belief, which can be neither confirmed nor repudiated. [...] The question whether laws will, in future, take a deterministic form, is quite irrelevant. All that we know is that we had to move far from the methods of classical physics, to free ourselves from the deep-rooted prejudices and habits of thought and to acquire new instruments of reasoning, and that, *at present*, our conception of the world is as it is and no other.

It is a very realistic and moderate view of science, free of extremes and false oppositions. Science is our most reliable and effective method of getting knowledge about the world. The knowledge is not absolute, however it is more and more close to the truth which can never be completely grasped. Living in the times of fast progress in science and participating in the construction of its fundamental theories, Infeld did not want to waste his time discussing problems that cannot be effectively solved. Science is effective, the philosophy of science is not. It was probably the reason why Infeld has paid so little attention to the philosophy of physics, although his scientific and popular works are full of philosophy. He was clearly aware that the results of scientific research very often forced a change in the philosophical view of problems which extended far beyond the restricted domain of science itself. Such questions as what the aim of science is or what is demanded of a theory which attempts to describe nature, exceed the bounds of physics but they are intimately related to it, since science forms the material from which they arise. Therefore, philosophical generalizations concerning science must be founded on scientific results.

Infeld emphasized the creative nature of science. He believed that it is not just a collection of laws, a catalogue of unrelated facts. It is a creation of the human mind, with its freely invented ideas and concepts. The aim of physical theories is to form a picture of reality and to establish its connection with the wide world of sense impressions. Infeld was interested in the history of the fundamental concepts of physics and he claimed that physics really began with the invention of mass, force and an inertial system. Those concepts, being all free inventions, led to the formulation of the mechanical point of view, which was replaced by the relativist and quantum physics at the beginning of our century. Infeld's contribution to this new physics was great; however, he was sure that in future it would also be replaced by new theories and concepts.

V. The last important issue worth mentioning in this short discussion of Infeld's worldview is the social role of science. During his lifetime, physics has radically changed the technological possibilities of humankind; particularly during the Second World War, when many powerful military technologies were invented and developed. Those technologies, of which the atomic bomb is whose a symbol and the most dangerous representative, endangered the existence of humankind. Infeld felt the responsibility of scientists for the future of our race and he was a secretary of Pugwash and an active member of other international organizations. In the last years of his life, he devoted much energy to the organization of physical education in Poland and to the Institute of Theoretical Physics in Warsaw.

SELECTED BIBLIOGRAPHY

1. (1934). *The World in Modern Science. Matter and Quanta*. London: Victor Gollancz Ltd.
2. (1938) (with A. Einstein). *The Evolution of Physics*. New York: Simon and Schuster.
3. (1980). *Quest. An Autobiography*. New York: Chelsea Publishing Company.

*Poznań Studies in the Philosophy
of the Sciences and the Humanities
2001, vol. 74, pp. 213-222*

Jan Płazowski

JERZY RAYSKI – A PHYSICIST AND A PHILOSOPHER OF PHYSICS

I. Jerzy Rayski (6.04.1917, Warsaw – 14.10.1993, Cracow) began studies in the Faculty of Medicine of the Jagiellonian University in Cracow in 1935. Having completed the first year of medical studies, he turned to studying physics, considered to be the most advanced field of the human knowledge of the Universe.

In 1944, in conspiracy, Rayski passed the final examination. From January of 1945 to February 1946, he held the position of a research assistant to Jan Weyssenhoff at the Department of Theoretical Physics at the Jagiellonian University. Then, until 1947 he was a senior assistant to Wojciech Rubinowicz at the Department of Theoretical Mechanics at Warsaw University. Then he presented his PhD thesis *On the Divergence Problem in the Theory of Quantized Fields*. The same year, his first publication *The Problem of Quantization of Higher Order Equations*, a letter to *The Physical Review*, was published. Both papers placed him in the mainstream of research in fundamental sciences.

From May 1948 to April 1949 he was on a scholarship in Zurich. He worked with Wolfgang Pauli, one of the most famous theoretical physicists of that time. Pauli, together with Jost, Luttinger, Rayski, Schaffroth and Villars who formed Pauli's team, investigated the method, suggested by Feynman, of eliminating ultraviolet divergences in quantum field theory.

Having returned from Zurich, Rayski together with Jan Rzewuski, organized a strong centre of investigation in the quantum field theory in Toruń. Rayski obtained his postdoctoral degree in January 1950 presenting the work *On Simultaneous Interaction of Several Fields and the Self-Energy Problem*. In June 1954, on recommendation of the Nicolas Copernicus University in Toruń, Rayski became associate professor. In October 1957, he started work in the Department of Theoretical Physics of the Jagiellonian University in Cracow.

In 1958, Rayski became full professor. In 1959 he took the Chair of the Department, and after the reorganization in 1978, he took the Chair of the Field Theory Department, which position he held till his retirement in

1987. He was twice elected Dean of the Faculty of Mathematics, Physics and Chemistry, first for the 1960-62 term, and then during the 600th anniversary of the Jagiellonian University.

II. Above all, Rayski was a physicist. He worked in those difficult and turbulent times when great ideas of contemporary physics were developed. The progress was neither easy nor straightforward. Physicists' emotions oscillated between euphoria and dismay. Rayski had full confidence in the prospects for science. As Einstein did, he preferred to "announce heresies", hoping that there was a real seed of progress in them.

Rayski was also a philosopher of science, in the special sense of a philosopher within science, not just an external commentator. There were several reasons for that.

First of all, it was his faith in science as a unique field of culture. On the one hand, he recognized the fundamental role of empirical observation and therefore the distinguished status of the observer. On the other hand, Rayski was a determined supporter of realism. Knowledge was for him an objective activity. No one can reduce the science to straightforward perceptions of our senses, so we have to accept the thesis stating the existence of other cognitive subjects. Why should we not accept the existence of external objects that are different from us? Rayski took this assumption as the necessary condition for starting scientific research of any kind, though he emphasized that it is an axiom, something that cannot be proved. It was an innovative position at a time when physics of isolated objects, possessing immanent features, which can be discovered in an experiment, was commonly subscribed to. Rayski intuitively used the modern system theory approach. He wrote:

> The price to be paid for a possibility of a realistic understanding of measurable parameters is a rejection of the concept of state as an immanent characteristic of the system. However, such resignation has nothing to do with a subjectivism [4, p. 117].

Further, Rayski denied understanding life as a phenomenon belonging to a different domain adverse to the law of physics.

> The modern scientific attitude consists in assuming that the life fits completely into the same framework of description which is valid for any other physical phenomena and is governed by exactly the same laws of physics, the differences being of a quantitative rather than qualitative character [4, p. 112].

It was not a postulate of reduction. Rayski recognized the great level of complexity characteristic of the living organisms, but he was convinced that fundamental features of life could be expressed in the same language, which so successfully works in physics and chemistry.

1. *Distinctiveness and manifest integrity* – many physical objects, like bound states with the total energy less than the sum of the rest energies of their constituents, possess intrinsic integrity, being a "whole" in a similar meaning as the living beings. For Rayski, a living organism is a kind of

gigantic super-complicated molecule. Its features like features of an ordinary molecule depend on the state of the medium surrounding it. The bound states do not exist without the environment. Resulting from such interactions is the energy exchange, which can lead to the growth of such a "living atom". Too large a growth leads to fission as the result of growing surface tension. In fact, the idea of heavy nuclei fission was borrowed by Bohr and Wheeler from biology. One can therefore speak about feedback between the explanatory platforms.

2. *Evolution of living creatures* – after fission, the molecule can grow again until the next fission. Under suitable conditions two different molecules can associate, form a metastable system and disintegrate quickly into two other molecules with an exchange of some of their constituent parts. Thus more stable molecules are produced.

3. *Apparent inconsistency with the law of entropy growth* – the second law of thermodynamics is a statistical rule and applies to isolated systems. The living organism decreases entropy in its interior at the cost of the large growth of entropy in its environment. Fundamental laws of physics are reversible but complex systems consisting of large numbers of particles are best described in statistical terms where irreversibility arises. That is nothing specific to the life process. The difference consists only in the fact that a living creature forms one object, though consisting of billions of atoms, which nevertheless form a bound state not a statistical ensemble. One can hope for the discovery of a precise, quantitative description of certain (rather superfluously chosen) aspects of life, not the whole organism.

4. *The capacity of living creatures to perceive* – to feel and, in its most advanced form, to have rich psychological life, consciousness, logical reasoning and even to produce knowledge and technology. The construction of cybernetic devices shows how numerous functions characteristic to life or intelligence can be performed by automata. It seems that even higher functions of organisms are not specific to life, but rather to the systems of high complexity. The common belief that automata only imitate intelligent behaviour is not obvious. Both in the case of an artificially constructed automaton and natural organism, even a human being close to us, psyche is only a hypothesis stated *per analogiam* to our own feelings. Such hypothesis cannot be proved, as we do not know any method to investigate psychological life of others. The acceptance of that fact forces us to conclude that similar systems should reveal similar features no matter what way they came to existence.

At any rate, the axioms of existence of an external world, independent of our "self" and of the existence of other human beings, whose experiences are similar to ours, are necessary for undertaking any research work. These axioms are not conventional, as we benefit from accepting them. The first is that the existence of the real external world independent of our "self" is the premise of any empirical knowledge, considered as a collective enterprise of humankind. The second identifies psychological phenomena with a special category of physical phenomena and thus completes the domains of science and philosophy.

How distant are these axioms from the narrowly understood "rationalism" of contemporary philosophy? Rayski's philosophy dwells in science, i.e. does not serve its criticism but its understanding of and search for new perspectives of development. It is therefore much closer to the contemporary system epistemology, being a kind of a bootstrap program.

III. In physics, Rayski was mostly interested in two fields. The first was the unification of physical theories. Some features of a field theory started to gain attention of researchers in the 1960s. They searched for a mechanism which would explain the large number fields and elementary particles. Physicists turned their interest to earlier theories describing the space-time in more dimensions, namely works by Kaluza (1921) and Klein (1926). They introduced a model of the field in five-dimensional space, i.e. the five-dimensional world-tube. L. De Witt (1963) and R. Kerner (1963) were keen to extend this unification to Yang-Mills field. Rayski presented Kaluza-Klein theories in the language of modern theory of fields and particles, suggesting further extension to a larger number of space-time dimensions.

IV. The other groups of topics were the problems of interpretations of quantum mechanics. Rayski's position was close to Bohm's and Heisenberg's. He rejected the simple ensemble interpretation. A probabilistic nature of predictions within the framework of the theory does not mean that we describe a genuinely statistical ensemble. Rayski gave his own interpretation claiming that the vector of state represents information about the quantum object available to the observer.

Both questions belong to the borderland of physics and philosophy. It is true that fundamental research in physics require the development of formalism, but not the language. The difference comes from metalogic implied by a logician and metaphysics used by a physicist. This is not a play on words but a completely different understanding of the world. Let us focus on the philosophical aspect of Rayski's interests.

The first group of problems resulted in deep thoughts concerning the actual state of physics and prospects for its development. According to Rayski, fundamental types of interactions should be reduced to four or five, though this does not mean reduction of the interaction terms. These could be expressed in the form of a non-linear function of classical Lagrangian. It would be especially interesting if one searched for a theory sufficiently regular and free from problems with convergence. Ultraviolet divergence results from neglecting the existence of a fundamental constant with dimension of length and the limitations it implies.

The two restrictive constants: c and h did not appear in physics by chance but through deep revolutionary changes in the theoretical system: relativity theory and quantum mechanics. Both required a new worldview. Concluding from history, a new constant should emerge in a similar way within a framework of a new theory.

Such a theory has to arise from a proper idea, which once formulated will be rather simple. According to that principle, Rayski tried to build new models. They possessed one common quality, i.e. were based on a non-linear Lagrangian. The new constant is a parameter separating its linear part. It obviously is of dimension of length and is non-trivially incorporated into the framework of the theory. An example can be: $L = 1/\lambda(1-\exp \lambda * L_0)$; where: L_0 is an ordinary Lagrangian describing the linear approximation.

The idea, though very simple, dramatically changes the foundations of observation and measurement. It breaks linearity and superposition rules, which have been incorporated into physics since Newton. The success of Newton's theory made us feel accustomed to those rules. Rayski thought, however, that superposition is not necessary to probabilistic interpretation and the emergence of new uncertainty relations, he considered interesting.

Theoretical physics includes phenomenological elements referring to the description of pure experiment, though it also gives opportunity of explaining many facts as consequences of more general and fundamental properties. Einstein theory brought gravitation and electromagnetism into the higher level. Both became rather interpretations of deeper and more fundamental ideas. Such unification is not a merely formal procedure but the discovery of a more fundamental hierarchy of real entity. Rayski believed Einstein's intuition; thus, he undertook attempts to impose geometrical character on the unification of gravitational and electro-magnetic, thus weak and strong, interactions. He wrote:

> We are still tapping in a complete darkness as regards a possible meaning of such concepts as isospin and strangeness (or hypercharge) encountered in nuclear physics. This branch of theoretical physics is still in an almost purely phenomenological stage of development. It remains to be hoped that, one day, it will be possible to raise it to a more advanced level of interpretation instead of mere description [4, p. 105].

The other problem, strictly related to philosophy, is the interpretation of quantum mechanics. Let us start the short review of Rayski's position on this issue from the very firm stand in the controversy on whether an interpretation of a theory is its internal unavoidable factor, or rather philosophical exegesis.

Let us recall Eddington: "the flesh of facts is strong, but the spirit of the theory is weak and thus it must quest for interpretation". Rayski considers the very term "interpretation of a physical theory" to be misleading. It seems to suggest an existence of a separate autonomous theory, which is interpreted or reviewed. He writes:

> This is wrong because there does not exist any physical theory without a physical interpretation incorporated into its framework. [...] The fact that the interpretational problems touch the deepest problems of philosophy is another matter, and should not be regarded as a vice but rather as a virtue of modern physics [4, p. 28].

It is the physicist's statement which professional philosophers would not dare to announce. The awareness of the deep and close relation between views of culture that are different but aim at the same entity is the real philosophy of science. The world is unity and, at least in the realm of ideas, this must be the fundamental premise. Without it, we are bound to go round in circles incessantly. Rayski, not without irony, criticizes those who fascinated with the abstract language restrain from apparently simple statements. He cites J. Bub: "By interpretation of a theory, I mean an account that shows in what respect the theory is related to preceding theories". "This is a shocking and completely confusing misuse of the term interpretation" [4, p. 29] – opposes Rayski. Quantum mechanics must have such interpretational link between the language of mathematics and experimental results.

Let us present shortly Rayski's position in some fundamental questions, often leading to much misunderstanding, bearing in mind that this was the discipline that brought him wide appreciation.

Does quantum mechanics refer primarily to single systems of the microworld, or exclusively to their ensembles?

The main reason for the belief that quantum description refers to an ensemble is the deep-rooted prejudice that probabilistic predictions can be verified only with the use of an ensemble, i.e. a large population of identical systems. It comes straight from the definition of probability. Applying this concept to a singular event seems ridiculous. Such a belief is, however, completely wrong, which can be exemplified with, say, a game of roulette. A casino's gains are the result of the probability theory, though it is not the number of games but rather the number of randomly betting players which guarantees the win. The matter lies not in the large number of events but in a large quantity of bets. A similar situation occurs in quantum mechanical predicting. Using quantum mechanics, the croupier can exactly predict the probability that the electron produces a track on a given element of the screen and accept bets against people. He can be sure to win even in the case of a single electron, provided that many people participate in the game and they bet at random. Thus, probabilistic prediction can refer to a single event – not necessarily to a statistical ensemble.

Is a final measurement something qualitatively different from an initial measurement, often referred to as a preparation of initial state? Have we to attribute a special significance to the circumstance that the acts of measurement have usually a finite duration? Is it necessary (or even possible) to require the repeatability of measurements?

According to many authors, the observer is active in the process of preparation of the initial state but passive during the final measurement. However, one may ask whether the initial process cut off the process from its past and the final measurement from its future. Is it really possible to make probabilistic pre-dictions and not retro-dictions?

The formalism of the theory is symmetric with respect to the time reversal and thus it is argued that irreversibility comes from the act of measurement, namely from the interaction of the micro-system with the macro-apparatus. That is the origin of the belief in the distinction of the preparation of an experiment and breaking the rule of repeatability through the unavoidable irreversibility. Moreover, the measurement lasts for some time, which breaks isolation of the system causing changes that cannot be neglected. According to quantum mechanics, however, only the non-commuting variables are subject to disturbance, not the ones represented by the commuting operator, including the multiple measurement of the same value. The rule of repeatability is the necessary assumption in any science.

Rayski thinks that misunderstanding lies in neglecting the difference between the gathering of information and the registering of it. Only the latter is irreversible. For instance, measuring the position of an electron is arresting it forever in a certain part of the film. Mixing up the two states results in erroneous conclusions. It resembles the story of a railway engineer who, in his letter to the Physical Society, pointed out Newton's mistake in the first law of dynamics, arguing that every train put into motion comes to a stop sooner or later.

How to determine (measure) experimentally the wave function? The background for the dispute on interpretation was the feeling, common among many physicists, that the Copenhagen interpretation breaks the postulate of scientific realism. This interpretation ascribes too much attention to the measurement process understood in a subjective way. Moreover, it leads to paradoxes like EPR that have never been satisfactorily solved. The problems disappear if we accept the interpretation that the measurement directly effects the state and thus one of the potentially or virtually possible opportunities comes to reality. This seems, however, to imply a certain kind of "hidden variables" and semi-classical description.

Rayski suggest a different approach. The wave function does not represent the state of the system, but the state of our knowledge of the system, and it is not merely psychological knowledge but the most complete and precise knowledge that we can obtain about from the physical system. It is neither subjectivism nor idealism.

Some questions became irrelevant due such interpretation. The question of whether the system is in a definite state if we do not perform any measurement turns into the question of whether we can obtain any information about the system if we do not get any information. The problem of determining the state also becomes unimportant as function ψ does not describe the state but our knowledge of the state and to this knowledge the limitations on the mutual measurement of non-commuting values apply. At the cost of suppressing the reality content in the wave function and regarding it as an "expectation function" one increases the reality of the concept of an observable. As they always possess a certain (known or unknown) value, the measurement becomes a passive act (measurement of pre-existing values) not a mysterious creation from Aristotelian potential to the act.

The above interpretation enables the revival of some related problems.

The first of them is the Einstein-Podolsky-Rosen paradox. If wave function does not represent the state, nothing prohibits the two particles from possessing really defined states. It is only our information about the remote particle that changes abruptly not the state of the particle. This, however, is nothing unusual.

The other question refers to the so-called quantum logic. Many prominent authors suggest that quantum phenomena do not comply with the two-valued logic. The necessity to quest for multi-valued logic seems to originate from the question: if the state is A, what is the probability of being B? Rayski transforms this question, according to his interpretation, into: if a measurement at a certain moment of a set of observables represented by A gave a value A_m what is the probability that the next measurement at the next moment of the set B gives the value B_m? All mystery disappears. The structure of reality is simple and logical though not necessarily intuitive. The evolution of our intuition proceeded through our direct contact with the macro-world, and the micro-world does not need to and is not a miniaturized copy of macro-world.

What do uncertainty relations mean exactly? What is the proper meaning of the time energy uncertainty with uncertainty within the framework of quantum mechanics?

The last problem is that of the so-called fourth component of the uncertainty relations. The complete analogy between the space-momentum relation $\Delta x_i * \Delta p_i = \frac{1}{2} \hbar$ and the corresponding time-energy relation $\Delta t_i * \Delta E_i = \frac{1}{2} \hbar$ is established once we realize that the former applies to the measurement of the position and momentum at exactly the same time. Similarly, the latter refers to the same place in space. Without this interpretational supplement the analogy does not hold. From the three dimensional point of view Δt means the uncertainty ΔE of the instant when the particle appears within the domain given by the hypersurface of the edge. The product of this uncertainty and the uncertainty of energy cannot be smaller than $\frac{1}{2} \hbar$. Thus, while the ordinary uncertainty relation refers to the initial conditions given on the space-like hyper-surface, the time energy relations holds for the initial conditions given for the time-like hyper-space.

Let us also mention some more general Rayski's remarks on epistemological question of quantum mechanics. They focus on the unclear and sometimes mistakenly analyzed concept of measurement. Rayski starts from the Schwinger analysis of measurement:

1. the object of a measurement is an ensemble
2. the result of the measurement is the pre-selected value a; which value of the set a', a'', ... the quantity A is able to assume depends on the apparatus being used.

Undoubtedly, there are devices called filters, which act in such a way. There are, however, also more subtle devices: analyzers. These analyzers constitute genuine measurements as they answer the question which set of possible values the quantity A possesses and which values are forbidden. An analyzer does not reject anything, it accepts any of the values a', a'', ... but at the same time it registers which one occurs when the system goes through the analyzer. With the use of analyzers we can perform a measurement even of a single object. The measurement performed on an ensemble means in fact an ensemble of measurements not a single act of measurement.

Contrary to the widespread belief in the statistical character of quantum mechanics, Rayski thinks that that the epistemological status of this theory is the opposite: "quantum mechanics constitutes the first example of a theory dealing successfully with single micro-objects whereas classical mechanics is a statistical theory because it deals only with expectation values, i.e. with averages over quantum phenomena" [4, p. 109].

Regarding the wave function as the information about the state solves, in addition, the difficulties connected with the problem of reduction of the wave packet. The discussion initialized by von Neuman analyzes the measurement procedure through the interaction between micro-object and qualitatively different macro-apparatus. Rayski calls such an approach a misunderstanding. One should not mix the apparatus used to perform the measurement with its physical description. These are completely different categories and placing them on the same level of understanding resembles the error of mixing up an element of a set with the set itself. The division of the world into the object of measurement and the apparatus cannot be avoided. The apparatus is an external agent with respect to the observed system. In a formal system, the Goedel theorems holds. Similarly, no conceptual system, including quantum mechanics, cannot be fully explained with the use of homogenous elements. This is another example of deep understanding of the situation in physics.

SELECTED BIBLIOGRAPHY

1. (1965). Unified Field Theory and Modern Physics. *Acta Physica Polonica* **27**, 89-97.
2. (1977). A Refined Born Approximation. *Journal of Computational and Applied Mathematics* **33**, 31.
3. (1978). Survey of Physical Theories from a Methodological Viewpoint. *Zeszyty Naukowe UJ, Prace Fizyczne*, zeszyt **14**.
4. (1979). Problems in Physical Interpretation of the Formalism of Quantum Mechanics. *Foundation of Physics* **9**, 217.
5. (1993). Toward a Unification of "Everything" with Gravitation. *International Journal of Theoretical Physics* **32**, 2125.
6. (1995). Evolution of Physical Ideas Towards Unification. *Zeszyty Naukowe UJ, Seria Fizyka*, zeszyt **XXXVIII**, Kraków.

Poznań Studies in the Philosophy
of the Sciences and the Humanities
2001, vol. 74, pp. 223-232

Józef Misiek

ZYGMUNT CHYLIŃSKI – PHYSICS AND PHILOSOPHY

I. Zygmunt Chyliński (18.08.1930, Cracow – 29.11.1994, Cracow) attended the Nowodworski Grammar School (partly in the underground schooling system). In 1948 he finished his secondary school and started studying physics and mathematics at the Jagiellonian University. He was granted an MSc degree in 1952. In the same year, he was employed at the University as Jan Weyssenhoff's assistant. In 1958, he became a doctor of sciences and in 1968 he received his postdoctoral degree.

Chyliński's academic output covers nearly all branches of physics – from the phenomenology of elementary particles to the problem of hydrogen's diffusion in metals. His main area of interest, however, was quantum mechanics. It was his philosophical reflections geared up with this theory that led him from narrowly understood physics to the domain of philosophy.

In the late 1960s, Chyliński took up a problem which preoccupied him for the rest of his lifetime. It was the problem of consistency of relativistic and quantum theories. The work on this problem brought about several publications which appeared both in philosophical and physical journals. In 1992, Chyliński published a book entitled *Kwanty i relatywistyka* [Quanta and Relativist Physics] [1.1], which was a sort of summary of the author's earlier achievements in this field. The present work is based mostly on that book.

After Chyliński's death a more technical version of his book was published in English [1.2].

II. For Chyliński, philosophy is not a set of established truths separated from physics by an insurmountable barrier. All the more so, it is not a set of metaphysical nonsense. Philosophy is treated rather as a method which a researcher must apply when s/he goes beyond the area of what is sufficiently examined and described, when the standard instruments used in his/her discipline fail to work. In particular, this concerns the situation when the researcher tries to improve the fundamental principles and/or basic concepts which prove inadequate to describe the examined reality.

It stands to reason that this role of philosophy is not Chyliński's invention. All great physicists trod this way. However, his undeniable contribution consists in the fact that in the times when neopositivism was commonly recognized as the only scientific philosophy, Chyliński was brave enough to present his own attitude irrespective of the damage it could cause to his academic career.

III. Despite the instrumentalism and operationalism that are so popular today, Chyliński adopted an realistic attitude to physical theories. This means that, according to him, a physical theory is something more than a mere abstract scheme that may be evaluated in one way only – by way of comparing with the results of experiments. For Chyliński, any theory was a means of describing the real world; the world that does not consist of facts but of real objects endowed with various properties and relations. This is why theories can be evaluated as inconsistent [1.1, p. 156] if the laws they postulate are not compatible with the laws governing the measurement of quantities described by these theories. Such an internal inconsistency is a serious drawback in the case of the theories which we would like to regard as fundamental. Let us stress once more: such an evaluation has nothing to do with inconsistency between theory and experiment.

Objects about which a theory speaks need not be directly observable, because the theory decides what is measurable.

> Obviously, as Einstein taught us, a theory must be adjusted to the possibilities of measurement, sometimes at the price of paradigms, but there is no pure observation. No one has ever seen force, and its hypothetical shape decides about the visible course of a classical particle [1.1, p. 91].

The thesis that a theory decides what is measurable refers to the special theory of relativity which posited – against the postulates of classical physics – that absolute simultaneity was unmeasurable. Unmeasurability does not signify here any limitations of human cognitive abilities but, just the contrary, it results from the recognition of a highly important and positive fact that there are no signals faster than light in nature.

IV. We may assume that Chyliński adopted a kind of operationalism in his philosophy. It must be underlined, however, that it was a specific operationalism which had nothing in common with the neopositivist tendency towards an instrumental treatment of concepts and theories in physics. According to the neopositivist conception of theories, their task consists in a faithful description of empirical facts and therefore all concepts referring to non-observable objects, properties or relations are merely intellectual constructs which serve the purpose of a concise presentation of facts. This is why neopositivism accepted Bridgman's operationalism which admitted such instrumental interpretation.

In physics, operationalism, it seems, derives from Einstein's analysis of the concept of simultaneity. It is widely known that Einstein rejected the concept of absolute simultaneity claiming that if there existed a limit to the velocity of signals, simultaneity must be relative. Einstein's whole argumentation, particularly that which appeared in popular works on the subject could evoke an impression that absolute simultaneity is unobservable and therefore it must be a metaphysical element in a physical theory. Here, however, we should be cautious because:

a) Einstein's experiments were only thought experiments;
b) Einstein did not analyze all possible experiments (since it was impossible), but only some of them.

Thus, Einstein's reasoning is based on certain assumptions of theoretical nature. One of them is the light principle, claiming that the speed of light depends neither on the speed of its source nor on the speed of the observer. Without it, Einstein's thought experiments do not yield any definite result. The second assumption is the principle of relativity. And the aim of this reasoning consists in showing that the above principles can be compatible if we give up the concept of absolute simultaneity. None of these principles can be derived from any real or thought measurement. Ingenuity of Einstein's reasoning lies in the fact that he derives these principles from Maxwell-Lorentz's electrodynamics, that is from the theory which he wants not so much to reject as to improve. Einstein aims rather at proving that all measurements, real or thought ones, must be compatible with these principles, hence they must take relativity of simultaneity into account.

All this has nothing to do with positivist operationalism which strives to prove that abstract physical concepts can be defined resorting to measurement operations. Defining is understood here in accord with the general instrumentalist tendency, i.e. as a means to show that a given concept does not refer to anything real save the very defining operations.

Such a claim implicitly assumes that measurement operations corresponding to different methods of measurement must lead to the same result. Such an implicit assumption may be justified, in some cases, if we assume the existence of appropriate laws of physics. This, in turn, shakes the foundations of the positivist operationalism since it aims at showing that laws of physics can be empirically tested because physical quantities that are bound by these laws can be measured without assuming any laws of physics. Einstein expressed it much more concisely in one of his later works: it is the theory that decides what can be measured. Chyliński fully accepts this view. Thus, we can state that a similarity between the discussed version of operationalism is illusory and it issues from a superficial understanding of Einstein's operationalism by neopositivists.

V. Chyliński's works shed a new light on the two classical approaches to nature of space-time: Newtonian conception of substantial space-time and Leibnizian conception of relational space-time. His attitude is not absolutely clear, since he uses the term "relational space" interchangeably with "attributive space", which refers to Descartes' conception. We can expect, however, that distinguishing between these two last conceptions is not necessary – at least for Chyliński – because they can both be treated as modifications of one anti-substantial conception.

In contradistinction to the views of most relativists, Chyliński's approach is anti-substantial. Moreover, he indicates that the conception of substantial space-time leads to the conclusion that the problem of one body in physics makes sense, while on the ground of anti-substantial conception it does not. According to the latter conception, it needs two or more bodies to make the problem possible to consider. On the other hand, however, it is known that in various physical theories a problem of one body can be posed. Therefore, we might conclude that physics denies anti-substantialism. Yet, Chyliński states that the so-called problem of one body in physics is always a problem of interaction between one body and an external field produced by another, infinitely heavy body. That is why, in fact, there is no problem of one body, and the above argument against anti-substantialism should be treated as unjustified.

This course of thinking leads to a conclusion that all physical theories are based on the assumption of the existence of an infinitely heavy body that determines the class of inertial systems. Such an assumption is acceptable as idealization which – due to certain peculiarities of Gallilean kinematics – may be compatible with non-relativist physics but in the case of relativist physics it brings about various difficulties. That is why the future theory that will make quantum and relativist physics compatible must abolish this idealization. Similarly, the beginning of relativist kinematics was connected with abolishing the idealization regarding the existence of infinitely fast signals.

Therefore, the future theory unifying quantum ideas and relativist ones must include not only the universal constants c (velocity of light) and h (Planck's constant) but also an additional parameter M (of the dimension of mass) representing the mass of a reference system. Thus, such a theory allows for three independent limiting cases (light velocity tends to infinity, Planck's constant tends to zero and M tends to infinity). In this way all fundamental theories known in physics can be obtained. This means that the theory postulated by Chyliński corresponds with the well-known basic theories. It stands to reason that it is a postulate of heuristic nature because the theory discussed here has not been developed yet and the postulate of correspondence is, in its nature, a plausible heuristic principle and not a proved theorem of physics.

VI. The principle of wave-particle dualism came into being at the beginning of quantum physics. While referring to quantum dualism Chyliński seems to speak of the dualism of physical quantities represented by non-commuting operators. He also claims that the uncertainty principles issue from duality. Such an understanding of dualism is in accord with the conception which N. Bohr called the principle of complementarity. We must add, however, that Bohr drew conclusions (on the basis of this principle) that were incompatible with realism, while Chyliński maintains that he did not have to reject realism. That is why we shall first briefly discuss Bohr's principle of complementarity and then indicate at which points it differs from quantum dualism as Chyliński understood it.

Bohr believes that the principle of complementarity is a discovery made in physics, whose importance goes far beyond this discipline. The principle states that the result of a measurement in quantum mechanics is relative in the sense that it depends on the whole measurement system that must embrace classical measuring instruments (i.e. such instruments which satisfy the laws of classical physics). That is why the state of a quantum system, independent of measurement, has no physical sense. It is only the results of definite measurements that have physical sense. Bohr used this formulation of the principle of complementarity in his discussion with Einstein to convince him that this principle is a kind of generalization of the principle of relativity. Just as the principle of relativity which requires that every physical measurement should refer to a definite inertial system, the principle of complementarity requires that every quantum measurement should refer to a definite measurement apparatus. And since – according to Bohr – the measurement apparatus is always classical and it is impossible to construct instruments that would measure two quantities represented by non-commuting operators, measuring of one quantity may exclude measuring another. Each of these measurements provide information about the system, relativized to the applied measurement apparatus but both these pieces of information are incompatible within the frames of any classical description.

The principle of complementarity, like the principle of relativity, is a heuristic principle and an each of them served as an instrument for the proper construction of quantum mechanics and special theory of relativity respectively. The difference between them is connected with the fact that at present the principle of relativity has a precise formulation while the principle of complementarity does not have it. This means that the principle of complementarity has not exhausted its heuristic power yet, and that its precise sense is not yet known.

The differentiation between the potentiality of quantum systems and their actualization achieved as a result of measurement is a consequence of the principle of complementarity. It means, among others, that the nature of quantum objects is completely different from that of classical objects because the measurement, according to quantum mechanics, not so much reveals the state of the system existing before the measurement, but rather the fact that the act of measurement creates new state of the system, usually different from that existing before the measurement. What is more, such formulation of the problem leads to a sharp dualism between quantum objects and classical measurement instruments. In effect, it is easiest to make the principle of complementarity compatible with instrumental understanding of quantum objects. In fact, the classical interpretation of quantum mechanics called the Copenhagen interpretation tends in this direction.

As it has turned out, the Copenhagen interpretation is sufficient to allow for application of quantum mechanics to describe a broad range of phenomena on the quantum level, and the results obtained in this way are amazingly concordant with measurement results. It ceases to be sufficient when we consider conceptual difficulties unsolved by this interpretation. Difficulties of this kind are particularly distinctly manifested when we assume that quantum mechanics is a fundamental theory, i.e. that it regards not only micro-objects but also macro-objects. Then, it turns out that, on the one hand, quantum mechanics is a generalization of classical mechanics, and, on the other hand, that the latter, which is its limiting case, must at the same time constitute its foundation as the theory of classical measurement instruments.

It seems that Chyliński had his own understanding of the principle of complementarity, which was closer to realism than to instrumentalism. This understanding is "interwoven" into his efforts aimed at developing a theory that would be a natural synthesis of quantum mechanics and the theory of relativity. Perhaps a thorough analysis of his works could reveal this understanding at least to some extent. The next chapter includes certain remarks on this subject.

VII. The problem of making quantum theories and relativist ones compatible occurred in physics immediately after the discovery of quantum mechanics. The field theory is usually recognized as a solution to this problem. Not all physicists, however, share this conviction. For instance, D. Bohm claims that relativist theories cannot be compatible with quantum theories because the former are local, linear and deterministic, while the latter are non-local, discreet and indeterministic. Chyliński expresses his approval of this view [1.1, p. 9], and he himself indicates that all applications of field theory are connected with the theory of perturbations, which is based on the adiabatic hypothesis. The latter is limited to the assumption that fields exert influence neither in the infinite past nor in the infinite future. This allows for the linearization of the initial field equations and application of perturbation calculus [1.1, p. 87].

This kind of approach is correct on the condition that it regards scattering states of two or more objects because only then can we assume that in the infinite past and future these objects are infinitely distant from one another. In the case of bound states such approach is theoretically unjustified and, in fact, it leads to various difficulties [1.1, p. 88].

Posing this diagnosis, Chyliński evaluates field theory as a provisional structure and tries to find a theory based on more solid foundations. Two observations constitute the starting point for his considerations:

1) Classical (i.e. nonrelativist) quantum mechanics allows for an examination and explication of an enormous range of phenomena. It is possible thanks to the fact that this theory allows for the separation of internal degrees of freedom of the system from the external ones.

2) Kinematics of the special theory of relativity is "unfavourable" for all interactions. On the classical level, it is expressed in the form of non-interaction theorems. On the level of quantum and relativist theories it evokes various difficulties connected with the inseparability of the internal and external degrees of freedom. For instance, in the theory derived from the Bethe-Salpeter's equation the internal state of the isolated complex system "must depend on the total four-momentum P no matter whether this system is in, scattering state or in bound one" [1.1, p. 59]. "In other words, the external state of the system $A_1 + A_2$ characterized by momentum P conditions its internal structure" [1.1, p. 60]. This means that the concept of internal structure cannot be maintained in this kind of theory. What is more, it is in accord with Landau and Lifszyc's general theorem that elementary particles must be treated as geometric points. No wonder that the concept of a form factor – necessary in experimental physics – has no firm theoretical basis.

Chyliński's main idea regarding the compatibility of quantum and relativist theories refers to Heisenberg's view that isolated quantum objects represent the hitherto unknown kind of reality called quantum potentiality. A mathematical expression of this potentiality can be found in operators which represent physical quantities in quantum mechanics. Potentiality is not directly accessible in measurement because measurement is connected with interaction of quantum potentiality with a classical measurement instrument. As a result of this interaction, quantum potentiality undergoes reduction which is mathematically described as projection of the quantum state on the measurement instrument's eigenstates. This view of Heisenberg was not approved of by those physicists who value definite equations (e.g. Schroedinger's equation) and do not favour philosophical speculations. Chyliński did otherwise. He accepted that

> Quantum mechanics discovered a two-level structure of the world connected with two kinds of "processes" [in time]. The problem is that the process of measurement [...] itself cannot be subject to Schroedinger's equation [1.1, p. 80].

In connection with the above Chyliński postulates (I omit here a thorough analysis of general situation in physics and the extremely interesting heuristic considerations that lead to this conclusion) that the isolated (and therefore non-observable) quantum systems evolve not "against the background" of Minkowski's space-time which he calls measurement space-time but "against the background" of the "internal spacetime of microworld" which is a Cartesian product of the space R_3 (representing the internal space of the system) and one-dimensional space R_1 (representing the internal time of the system). At the same time, however, he presents a heuristic reasoning which shows that the results of measurements conducted on such a quantum system are in accord with relativist symmetry because they are obtained with the use of macroscopic measurement apparatus which is subject to this symmetry.

This hypothesis, modestly called by Chyliński the hypothesis of geometry of R provides a new point of view on various problems of contemporary physics. To illustrate this, we shall choose just two of them.

The first one is the problem of inseparability of internal and external degrees of freedom. R hypothesis solves this problem and therefore creates a possibility of quantum-relativist description of bound states [1.1, p. 61, 62].

The second problem regards Landau-Peierls's principle of uncertainty which states that the product of uncertainty of momentum and of time must be higher than the product of Planck's constant and light velocity. This principle is both of quantum and relativist nature because both universal constants: Planck's constant and light velocity occur in it [1.1, p. 84]. Nevertheless, its physical sense is not quite clear. In connection with this principle, Landau himself postulated that the future theory, being both quantum and relativist, must reject temporary-local equations of movement [1.1, p. 85]. Chyliński supplements this principle with his own [1.1, p. 74, formula 10.3], which regards the uncertainty of the zero-th component of the four-momentum (which multiplied by time uncertainty must be higher than the product of Planck's constant and light velocity). According to Chyliński both these principles acquire a clear physical sense if they are referred to internal variables of an isolated quantum system.

SELECTED BIBLIOGRAPHY

1. Books:

1. (1992). *Kwanty a relatywistyka czyli relacjonizm a relatywizm* [Quanta and Relativist Physics or Relationism and Relativism]. Kraków.

2. Papers:

1. (1965). Uncertainty Relation Between Time and Energy. *Acta Physica Polonica* **28**, 631-638.
2. (1965). Zagadnienie lokalizacji czasoprzestrzennej w świetle zasad nieoznaczoności [The Problem of Spatio-temporal Localization in the Light of Uncertainty Principles]. *Studia Filozoficzne* **43**, 47-65.
3. (1966). The Nature of Lorentz Invariants. *Acta Physica Polonica* **30**, 293-321.
4. (1966). Red Shift in Gravitational Field. *Zeszyty naukowe UJ* **121**, 31-36.
5. (1966). Two-Body Problem. *Zeszyty naukowe UJ* **122**, 19-30.
6. (1966). Localization Problem and Space-Time Continuum. *Studia Filozoficzne* 3, 19-41.
7. (1967). The Koenig Theorem and Relativity. *Acta Physica Polonica* **32**, 3-20.
8. (1967). High-energy Collisions of Nonrelativistic Composite Systems. *Acta Physica Polonica* **32**, 839-853.
9. (1968). Analysis of Measurement in Macro- and Microphysics. *Nukleonika* **13**, 23-28.
10. (1968). Czasoprzestrzeń jako kontinuum graniczne [Space-time as Bound Continuum]. *Studia Filozoficzne* 54-55, 78-87.
11. (1972). Wave Functions of Fast-Moving Two-Body Systems. *Acta Physica Polonica* **B5**, 575-583.
12. (1975). Non-Equivalent Hamiltonians and Relativistic Symmetry of Mechanics. *Acta Physica Polonica* **B6**, 387-395.
13. (1977). Czy symetria relatywistyczna może być uwarunkowana? [Can Relativist Symmetry be Conditioned?]. *Studia Filozoficzne* 135, 39-54.
14. (1978/9). Podstawowe modele teoretyczne fizyki i operacjonizm [Foundamental Theoretical Models of Physics and Operationalism]. In: *Zagadnienia Filozoficzne w Nauce* I, Kraków.
15. (1980). Rola symetrii czasoprzestrzennej w światopoglądzie [The Role of Spatio-temporal Symmetry in the Worldview]. In: *Nauka–Religia–Dzieje. Seminarium w Castel Gandolfo. 16 sierpień 1980* [Science–Religion–History. Seminar in Castel Gandolfo. 16th August, 1980]. Roma: Abilque, 150-157.
16. (1982). Bound States and Lorentz-Poincaré Symmetry. *Acta Physica Polonica* **B13**, 625-665.
17. (1983). Operationalism and the Fundamental Models of Physics. In: *Philosophy in Science* 1, Tucson, 149-163.
18. (1984). Experimental Foundations of Relational Spacetime. *Acta Physica Polonica* **A65**, 369-375.
19. (1985). Internal Symmetry of Quantum Systems and Vertices of Composite Particles. *Physical Review* **A32**, 764-774.
20. (1987). Dilatation of Lifetimes of Micro-Clocks and Quantum Nonlocality. *Acta Physica Polonica* **B18**, 1165-1178.
21. (1987). O tak zwanych twierdzeniach limitacyjnych [On the So-called Limitative Theorems]. *Zagadnienia Naukoznawstwa* **91-92**, 401-410.
22. (1988). Relacyjny czas fizyki [Relational Time of Physics]. *Studia Filozoficzne* 267, 39-53.
23. (1988). A Quantum-Relativistic Puzzle of the Meso-Atom Decay. *Physics Letters* **A134**, 152-160.

24. (1992). The Measurement of Red-Shift and Flat Spacetime. *Acta Physica Polonica* **B23**, 745-757.
25. (1993). Gravity and Non-Extensive Nature of Mass. *Acta Physica Polonica* **B24**, 1475-1479.
26. (1995). Relationism of Quantum Physics. *Acta Physica Polonica* **B26**, 1547-1665.

Poznań Studies in the Philosophy
of the Sciences and the Humanities
2001, vol. 74, pp. 233-238

Małgorzata Czarnocka

GRZEGORZ BIAŁKOWSKI – SCIENCE AND ITS SUBJECT

I. Grzegorz Białkowski (8.12.1932, Warsaw – 29.06.1989, Warsaw) was one of the most distinguished Polish elementary particle physicists. Since 1971, he was a professor of Warsaw University and Head of the High Energy Theory Department in the Institute of Theoretical Physics. He was also a poet (five volumes of his poetry were published) and the president of Warsaw University (since 1985).

II. Białkowski's philosophical views are composed of two elements, inconsistent with each other. One element is close to the standards of metascientific opinions that are widespread among physicists: close, but not identical, because of its epistemological maximalism and related metaphysical claims. By contrast, the other element of his views pictures an exceptional image of science. Notwithstanding its incompleteness and sketchiness, this image determines a clear and uniform basis of a model of science, since it lays out a definite perspective (by specifying the conceptual apparatus and problems to be posed) of apprehending science. Białkowski's idea of science differs essentially from present-day conceptions of science. First, its differs from current philosophical views of physicists, or, to be more precise, it differs from the metascientific opinions shared by the physicists community. Second, it differs from theories of science constructed by professional philosophers.

Białkowski's philosophical views can be adequately specified as romantic ones, in a rather loose sense of the term. The term refers to his conservative ideas (that is, those viewed as outdated in the philosophy of science, of only historical significance for it present-day state) as well as ideas that are innovative with respect to the standards of present-day philosophizing about science.

In Białkowski's conception physics does not solve puzzles nor explore natural phenomena. It is inspired by a deeper and nobler goal, namely, an everlasting fundamental intellectual (and, perhaps, also emotional) goal to view the world as an unity, to discover few fundamentals of the nature which are presented to the cognitive subject in the form of the vast variety of phenomena. In Białkowski's opinion, the most adequate notion for determining the quest of physics is the ancient notion *arche*. Białkowski claims that elementary particles and, on a deeper level, quarks are contemporary *arche*. Thus, he admits physics to be the basic science, the most appropriate one to disclose the basis of all reality, its deepest level, in an ontological sense, and not relativized to the state of scientific knowledge. Hence, he reserves for physics a distinguished place of metaphysics, and not just a special science. This return to pre-Aristotelian philosophy of nature is a conservative element of his views.

III. Two of Białkowski's broad ideas seem to be of special importance. Their value consists in that they outline an entirely new inquiry into science. Both ideas inspired an entirely new way of grasping science, different from those prevailing in the contemporary thinking about science – in philosophy and beyond it as well. They initiate completely new discussions on science – carried in new conceptual frameworks, embracing entirely new questions, and, as a result, offering an image of science constructed from an astonishing perspective, even rebellious one. The ideas are partial and incomplete. Despite their partiality and incompleteness, they can – if appropriately extended – be a basis of radical new models of science. In short, Białkowski claims, first, that a specific human being (not reduced to the Cartesian cognitive subject, that is to a pure rational mind) element of acquiring scientific knowledge and, second, that science resembles art in some essential respects.

In the centre of his model of science, Białkowski situated a specific *human* cognitive subject. His notion of the cognitive subject is much richer than that notoriously adopted in epistemology since Descartes. It differs significantly from the modern notions of cognitive subject elaborated in the philosophy of science as well. In Białkowski's view, being a human being (at the various levels of human existence) determines forms of physics (and of all science), its achievements, the ways in which sciences develops. For Białkowski, human goals and cognitive yearning, human biological nature, human prephilosophical ideas (especially metaphysical ones), emotions, aesthetic values accepted by human beings, and – last but not least – their individual ways of thinking play a role in scientific cognition. Białkowski distinguishes three types of subject-loading of scientific knowledge: social, specific, and related to the individual features of the cognitive subject. Białkowski does not question the social nor species-related constitution of the cognitive subject. However, he focuses his considerations on the subjective level of the cognitive subject constitution. While the social and species-related level are admitted by at least some contemporary philosophies of science (e.g. those grounded in Marxism, those drawing on the so-called strong programme in the sociology of knowledge, and from evolutionary theories of knowledge) the influence of the subjective level is much more controversial. Commonly, it is neglected as occurring only in nonveridical cases of knowledge acquisition, i.e. in nonveridical sense perceptions. However, in Białkowski's image of science the individuality of the human being, his/her mental specificity, all that occurs in the human internal psychological sphere are essential and non-eliminable (e.g. by means of reduction to the human biological nature) elements of knowledge acquisition. In Białkowski's opinion, not only forms but also contents of scientific knowledge are loaded by human subjective spheres. He points to the phenomenon of the so-called parallel discoveries as the most salient historical evidence of subjective loading. However, it is an extremely complex matter to show on the basis of some historical evidence that the explorer's subjective spheres share in the acquisition of scientific knowledge. It requires to study in detail the histories of the lives of authors of parallel discoveries. In particular, it requires taking into account all elements forming the discoverer's personalities.

IV. Białkowski reveals subjective elements of knowledge among other things in criteria of scientific knowledge, that is, in general, the criteria of admitting scientific knowledge, or of testing it. He specifies nine criteria used in scientific practice. Most of them embrace non-objectivizable factors. Thus, such subjective criteria introduce subjective factors to scientific knowledge. Białkowski modifies the criterion of intersubjectivity. In the version widely accepted in the philosophy of science, this criterion guarantees the complete equivalence of all cognitive subjects with respect to cognition and, in result, to scientific knowledge. The criterion of intersubjectivity in the version adopted in the philosophy of science rejects any cognitive privilege of any subject. On the contrary, Białkowski rejects the total (i.e. unconditional) democracy of all cognitive subjects. He claims that the criterion of intersubjectivity obliges one "to refuse (neglect) all explanations referring to some features of an individual person or even of a group of persons if their privileged status is not given" [1.3, p. 385]. If there are privileged cognitive subjects then some features of individual subjects have (in some conditions at least) an epistemic value. Thus, Białkowski's version of the intersubjectivity criterion admits subjective factors in scientific knowledge.

Białkowski formulates the criterion of simplicity in a nonstandard way, different from versions reconstructed in the philosophy of science. Namely, he applies the notion of imagination. A theory is simple if it is based on physical assumptions easily attainable by the individual imagination of the cognitive subject. Imagination is subjective. It can be quite different for any two individual scientists, even trained in comprehending abstract assumptions by the same methods. The criterion of simplicity in Białkowski's version embraces obviously subjective elements.

The criterion of convenience inclines scientists to choose a known, already practised set of forms of representations. Choosing this criterion, physicists frequently apply continuous functions. As a result of this they can apply perfectly known differential calculus [1.3, p. 388-389]. Although the criterion of convenience refers to the notion of representation, it is of fully pragmatic type. The very essence of this criterion ignores – despite Białkowski's reference to representation – any idea of reliable representation of the investigated object. However, it would be not so if Białkowski had been introduced an additional condition for allowed forms of knowledge. In any case, the subjective impression of convenience is more important in science than the tendency to construct reliable representations, that is representations corresponding (but not uniquely copying) in some loose sense to the explored domain (but not copying it uniquely).

In Białkowski's image of science, the criterion of beauty plays a significant role. The criterion is comprehended in a special way as it does not reduce itself to another criteria, e.g. to the criterion of simplicity. Białkowski claims that the estimations of theories, models, etc., resembles experiencing works of art. In estimating a cognitive result any cognitive subject takes into account (among other things) its aesthetic value, especially its beauty. The beauty of scientific cognitive results renders the beauty of the nature itself as well as the beauty of its symbolic coding.

In Białkowski's view, the aesthetic experience in science embraces such elements as the admiration at the depth and originality of the very idea, the impression of dramatizing the leap from the given evidence to the final cognitive result, the feeling of the irrefutability of the result and of its ingenious character. An extensiveness of consequences is also an element of the beauty in science [1.3, p. 391]. Scientific beauty embraces the ingeniousness and novelty of the scientific idea. Beauty ideas are beautiful if they effect our imagination especially strongly [1.3, p. 389]. The beauty of scientific works is a product of human creation, of coding the nature by means of a symbolic code and decoding it. It is, however, also a product of the nature itself, of its structures and symmetries.

Białkowski rejects the idea of the conventional origins of criteria. He indicates the origins of criteria in the nature of subject behaviour, in the structure of psychical powers, and of human material, structural and functional unity with human environment, and in the consciousness of this unity [2.7]. Psychological powers impact on the criteria especially strongly. For instance, the criterion of conservativeness origins in the way human imagination functions, especially, the limited restricted character of the imagination. The man chooses to use those explanations which add as little as possible to the knowledge already accepted. The criterion of convenience has similar origins [1.3, pp. 385-386, 388].

The source of the criterion of beauty inheres in the subject mental and emotional sphere, in the subject impression of communicating with mystery. Mystery is necessary for internal mental equivalence of the subject. "Mystery generates an internal tension in a man enabling him to be creative" [2.6, p. 25].

Białkowski rejects the isolated mind as representing the cognitive subject. Instead of this, he sketches the picture of the multidimensional subject. In his construction, the subject is connected with his space-time environment, is determined by various social links (chains), and is conditioned by biological and psychological needs as well as by his emotional system. This notion of cognitive subject generates essential changes in the image of science in comparison with the hitherto constructed epistemological models.

Contemporary studies on science are founded on the assumption of the peculiarity of science. According to it, science is a phenomenon which is autonomous and incomparable with other specialized branches of culture, above all with art. Białkowski rejects this assumption, pointing to aspects of science which resemble relevant aspects of art. Science as well as art embrace subjective elements. He claims that cognitive scientific results as well as works of art offer information on their subjects. Modern art tends, as Białkowski maintains, to minimize subjective elements and to avoid the burden of art creators. In this, it resembles a science in which tendencies to neglect cognitive subjects have been transformed into the dogma of science objectivity [2.6]. Science as well as art refer to the category of beauty to estimate results. In Białkowski's view, objects created in artistic processes are, in fact, objects of a metaphysical thought experience. Cognitive scientific results are of a similar type. They do not take into account all real conditions either. Thus, they refer to a possible world. Contents of works of art are always presented in a coded form. Their comprehension consists mainly in decoding their symbolic forms. This process resembles the so-called interpretation of cognitive results.

V. A general outstanding value of Białkowski's views presented above is their controversial character. Białkowski neglects – by proposing a constructive counterproposal – some main basic elements of the contemporary models of science and challenges aspects of the image of science. In Białkowski's view, the subject of science is creative, follows among other things by (to be guided, to be prompted, to be influenced by something) subjective agents (e.g. beauty, emotions, imaginary). Scientific cognitive results resemble works of art. Both are products of human imagination, aesthetic feelings, emotions, metaphysical quest, the symbolic coding of a world. Therefore, Białkowski claims that the predominant image in which science is reduced to algorithmic operations, calculations, and precise measuring is completely erroneous.

SELECTED BIBLIOGRAPHY

1. Books:

1. (1980). *Stare i nowe drogi fizyki.* T. I: *U źródeł fizyki współczesnej* [Old and New Ways of Physics. Vol. I: At the Sources of Modern Physics]. Warszawa: Wiedza Powszechna.
2. (1983). *Stare i nowe drogi fizyki.* T. II: *Fizyka XX wieku* [Old and New Ways of Physics. Vol. II: 20th-century Physics]. Warszawa: Wiedza Powszechna.
3. (1985). *Stare i nowe drogi fizyki.* T. III: *Fizyka dnia dzisiejszego* [Old and New Ways of Physics. Vol. III: Present-day Physics]. Warszawa: Wiedza Powszechna.

2. Papers:

1. (1975). Teoria naukowa a eksperyment artystyczny [Scientific Theory and Artistic Experiment]. *Studia Filozoficzne* **9**, 59-75.
2. (1978). Cognitive and Aesthetic Values in Artistic and Scientific Work. *Dialectics and Humanism* **5** , 2, 39-52.
3. (1980). Świat nauki a świat sztuki [The World of Science and the World of Art.]. *Życie Literackie* **30**, 29, 1-17.
4. (1980). Twórczość pseudonaukowa [Pseudoscientific Creation]. *Problemy* **12**, 12(417), 12-16.
5. (1981). Subjective Determinants of the Process of Cognition in Physics. *Science of Science* **3**, 261-278.
6. (1981b). Uwarunkowania podmiotowe procesu poznawczego w fizyce [Subject Conditioning of the Cognitive Process in Physics]. *Zagadnienia Naukoznawstwa* **17**, 1-2, 13-27.
7. (1983). Rola wyobraźni i emocji w nauce i w sztuce [The Role of Imaginary and of Emotions in Science and in Art.]. *Problemy* 4(441), 42-47.

APPENDIX

POLISH PHILOSOPHY OF SCIENCE.
THREE COMMENTARIES

Poznań Studies in the Philosophy
of the Sciences and the Humanities
2001, vol. 74, pp. 241-249

Kazimierz Ajdukiewicz

LOGISTIC ANTI-IRRATIONALISM IN POLAND*

The present paper is an account of the Polish schools of thought associated with the orientation represented by the so-called Vienna Circle. I am presenting it here at the invitation of the conference organizers. It is to be merely a rough sketch, designed to provide some introductory orientation in the field, and it is therefore going to be cursory and incomplete.

To start with, let us establish which Polish thinkers will be included in the group thought to be close to the Vienna Circle. There are no unreserved supporters of the Vienna Circle in Poland; I do not know of any philosopher who has accepted and adopted the actual contents of the theses propounded by the Vienna Circle. At best, the affinity between some Polish philosophers and the Vienna Circle consists in a similarity of the basic methodological attitude and the kinship between the issues under investigation. The hallmarks of the methodological attitude include: *anti-irrationalism*, i.e. the postulate stating that only such propositions can be acknowledged which are justified in a way that can be verified, and *linguistic precision*. Apart from those two hallmarks, one should also stress the third element, i.e. accepting the logistic conceptual apparatus and the powerful influence of symbolic logic. As far as the scope of the investigation is concerned, priority is given to issues which have scientific cognition as their object, and thus to issues in the so-called metatheoretical research. This involves an interest in semantics, rooted in the belief that cognition itself can be investigated only through the consideration of its verbal expression. Closely connected with the above are the deliberations on the foundations of science, i.e. not so much a metatheoretical as an intratheoretical investigation of the fundamental areas of specific sciences, especially deductive sciences.

The methodological hallmarks and the range of issues outlined above are characteristic both of the philosophy of the Vienna Circle and of the research conducted by a large circle of Polish philosophers. Hence, those

---

* Originally published in Polish as "Logistyczny antyirracjonalizm w Polsce" in *Przegląd Filozoficzny* **37**, 4 (1934), 399-408.

who research the relevant issues in accordance with the methodological criteria listed above will be treated as being close to the Vienna Circle, and they are the subject of this paper.

Before I look at the present situation, however, I would like to embark on a brief historical tour. The anti-irrationalist and antimetaphysical tendencies are hardly anything new in Poland. Jan Śniadecki, a leading Polish philosopher and mathematician of the late 18th century and the early 19th century, was one of the most ardent champions of anti-irrationalism and a dedicated opponent of metaphysical phantasy. He denounced first Kant and then Romanticism, and he can be said to be a forerunner of Comte's positivist theory of cognition. He was not alone in his combat against the irrationalism of Romanticism. He was assisted by his brother Jędrzej Śniadecki, by the philosopher Przeczytański, and by others; all of them opposed – in the spirit of positivism – the increasingly influential irrationalist metaphysics of the Polish romantics. In the latter half of the 19th century the irrationalist orientation lost its momentum, and positivist started to dominate in Poland. Its representatives, who were mainly influenced by Mill and Spencer, included Julian Ochorowicz and Aleksander Świętochowski. It is under their influence that a new orientation, known in the history of Polish philosophy as "Warsaw positivism" came into being, with anti-irrationalism as its principal characteristic feature.

In today's Poland, however, anti-irrationalist trends with a logistic flavour are not directly descended from the Warsaw positivism. The foundations for expanding this orientation were laid by someone else, namely by Kazimierz Twardowski and Jan Łukasiewicz. Twardowski was a disciple of a well-known Austrian philosopher, Franz Brentano, who wrote the following memorable sentence in his postdoctoral dissertation of 1866: "Vera philosophiae methodus nulla alia nisi scientiae naturalis est" (F. Brentano, *Über die Zukunft der Philosophie,* Leipzig 1929, p. 136). Twardowski remained faithful to his master's anti-irrationalism. After a brief period when he held the position of associate professor in Vienna, Twardowski started teaching at Lvov University as full professor in 1895. Since then, he was active as a professor of philosophy for 30 years and educated dozens of his followers. Twardowski himself did not belong to the anti-irrationalist orientation with a logistic slant, because logistics did not influence him in any major way. However, he can be credited with preparing the ground for logistics, because – both in his writings and his teaching – he fervently combated conceptual vagueness and raised his disciples in the spirit of the strictest possible adherence to scientific methods. Also, Twardowski's interests were mostly connected with the fields in philosophic logic which are usually referred to as the science of the concept and the science of judgement. His writings such as *Zur Lehre vom Inhalt und Gegenstand der Vorstellungen, Über begriffliche Vorstellungen* and others have prepared a conceptual apparatus which was used by further metatheoretical research and especially semantic research as a solid foundation to be built on. Soon, Jan Łukasiewicz, one of Twardowski's oldest and most distinguished disciples, started teaching. Łukasiewicz, who received both philosophical and mathematical education, discovered logistic for Poland. It may well be that Twardowski played a part in that as well, because it was Twardowski who talked about the algebra of logic in his university lectures in the academic year 1899-1900. Since 1906, Łukasiewicz started teaching logistic to his own as well as Twardowski's students, first as associate professor and then as full professor of the University of Lvov. Under Twardowski's and Łukasiewicz's influence a new generation was raised, whose members studied philosophy in the spirit of anti-irrationalism flavoured with logistic shortly before the [first world] war. This generation included philosophers such as Zygmunt Zawirski, Tadeusz Kotarbiński, Stanisław Kaczorowski, Tadeusz Czeżowski and Kazimierz Ajdukiewicz. They were joined by Stanisław Leśniewski, who had studied in Germany under the supervision of Cornelius and others and who supplemented his study of philosophy in Lvov, where he was first exposed to logistic, thanks to Łukasiewicz. After the war, Twardowski and Łukasiewicz's disciples were dispersed all over Poland and took chairs of philosophy in the newly founded universities in Warsaw, Vilna and Poznań. Łukasiewicz, Leśniewski and Kotarbiński went to Warsaw, Czeżowski moved to Vilnius, Zawirski settled in Poznań, while Ajdukiewicz remained in Lvov, apart from the two-year interlude during which he taught in Warsaw. The Warsaw group has produced numerous outstanding young scholars, one the first of which was Alfred Tarski. Before long, Tarski equalled his masters, Łukasiewicz and Leśniewski, and became one of the leading celebrities of the so-called Warsaw School. As a side remark, let me add that Twardowski's anti-irrationalism is also discernible in the writings of those of his disciples whose research was not affected by logistic. This concerns primarily Władysław Witwicki, a Warsaw professor of psychology.

Aside from this logistic and anti-irrationalist group, descending from the Lvov School, one should also mention another scholar exhibiting a similar intellectual attitude, who – however – followed a different path in his scholarly development. I am referring to Leon Chwistek, a disciple of Zaremba, a Cracow mathematician. For many years, Chwistek was associate professor at the Jagiellonian University in Cracow, and for four years now he has been professor of logic at Lvov University. During a certain period, Cracow played the role of the second most important centre of logistic in Poland. Next to Chwistek, professor Sleszyński, a logician, was active in Cracow as well. Zaremba, an outstanding mathematician was also concerned with logistic and investigated the issue of foundations in the spirit of logistic. Professor Wilkosz and associate professor Nikodym also deserve to be listed here.

Following this historical introduction, I would like to proceed now to discussing the most important scholarly accomplishments of the researchers I mentioned above. Łukasiewicz is primarily concerned with the theory of deduction and the history of logic. The theory of deduction became a laboratory specimen for him, because it constituted a relatively simple material on which to test various methodological properties of deduction systems. He built a range of various axiomatic systems of the theory of deduction, which can be viewed as masterpieces of deductive insightfulness and scientific strictness. Independently of Post and Bernays, he discovered a method of proving the logical non-contradiction of axiomatic systems. In his further research, he also proved the independence and completeness of the axioms of the theory of deduction. He found the method of proving the independence of axioms by using philosophical problems as a point of departure. This is because he also strove to justify the indeterminism of future events. Those efforts led him to discover many-valued logic, and at the same time allowed him to conduct the proof of the independence of the axioms of the deduction theory. Łukasiewicz has authored numerous systems of many-valued logic, in which propositions can take on values other than "false" and "true"; in fact, they can take on an infinite number of logical values. Those many-valued types of logic provided a framework within which attempts were made to construct theories of deduction that would be different from the ordinary theory, eg. Lewis's theory of strict implication or Brouwer's logic in Heyting's interpretation. In the earlier period of his research, Łukasiewicz was concerned with probability, but he interpreted it from the point of view of two-valued logic. Of particular interest is Łukasiewicz's research devoted to the history of logic, which unfortunately has only partially been published. It is in that field that he combines his great competence in logic with thorough knowledge of ancient and medieval sources. He has contributed greatly to our knowledge of Aristotle's teachings on the principles of contradiction and excluded middle, and particularly to our knowledge and understanding of the logic of the Stoics. Łukasiewicz discovered the germs of the theory of deduction in the Stoics, and he considers their logic to be much close to contemporary logic than that of Artistotle.

Unlike Łukasiewicz, who limited his research to the theory of deduction, Leśniewski extended his activities to the whole field of mathematical logic. He constructed a comprehensive system of the foundations of mathematics, which matches the scope of Whitehead and Russell's system, but surpasses it in terms of its precision. The system is composed of three parts. Firstly, it comprises the so-called protothetics, which covers the theory of deduction including the so-called theory of the apparent variable, and which is based on only one axiom with the single primary term. The second part is constituted by the so-called ontology. This name was picked not only because the only constant that its only axiom contains is "is" (έστι), but also because of the author's conviction that – as the most general theory of objects – it is an implementation of Artistotle's *prote philosophia* programme. In Leśniewski's system, the role of his ontology is similar to that played by the theory of classes and relations in Whitehead's and Russell's system, but there are important differences. The so-called mereology is the third part. It can be described as a theory of collective aggregates. It deals with the same subject that Whitehead addresses, albeit in a rather sketchy manner, namely the theory of events. Leśniewski goes to considerable lengths to formulate the directives of his system with greatest possible precision and strictness. For the first time, directives for defining a certain deductive systems were formulated there with complete precision. Frege's influence can be felt in this. In order to avoid antinomies, Leśniewski applied greatest possible accuracy in expanding his theory of types, which he calls the theory of semantic categories. It roughly corresponds to a simplified theory of types, but is richer. In contrast with other similar attempts of this kind, which always contain something artificial, Leśniewski's theory of semantic categories is completely natural and fully concordant with intuition. It was this research, the aim of which was to resolve all the known antinomies, that drove Leśniewski towards semantic problems. The results that he achieved in his research on extensional functions are among the most profound and ingenious in the field of logical semantics so far. It should be added that the results that Leśniewski accomplished in his efforts to achieve a precise formulation of his system's directive have profound significance for philosophy in general. A characteristic quality of Leśniewski's general approach as a researcher that should be emphasized is his hostile attitude to any formalism and conventionalism. He believes in the unconditional truth of certain assumptions and the objective worth of certain kinds of reasoning, and he refuses to view his logical system as a play on concepts, but as a system designed for cognizing reality..

Alfred Tarski, a disciple of both of the scholars presented above, comfortably occupies the same scholarly heights that they do. His first major accomplishment was to define all the truth functors by means of the equivalency symbol, as the only primary term (aside from a quantifier).

This allowed Leśniewski to limit his work to the equivalency symbol only, as the only primary term, when constructing his protothetics. Generally, it should be stressed that Tarski co-operated both with Leśniewski and Łukasiewicz in their research through his numerous detailed results. Later on, Tarski devoted himself to the issues of foundations [of mathematics]. The results that he achieved in this field and that are worth mentioning include a number of definitions of finite sets, which he constructed in collaboration with Lindenbaum, then his geometrical system, where the concept of the sphere rather than that of the point is used as the primary term, and many others. In the last period of his research activity, Tarski succeeded in constructing a strictly deductive system of metatheory, i.e. he made deductive systems into the object of strictly deductive investigations. Finally, in collaboration with Kuratowski, Tarski discovered a method which allows one to solve problems specific to a theory under investigation on the basis of metatheoretical deliberations, that is, for example, to provide simple solution to mathematical issues on the basis of metamathematics wherever the solutions attempted by purely mathematical means encountered serious problems. This method met with a very lively response from mathematicians, whom it convinced of the usefulness of logistic research for mathematics. In his last major publication, which deals with the concept of truth, Tarski demonstrates that for highly formalized languages of the finite order it is possible to define this concept by means of purely morphological terms, i.e. terms which only refer to the external forms of expressions and the relations between them. Furthermore, this publication demonstrates that the semantics of every formalized language of any order can be harnessed in a deductive theory equipped with its own axioms and primary terms, based solely on the morphology of the language in question.

A group of younger logicians have clustered around Łukasiewicz, Leśniewski and Tarski, among which Lindenbaum, whom I already mentioned, as well as Wajsberg, Jaśkowski, Sobociński and others have accomplished valuable results.

The issues of formal logic and methodology were also of concern to Kazimierz Ajdukiewicz, whose scholarly efforts are however mainly directed towards putting the results of logistic research to philosophical use. He was apparently the first Polish scholar to formulate (under Hilbert's influence) the idea of strictly formalized deductive science based on structural directives of reasoning (i.e. those which abstract away from the meaning of words). Guided by this idea, he strove to define the meaning of words as a logical structure of certain relations obtaining between expressions of a given language, which – alongside with sounds – are necessary for providing a description of that language. Those relations are based on the so-called linguistic directives, which lay down certain specified behaviour with respect to recognising certain sentences as necessary prerequisites for the correct use of expressions. It is on this interpretation of the meaning of expressions that radical conventionalism, which Ajdukiewicz represents, is based. The guiding principle of this radical conventionalism is that the scientific image of the world is conventional down to minute details, and can be changed through appropriate changes in the conceptual apparatus (which is constituted by the meanings of words in a given language), and that each of those scientific images of the world has equal rights to claim recognition as the "real" image.

T. Kotarbiński was another scholar who, while not engaging actively in logistic research, nevertheless remained profoundly influenced by this research. Kotarbiński is one of the most influential philosophers of contemporary Poland and he has greatly contributed to boosting and then preserving the influence of logistic on the Polish philosophy. His point of departure for philosophical work were ethical and social issues. His early work contains numerous contributions to the general theory of action, which he termed "praxiology". In that early period, he also published his writings devoted to the issue of the indeterminacy of future events, which – as mentioned earlier – led Łukasiewicz to constructing his many-valued logic. Later, Kotarbiński dedicated himself to the analysis of concepts inherited from the philosophical tradition, and though this analysis he reached solutions to some of the relevant issues posed by this tradition. For example, the so-called reism offers a solution to the problem of ontological categories. In Kotarbiński's view, there is only one ontological category, namely the category of things. Names of properties, relations, classes, universals, immanent contents, psychological acts, etc., are not in reality names of any objects. The sentences which contain those apparent names are only abbreviations of those expressions which only contain names of things. Kotarbiński's reism gets sharper in his statement that every thing is a body, i.e. – to use Kotarbiński's term – in the somatism thesis. The negative attitude of reism to anything that is not a thing implies a denial of the existence of mental qualities, and for Kotarbiński this denial offers a radical resolution of the dispute between idealism and realism in favour of realism.

Kotarbiński, who has been active in Warsaw for sixteen years, has attracted numerous followers who share essentially the same philosophical orientation. They are concerned with semantic issues (Maria Ossowska), the theory of physical cognition (D. Sztejnbarg, Wundheiler, Poznański), as well as methodological problems. Among those who deal with methodology, Janina Hossissonówna deserves to be singled out for her work devoted to the issues of induction. J. Drewnowski, Kotarbiński's disciple, is exploring his own philosophical paths, aiming to construct a philosophical system.

Z. Zawirski's research in Poznań and T. Czeżowski's work in Vilna show a certain affinity with the Warsaw School. Zawirski is mainly concerned with the theory of physical cognition. But his work is primarily of informative value. This year, Zawirski's publication "O rozwoju pojęcia czasu" [On the Development of the Concept of Time] won the first prize – on a par with G. Giorgi's contribution – in a contest organized by the *Scientia* journal as a tribute to E. Rignano. Recently, Zawirski devoted his attention to the application of Łukasiewicz's many-valued logic in the quantum theory and the probability theory. Czeżowski is mainly interested in certain fine issues in logic.

Next, we should not omit to mention a number of younger scholars who studied in Lvov after Łukasiewicz had moved to Warsaw. They are partly Twardowski's and partly Ajukiewicz's and Ingarden's disciples. Some of them are not influenced by logistic and work with methodological and logical issues (Dąmbska, Kokoszyńska, Łuszczewska, Mehlberg, Schmierer).

As already mentioned, the progress in Chwistek's work is not descended from the research conducted in the Lvov-Warsaw community. Chwistek's major influences include Poincaré, Mach, Mill, Russell and Einstein. His field of research is not limited just to logic and philosophy. As a painter and art theorist, he used to be one of the leading representatives of the Polish futurism. He started his work in logic by formulating a critique of Russell's system, particularly of his concept of class. He demonstrated that, if one accepts the axiom of reducibility, one can reconstruct Richard's antinomy on the basis of Russell's theory of types. Therefore, Chwistek rejected the axiom of reducibility and attempted to provide some groundwork for mathematics by using the so-called pure theory of types, which he called the theory of constructive types. In the course of this work, Chwistek was the first to observe that the concept of class can be replaced with the concept of a sentence function. As early as in 1921, i.e. before Ramsey, Chwistek hit upon the conception of the simplified theory of types, under which all sentence functions sharing the same argument belong to the same logical type. In a later period, Chwistek took up the idea of replacing the logical grounding of mathematics with a metalogical one. Since 1928, this conception has been the subject of a number of his publications, and he has finally consolidated those contributions into a system known as "rational metamathematics" in a book co-authored by his followers, Hetper and Herzberg. The system is described by the authors as follows: "En se basant sur la métamathématique rationelle on peut reconstruire le cantorisme entier, sans faire appel aux objets inconstruisibles et, par conséquent, on peut écarter complètement les hypothèses d'existence telles que l'axiome du choix". The overriding idea of this system is to "remove metaphysics from the fundaments of mathematics by adopting a nominalist and radically formalist position". Chwistek's engaged in philosophical deliberations about the problem of reality and came to the conclusion that there is a range of concepts of reality which we use depending on the attitude that we adopt in various real-life situations. There are at least four concepts of reality, with the following corresponding types: sensory reality, representational reality, daily-life reality, and physical reality. None of the four realities is privileged over another. Those various concepts of reality contain nothing conventional, but – depending on our current disposition – we must apply one of them. Chwistek's writings, particularly those which go beyond the limits of logistic, are characterized by considerable linguistic latitude, in stark contrast to the Lvov-Warsaw School.

Similarly to Chwistek, B. Bornstein of Warsaw and A. Wiegner of Poznań did not belong to the Lvov-Warsaw School, but their work shows some influence of logistics.

The above was a very sketchy overview of the present Polish anti-irrationalist trends infused with logistics. It should be pointed out that the percentage of philosophers who conduct their research in the spirit of logistics is much higher in Poland than in other countries. As was stated above, this powerful anti-irrationalist movement gained ground mainly due to the influence of Franz Brentano's Austrian School, which was also the source of some scholastic elements. The logistic element, which was appended to anti-irrationalism, so ardently championed by Twardowski, was primarily the influence of B. Russell, G. Frege, and then E. Schröder. Apart from this decisive influence, Hume, Leibniz, Mill, Spencer, Poincaré, Duhem, Hilbert, Einstein and others have all played an important role as sources of inspiration. All those influences only served to provide a stimulus to the Polish thought, because it soon found its own paths; by following those paths, it has achieved independent and – I dare suggest – worthy results.

Translated by Piotr Kwieciński

*Poznań Studies in the Philosophy
of the Sciences and the Humanities
2001, vol. 74, pp. 251-266*

Klemens Szaniawski

PHILOSOPHY OF SCIENCE IN POLAND[*]

The reader is warned that this is not a systematic presentation of the development of philosophical thought in postwar Poland. There is no attempt to cover all the disciplines that come under the name of philosophy. For instance, ethics and aesthetics, areas in which such distinguished writers as Maria Ossowska and Władysław Tatarkiewicz have been active, are omitted. Also omitted are the history of philosophy and some trends in systematic philosophical thought, e.g. phenomenology, where the name of Roman Ingarden comes first.

There is no value judgement implied in these and other omissions. They are explained by the fact that the task which I have set myself is simply to present some developments in my own field of research, i.e. the philosophy (or general methodology) of science. In addition, the selection is guided by the tradition established in Polish philosophical thought in the period between the wars (1918-1939), which found continuation in the years after 1945. A few words of comment on this trend in Polish philosophy will perhaps be useful as an introduction to the present essay.

No single appellation is used to denote this school of thought, which is connected primarily with the names of Kazimierz Ajdukiewicz (1890-1963) and Tadeusz Kotarbiński (1886-1982). Ajdukiewicz himself used to call it "logistic anti-irrationalism" in order to stress its association with the development of modern logic and also its polemical character directed against irrationalism of any kind. Sometimes the expression the Lwów-Warsaw School was used, after the two main academic centres where this style of philosophy originated. H. Skolimowski uses the term "analytic philosophy in Poland", which has the disadvantage of overemphasizing the similarities with the British school known under the same name. For the purpose of the present exposition, I shall use – in a purely conventional way – the term "Polish positivism."

[*] Reprinted from: J. Barr (Ed.) (1980). *Handbook of World Philosophy*. Westport: Greenwood Press, pp. 281-295. Published on the permission of Greenwood Press. According to the author's will, the original Polish spelling of the name of the city of Lvov, i.e. "Lwów", is preserved throughout the article.

The principal characteristics of this school were (1) a critical attitude towards the traditional language of philosophy; and (2) the use of newly discovered logical tools in philosophical analysis.

The critical component constituted a similarity with the philosophy of the *Wiener Kreis*. At the same time, the Polish positivists differed from their Viennese colleagues as to the nature of this criticism. They did believe that the language of philosophy was in need of radical reform, aiming at eliminating ambiguity and vagueness from its basic vocabulary. In this way, it was hoped, a number of spurious problems would disappear, together with the misunderstandings they had been based on. On the other hand, and contrary to the *Wiener Kreis* doctrine, the Polish positivists did not believe philosophy's fundamental problems to be spurious. In trying to solve them, they adopted the realist attitude in ontology and the empiricist attitude in theory of knowledge.

The application of logical tools to the articulation and solution of philosophical problems, now a common enough procedure, was at the time relatively new. Modern logic itself was in an early stage of development; its potentialities were a matter of conjecture rather than of firm knowledge. Stimulation, therefore, was mutual since the demands of philosophy led to discoveries in logic. One striking example is the work of Ajdukiewicz who contributed as much to logic as to philosophy, while trying to reformulate classical problems in such a way as to render them capable of a strict solution. It is also well known that the discovery by Jan Łukasiewicz of many-valued logical systems (in 1920) was motivated by philosophical considerations concerning the indeterminacy of future events.

The combination of logical and philosophical inquiry in Polish positivism is the result of the above-mentioned tendency to reform the language of philosophy. It also stems from the fact that during the prewar years Warsaw was one of the most active centres of research in formal logic. It suffices to mention the names of Jan Łukasiewicz (1878-1956), Stanisław Leśniewski (1886-1939), and Alfred Tarski (1901-1980). The first of them, one of the founders of contemporary logic, was also a philosopher *par excellence*. He wrote on such topics as determinism, causality, limits of science, and the concept of reasoning. Leśniewski expressed his nominalistic beliefs in the form of logical systems: "ontology" and "mereology." Tarski's contribution to the theory of truth is widely known, as well as his pioneering work, together with Łukasiewicz, on philosophical aspects of many-valued logical systems.

These facts are stressed here because the close ties that existed in Poland between logic and philosophy survived World War II and influenced at least some developments in postwar philosophical thought. Indeed, the dividing line between logic and the philosophy of science became blurred and, as a result, a lot of philosophical work went under the name of logic, conceived in such a way as to cover more than just the logical calculus.

The continuation of the prewar trend outlined above was assured by the fact that, in spite of heavy personal losses in the period 1939-1945, its principal representatives were active for several years after 1945. One such representative, T. Kotarbiński (1886-1982), is one of the best known Polish philosophers today. Until his retirement in 1960, he was a professor at Warsaw University and also occupied important positions in academic life. From 1957 until 1963, he was president of the Polish Academy of Sciences. During the postwar period, he was president of the Polish Philosophical Society until 1977, when he refused reelection for reasons of health.

Kotarbiński's epistemological and methodological views are expressed in *Gnosiology: The Scientific Approach to the Theory of Knowledge*. Written as a handbook for philosophy students, the book is actually a systematic exposition of a *Weltanschauung* in which the doctrine of reism, or concretism, plays a prominent role. Reism, in its ontological version, asserts the existence of material objects ("things") and of such objects only. As a program for discursive language, it postulates reducibility (in principle) of all sentences containing names of properties, relations, processes, and the like, to sentences which contain only names of things. The postulate was intended to act as an extremely sharp version of Occam's razor, eliminating pseudo-entities which are the cause of fruitless philosophical argument.

The implementation of this program turned out to be more difficult than had been expected. Large parts of Kotarbiński's postwar writings have been devoted to overcoming those difficulties. Probably the most serious of them was that the standard concept of set, used in mathematics, cannot be interpreted in the postulated manner. Sets, in the mathematical sense of the word, are not "things," i.e. material aggregates of their elements. (An exposition of the reason for that is to be found in Quine's well-known book *From a Logical Point of View.*) Sets must be treated as abstract entities, a fact which contradicts the existential tenet of reism. Moreover, the language of microphysics does not lend itself easily to the postulated translation. The concept of "thing" is well adapted to the world of macro-objects but seems hardly adequate to deal with subatomic phenomena. Kotarbiński, being aware of these objections, devoted part of his efforts to outlining possible solutions.

The part of Kotarbiński's work which is most original and interesting in the present context is the philosophical discipline that he created, called "praxiology" or "general methodology." Foundations for this discipline were laid as far back as 1913 when he published a series of articles under the title *Szkice praktyczne* [Practical Essays]. But it was only in 1955 that the monograph *Praxiology. An Introduction to the Science of Efficient Action* appeared, in which the subject was treated exhaustively and systematically.

Briefly speaking, praxiology is the general study of human action, from the point of view of its efficiency. The term "action" is taken here in its broadest meaning, including mental activity. Praxiology's objective is first to analyze the meaning of basic terms used to describe goal-directed actions. Such concepts as "goal", "agent", "material", and "instrument" are analyzed and redefined to form a consistent framework for a general theory of efficient action. Classic philosophical problems arise in this connection, e.g. those concerning the nature of causal relation (presupposed in the statement that X's action brought about a certain state of things) or the responsibility of the agent for the results of what he did.

This conceptual framework makes it possible to formulate principles of efficient action. They are of a very general nature, and in this respect they differ from the results obtained by the numerous specific disciplines that came into existence during the last decades, such as the theory of games, operations research, and theory of decision-making. In addition, while those disciplines build up formal models of rational behaviour, praxiology's results (with the exception of a very few attempts to formalize certain concepts) are expressed in natural language sharpened by a number of terminological conventions.

The difference in approach may be exemplified by the case of rational behaviour in conflict. It is well known how the theory of games reduces the problem to the choice of strategy from a given set, all the valuations having been summarized by a single numerical function (so-called utility). Praxiology, on the other hand, looks for generalizations arising out of observation of human practice. For instance, one such guiding principle, obviously inspired by analysis of military battles, recommends concentrating the effort on disabling that part of the opponent's forces on which the behaviour of other parts depends (such as the commanding unit or the nervous system).

The main interest of praxiology is in *homo faber* as such. The pioneering work of Kotarbiński, who created this branch of philosophical investigation, has been one of the major components in the development of philosophy in postwar Poland. It soon ceased to be an individual effort. A centre for praxiological research has been created in the Polish Academy of Sciences, where the late J. Zieleniewski (as well as W. Kieżun, T. Pszczołowski, W. Gasparski, *et al.*) have been active. A quarterly, *Prakseologia* [Praxiology] appeared regularly for several years.

Another philosophical school owes its origin to Ajdukiewicz. Ajdukiewicz became known internationally in the 1930s, mainly through a series of articles he published in *Erkenntnis*. In these articles, he expounded his views on the relation between language and cognition. Starting from an analysis of the concept of meaning, he obtained a system which he called "radical conventionalism." Its principal tenet was that all knowledge (not only some part of it, as Poincaré and Le Roy used to assert) depends in an essential way on the choice of language.

In his postwar publications, Ajdukiewicz repudiated radical conventionalism (because of its unrealistic conception of language) and expressed some scepticism as to the correctness of such sweeping generalizations. Instead, he concentrated on more specific problems in the philosophy of science. For this discipline he formulated a program of "methodology that aims at understanding."

The program is an attempt to solve the old dilemma concerning the interpretation of methodological statements. If, on the one hand, they are assumed to be nothing more than descriptions of general patterns of behaviour, reflecting the actual practice in science, then the question *"quid juris?"* is left unanswered. In other words, the philosopher avoids the problem of justification. If, on the other hand, methodological statements have normative meaning, then the philosopher is open to the objection that he places himself "above science," since he tells scientists how they should do their job. The dangers in such a ruling on science are easy enough to see.

Ajdukiewicz's solution of the dilemma can be outlined as follows. Any method is a prescription for acting in a certain way, and the reason we comply with the prescription is that we thereby hope to achieve something. Hence, the fundamental characteristic of any method is its efficiency with respect to a given objective. The argument for adopting a method is thus relative to the purpose the method is intended to serve and consists in showing that the method is efficient enough with respect to this purpose, or that it is the most efficient in a class of methods available. Efficiency, in turn, may be defined in various ways, depending on the context, but the main ingredient of the concept is the probability of achieving the purpose by the application of the method.

If this argument is agreed upon, the adoption of a method in science may, in principle, be justified along these lines. Such a procedure presupposes that the objective is unequivocally defined. Quite often, however, the scientist himself is not clearly aware of the purpose the method is to serve. In such cases, a hypothetical justification is possible of the following type: if this is the purpose the scientist had in mind, then the method is good in terms of its efficiency with respect to the purpose.

Science can then be understood as a rational pursuit of possibly hypothetical goals.

This was the task Ajdukiewicz and his followers set themselves. Out of the results obtained, only a few can be mentioned here. Ajdukiewicz himself remained faithful to his old interest in semantics. He continued to investigate the problem of meaning and the way the conceptual framework of an empirical theory is built up – hence, his work on definitions, on analyticity, and on interpreting axiomatic systems. He also wrote on the logical foundations of measurement (see his *Pragmatic Logic*, published posthumously) and was one of the first to justify rules of non-deductive inference in terms of a non-negative balance of losses and gains in the long run. (See his paper on fallible methods of inference, reprinted in *Twenty Five Years of Logical Methodology in Poland.*) His essay on contradiction and change (1948) played a decisive role in the discussion concerning the alleged conflict between dialectics and formal logic. The Eleatic paradox of the flying arrow had been used to argue that movement implies contradiction; on the other hand, formal logic condemns contradiction as an inexcusable fallacy. Ajdukiewicz has shown conclusively that the argument is not valid since it rests on an error (in the definition of "being at rest") eliminated in the 19th century thanks to the mathematical definition of continuity.

The programme outlined by Ajdukiewicz was carried out by a number of people, including, first of all, the older generation of philosophers who had already been active before the war. One of these, S. Łuszczewska-Romahnowa, has studied the concept of classification, interpreting it in terms of distance function (see: *Twentv Five Years of Logical Methodology in Poland*). I. Dąmbska analyzed the role of models in science, "Le concept de modèle et son role dans les sciences." T. Czeżowski defined in a new way the types of reasoning in science, modifying the well-known classification of reasonings by J. Łukasiewicz; he also wrote on testability in empirical sciences (see *Twenty Five Years of Logical Methodology in Poland*). J. Kotarbińska wrote on definitions (so-called deictic definitions in particular), on the concept of sign, and the falsificationist view of hypothesis testing (see her papers in *Twenty Five Years of Logical Methodology in Poland.*) M. Kokoszyńska worked mainly on the problem of analyticity. Since considerable progress has been achieved in this field since World War II, it s described in more detail here.

It may be said that the initial stimulus came from Ajdukiewicz who forcefully presented the problem and showed its philosophical implications. (See his paper, "The Problem of Justifying Analytic Sentences," reprinted in *Twenty Five Years of Logical Methodology in Poland.*) He questioned the almost universally accepted view that definitions and other terminological postulates are "true by convention", i.e., in a way that does not depend on experience. As regards non-definitional postulates, such as the so-called reduction sentences introduced by Carnap, this had been noticed early enough. It had been shown, for instance, that two or more reduction sentences specifying the meaning of a term, say *T*, may logically imply a non-tautological sentence in which the term *T* does not appear. We then have a paradoxical consequence: truth by convention would imply empirical truth, or, even worse, it would imply empirical falsehood, in which case truth by convention would be no truth at all.

A considerable amount of research was devoted to answering this question. One answer consisted in saying that analytic sentences depend upon experience as to their meaningfulness (but not truth-value), whereas synthetic sentences depend upon experience in both ways.

There have been attempts to "split up" meaning postulates of a given language into a synthetic (factual) component and an analytic (definitional) one. In order to do that, both the language and its semantic properties had to be specified. Explicit consideration of relations between language and the reality it is intended to describe (its "model") is characteristic of such an approach to the classical problem of analytic sentences. The tools used in this investigation therefore belong to the logical theory of models.

Since it is impossible to enumerate here all the results obtained in this field, only some of their authors are mentioned: M. Kokoszyńska, L. Borkowski, Z. Czerwiński, A. Nowaczyk, M. Przełęcki, and R. Wójcicki. The last two authors gave an excellent summary of this work, up to 1969, in "The Problem of Analyticity", reprinted in *Twenty Five Years of Logical Methodology in Poland.*

More generally, the conceptual structure of empirical theories has received considerable attention from the Polish philosophers of science. *The Logic of Empirical Theories* by M. Przełęcki can be quoted as a typical example. This book analyzes theories, both in terms of their syntactic properties and of their empirical interpretation. In the monograph by R. Wójcicki, *Metodologia formalna nauk empirycznych* [Formal Methodology of Empirical Sciences], we also find an attempt to describe the semantic aspects of empirical theories. Wójcicki in particular tries to define the relations between the language of a theory and its scope in such a way as to make it possible to say that the theory is approximately true. Although the concept of approximate truth seems necessary for a realistic description of science, it was seldom investigated by means of modern logic. Wójcicki's book is largely devoted to that purpose.

Here it ought to be mentioned that a number of Polish philosophers and logicians were concerned with problems of semiotics. This is not surprising, if one remembers that emphasis on language was characteristic of the Lwów-Warsaw School in philosophy. Thus, for instance, Kotarbiński's reism was to a large extent a conception of language, and Ajdukiewicz obtained his famous results when he tried to define the concept of meaning.

The postwar continuation of semiotic studies is due to many people. Most active was J. Pelc, professor of logical semiotics at Warsaw University. His main interest is in the logical structure of natural language; see his *Studies in Functional Logical Semiotics of Natural Language*. Pelc has also studied the history of semiotics and, as a result, has published several review papers and edited the anthology *Semiotyka polska 1894-1969* [Polish Semiotics 1894-1969]. He is the editor of a serial publication called *Studia Semiotyczne* [Semiotic Studies], eight volumes of which appeared in the years 1970-1977.

A large part of semiotic studies in Poland has been devoted to the analysis of natural language in terms of formal logic. Thus, B. Stanosz showed how to define the concept of property in set-theoretic terms (which is important if we want to eliminate intentionality from natural language). W. Marciszewski has investigated the semantic structure of a coherent text, and R. Suszko has analyzed semantic antinomies. L. Koj has published a monograph, *Semantyka a pragmatyka* [Semantics and Pragmatics], on the relations between semantics on the one hand and linguistics and psychology on the other. An approach to semantics from the Marxist point of view is represented by A. Schaff's book, *Introduction to Semantics*.

Let us now return to the description of how Ajdukiewicz's programme worked in the philosophy of science. A group of problems concerns the validity of scientific generalizations. Under which conditions is it justified to accept them as true? Philosophically, this is, more or less, the age-old problem of induction.

Attempts to solve it in the spirit of Adjukiewicz's program take as their starting point an analysis of scientific practice. In this case, the paradigm is provided by the so-called statistical inference, where the relation between premise and conclusion is of a probabilistic nature. The acceptance of a conclusion according to a method of this type is justifiable relative to a hypothetical goal, such as minimization of error, maximization of information, and some compromise between these two. Research along these lines has been conducted by Z. Czerwiński (e.g. "On the Relation of Statistical Inference to Traditional Induction and Deduction"), K. Szaniawski (e.g. "Types of Information and Their Role in the Methodology of Science" and the papers reprinted in *Twenty Five Years of Logical Methodology in Poland*), and H. Mortimer who also studied probabilistic aspects of the language of scientific theories (e.g. the article on probabilistic definition of genotype in *Twenty Five Years of Logical Methodology in Poland*).

A different approach to the problem of induction is seen in H. Greniewski (1903-1972). Greniewski adopted a cybernetician's point of view, and he stated that the purpose of Mill's methods was to find out how a relatively isolated system (a "black box") works, i.e. what is the relation between the possible states of its inputs and those of its outputs, see: *Elementy logiki indukcji* [Elements of the Logic of Induction]. Earlier, in 1947, J. Łoś provided a formal analysis of Mill's methods; his paper is reprinted in *Twenty Five Years of Logical Methodology in Poland*.

Before closing this account of how the program for the philosophy of science has been put into effect, one ought to mention the work that has been done on the concept of "problem" or "question". Here, again, the initiative goes back to an early (1934) paper by Ajdukiewicz, in which foundations were laid for the logic of questions and basic types of questions were distinguished. T. Kubiński published extensively on formal theory of questions (e.g. "An Essay in Logic of Questions" and "Logic of Questions"), while J. Giedymin in *Problemy, założenia, rozstrzygnięcia* [Problems, Assumptions, Solutions] investigated the types of questions (also, their presuppositions, the existence of solutions, and the like) in the social sciences and history.

A number of results just described appeared in the philosophical journals published in Poland: *Studia Logica* (in foreign languages, since 1974 exclusively in English), *Studia Filozoficzne* [Philosophical Studies], and *Dialectics and Humanism* (in English).

A number of philosophically significant contributions in Poland have come from disciplines other than philosophy. Sociology provides one instance. S. Ossowski (1897-1963), himself a former student of Kotarbiński, is the author of a penetrating analysis of the way in which the study of social phenomena differs from that of natural ones: *O osobliwościach nauk społecznych* [On the Peculiarities of the Social Sciences]. In his other works, Ossowski applied the exacting standards set by the Lwów-Warsaw School to analyzing the meaning of emotionally loaded concepts that express the state of social consciousness. Ossowski created a school in the philosophy of the social sciences. A typical example is the collection of essays *Understanding and Prediction* by S. Nowak, his student and collaborator.

Another example of philosophical inquiry into the foundations of a discipline is to be found in the writings of O. Lange, the eminent economist (1904-1965). Looking for general principles that would make it possible to justify normative statements made in economics, he found them in a philosophical doctrine, namely, Kotarbiński's praxiology. According to Lange, economic statements, if they are correct, should be derivable from the principles governing efficient action. Lange was one of the leading Marxist thinkers in the field of economic theory. However, he did not consider Kotarbiński's philosophy to be foreign to his convictions.

The appearance of Marxism was probably the most important factor shaping the development of philosophy in postwar Poland. The word "appearance" is used here because before 1939 very few academic philosophers in Poland professed to be Marxists. After the fundamental changes in the political and social structure of Poland, Marxism became the officially acknowledged ideology. This change was, of course, not without consequences for philosophy as one of the subjects taught at the Polish universities.

The process that assured Marxism its present position in Polish philosophy was neither short nor simple. It had its dramatic periods, and the arguments used were not always of a purely theoretical nature. Any fair account of that process in all its complexity would have to place it against a wider historical background. Since such an account is impossible within the limits of this exposition, a summary description of the present situation must suffice.

As a subject for teaching, Marxist philosophy belongs to practically all the curricula of higher education in Poland. The majority of publications in philosophy represent the Marxist school of thinking. The policy of appointments to academic positions is to give preference to candidates who have Marxist views. Philosophical journals are run on Marxist lines (although non-Marxist articles do appear in them). These are but a few examples of Marxism's dominant role in the philosophical life in Poland.

The principle applies to all domains of philosophy: ontology, theory of knowledge, ethics, aesthetics, but in varying degrees, depending on the importance of a commitment to a philosophical doctrine for solving problems of a given type. Philosophy of science, with the exception of its most general problems, is relatively neutral in this sense.

It can be said that in the philosophy of science there was an exchange of ideas between Marxist philosophers and those who drew inspiration from the Polish positivistic tradition. The Marxist thinkers, while retaining their ontological views, *viz.*, those of dialectical materialism, took over from "classical" philosophy of science some of its concepts and methods. In particular, they used in their analysis of scientific cognition the logico-mathematical tools that had been created in a rather different philosophical context. At the same time, it must be stressed that a philosophy of science developed on Marxist principles was openly opposed to logical positivism and to similar views expressed by the Lwów-Warsaw School in philosophy.

A good example is provided by the writings of a group of philosophers from Poznań University (J. Kmita, L. Nowak, J. Such, and J. Topolski are the most prominent representatives). The theoretical position of the Poznań group can be outlined as follows. First, we have to reject the current interpretation of Marx's views on the nature of scientific knowledge. According to that interpretation, Marx maintained that abstract statements, constituting scientific theories, are obtained by means of inductive generalization. This, however, is inconsistent with the fact that Marx used to distinguish between essential and inessential factors in opposition to the positivist view which replaces the concept of essence by that of repeatability.

The genuine position of Marx will then have to be reconstructed from his scientific practice, i.e. from the method he used in building up his economic theory. A careful analysis of *Capital* shows that Marx proceeded in a much more complex way than by simple induction. He started by formulating a law (for instance, the law of value) in a highly idealized form, i.e. under assumptions which are patently untrue: that capitalists are directed by one goal only (maximizing the surplus value), that the commercial profit on the goods produced equals zero, and so forth. In this form, the law is simple and shows the essential relation between value and price.

The next stage consists in successive concretizations of the law. The simplifying assumptions are removed, one by one, which brings the law closer to the phenomena it intends to describe. The different factors, first eliminated by the simplifying assumptions and then reintroduced, have different degrees of significance. In its basic form, the law deals only with what is essential. However, this essential regularity is obscured by the intervention of less important factors. Hence, it is necessary to abstract essential connections from their real life context.

According to the Poznań group, the method of abstraction and concretization constitutes the central element of Marxist philosophy of science, as opposed both to positivism in its many variants and to the hypotheticism of Karl Popper. Its description and development are to be found in the numerous publications by L. Nowak, e.g. *Zasady marksistowskiej filozofii nauki* [Principles of Marxist Philosophy of Science] and his collaborators.

Another classical problem which the Poznań group took up concerns the foundations of the humanities. What is the nature of explanation in the social and historical sciences? The answer provided by J. Kmita – *Z metodologicznych problemów interpretacji humanistycznej* [On Methodological Problems of Humanistic Interpretation] – is, to some extent, based on Ajdukiewicz's methodological ideas. In agreement with Ajdukiewicz,

Kmita wants to make explanation depend on the hypothetical goal which the agent presumably wanted to achieve. But he goes further in trying to bring out the structure of reasoning that leads from the postulated goal to the act to be explained. Unless this reasoning is a correct deduction, we cannot speak of explanation in the strict sense of the word. In practice, the reasoning is usually enthymematic, so that its reconstruction must supply the missing premises which, of course, are a highly controversial matter.

According to Kmita, the assumption tacitly made in explanations of this type is that the agent is a rational being, in the sense that s/he so chooses his/her actions as to bring about the state of things which for him/her has the highest value. In spite of its almost tautological character, the assumption of rationality is not the only candidate to play the role of general principle in explanations of human behaviour. Thus, Kmita argues against the positivistic tendency to link goals with acts by means of some special psychological laws. Humanistic interpretation of human acts (this is the name he gives to the procedure just outlined) is empirically testable in just the same way as are explanations offered by the sciences.

It ought to be clear from the above examples that the interpretation of Marxism by the philosophers from the Poznań group aims at retaining the spirit, rather than the letter, of the original sources. In an article defining their theoretical position ["Against the False Alternatives"], Kmita, Nowak, and Topolski express the belief that

> Marxist philosophy is a coherent system rich enough to compete with the remaining systems and win the competition. Upon the condition of course, namely, that Marxism will be given a clear, precise form and thus enabled to solve contemporary problems.

The article was published in the journal *Poznań Studies in the Philosophy of the Sciences and the Humanities*, which has appeared since 1975 and is edited by philosophers from the Poznań group. Chronologically, the Poznań group was formed relatively late.

Considerable work has been done in Poland on the philosophical and methodological problems of physics. Z. Augustynek specializes in analyzing the concept of time. For instance, in his "Three Studies in the Philosophy of Space and Time", he discusses the topology of time as well as the relation of symmetry properties to other space-time properties. I. Szumilewicz has published on the direction of the flow of time, *O kierunku upływu czasu* [On the Direction of the Flow of Time], and on the nature of our cognition of the universe. S. Amsterdamski has analyzed certain basic concepts of science, e.g. that of probability, in "O obiektywnych inter-pretacjach pojęcia prawdopodobieństwa" [On the Objective Interpretations of the Concept of Probability]. H. Eilstein, in her important essay on the concept of matter, "Przyczynki do koncepcji materii jako bytu fizycznego" [Towards the Concept of Matter as Physical Being], discusses the possibility of reducing in a non-mechanistic way all natural sciences to physics. It ought to be said here that H. Eilstein, for several years the editor of the journal *Studia Filozoficzne*, initiated many studies in the philosophy of science from the Marxist point of view; this trend owes a great deal to her activity.

The work of W. Krajewski also concerns the philosophical foundations of physics. Out of his numerous publications, only his recent *Correspondence Principle and the Growth of Science*, published in English, will be mentioned here. The main problem taken up in this book is the development and application of Bohr's principle of correspondence to an analysis of development in science. The book is also an excellent source of information on the results obtained in Marxist philosophy of science. Krajewski describes his own philosophical position in the following words:

> I associate basic ideas of dialectical materialism (liberated from dogmatism and Hegelian phraseology) with logico-methodological achievements of the analytical philosophy of our century (liberated from positivistic narrowness). And first of all, the methods used in the sciences and the history of knowledge must be seriously examined" (p. x).

This attitude is characteristic of the relations between Marxist philosophy of science and the traditions of Polish philosophy.

Problems of the evolution of scientific knowledge occupy a prominent place in the philosophical thinking of the last decades. S. Amsterdamski takes them up in his recent book *Between Experience and Metaphysics: Philosophical Problems of the Evolution of Science*. One of the main assertions of this book is that Kuhn's representation of the growth of science ought to be modified. The change of paradigm does not, as a rule, affect those convictions which serve as the basis for overcoming the crisis. Therefore, "revolutions in science" are in most cases local rather than global.

An attempt to describe formally the process of development of science, in the spirit of Marxism, is found in R. Suszko ("Formal Logic and the Development of Knowledge"). For this purpose, he uses "diachronic logic", i.e. the logic that takes time expressly into account. According to Suszko, knowledge is an epistemological relation between the subject (the language he uses) and the object (the model of the language). Changes in knowledge affect either the set of sentences asserted by the subject (evolution) or the language and its model (revolution).

Here we are concerned with the philosophy of science proper. However, certain Marxist contributions on the borderline between the philosophy of science and epistemology or ontology are sufficiently important to mention here. B. Wolniewicz provides a new interpretation of Wittgenstein's early philosophy; he also tries to elucidate basic epistemological ideas of dialectical materialism. C. Nowiński attempts to show (in "Biologie, théories du développement et dialectique") that dialectical methods are used on an increasingly large scale in modern biology and in other sciences analyzing development. In particular, he interprets the work of Jean Piaget in terms of dialectics. According to Nowiński, the relation is reciprocal: progress in biology influences the elaboration of dialectical methods.

O. Lange, in "An Essay in Logic of Questions", has shown the application of cybernetical concepts to the elucidation and proof of certain Marxist tenets. He analyzes the way in which the functioning of any system, biological or technical, depends on the behaviour of its subsystems ("parts") and on the network of relations between them, i.e. the structure of the whole. Lange considered this to be a refutation of mechanism since it proves that the behaviour of the parts is not enough to account for the behaviour of the whole. Moreover, the mathematical theory serves as an argument against finalism. It turns out that there is no need to postulate some mysterious "force" in order to explain the phenomenon of adaptation to changing environment. The theory of ergodic processes is enough for that purpose.

No account of Marxist philosophy in Poland would be complete without mention of the work of A. Schaff. He is one of the initiators and main representatives of this philosophical school in postwar Poland. His writings are rather widely known through many translations. Schaff is concerned with classical problems of philosophy rather than with the philosophy of science proper. However, in his epistemological works, like *Geschichte und Wahrheit* or *Theorie der Wahrheit. Versuch einer marxistischen Analyse*, he discusses problems of validation of assertions made in science and historiography. Schaff's writings are strongly polemical, directed against any brand of what he considers idealism, especially neopositivism. He has had considerable influence on the development of Marxism in Poland, particularly in the first two decades.

In summary, it must be stressed that the philosophy of science, together with logical semiotics, traditionally occupied a prominent place in Polish philosophical thought. The tradition had much in common with logical positivism, but it had its own distinctive features. I call this trend, for lack of a better word, Polish positivism or the Lwów-Warsaw School. After 1945, Marxism was another factor that began to influence methodological studies. Although Marxism was and is explicitly opposed to positivism in any form, its analysis of scientific knowledge has very often been conducted in a way closely parallel to the Lwów-Warsaw School style of thinking. This phenomenon is worth noting.

Contemporary philosophy of science is preoccupied with the formal structure of scientific methods and scientific theories – hence the use of logical and mathematical language, sometimes specially developed for the purpose. This tendency is in harmony with the Polish tradition of philosophizing. The last statement can be substantiated by referring the reader to the prewar publications of the Lwów-Warsaw School and to *Twenty Five Years of Logical Methodology in Poland*, an anthology of methodological writings in Poland covering the postwar period up to 1974. In 1974, an international conference was held in Warsaw devoted to formal methods in the methodology of the empirical sciences. Its recently published proceedings *Formal Methods in the Methodology of Empirical Sciences*) provide a source of information on the present state of the philosophy of science in Poland.

REFERENCES

Ajdukiewicz, K. (1974). *Pragmatic Logic*, transl. by O. Wojtasiewicz. Boston and Dordrecht: Reidel.

Ajdukiewicz, K. (1978). *The Scientific World-Perspective And Other Essays. 1931-1963*, ed. and introduction of J. Giedymin. Dordrecht: Reidel.

Amsterdamski, S. (1975). *Between Experience and Metaphysics: Philosophical Problems of the Evolution of Science*, transl. by P. Michałowski. Boston and Dordrecht: Reidel.

Amsterdamski, S. (1964). O obiektywnych interpretacjach pojęcia prawdopodobieństwa [On the Objective Interpretations of the Concept of Probability]. In: S. Amsterdamski, Z. Augustynek and W. Mejbaum (Eds.). *Prawo, konieczność, prawdopodobieństwo* [Laws, Necessity and Probability]. Warszawa: Książka i Wiedza, pp.1-124.

Augustynek, Z. (1967). Three Studies in the Philosophy of Space and Time. In: R.S. Cohen and M.W. Wartofsky (Eds.). *Boston Studies in the Philosophy of Science* **III**. Dordrecht: Reidel, pp. 447-466.

Czerwiński, Z. (1958). On the Relation of Statistical Inference to Traditional Induction and Deduction. *Studia Logica* **8**, 242-263.

Dąmbska, I. (1959). Le concept de modéle et son role dans les sciences. *Revue de Synthése* **80**, 13/14, 39-51.

Eilstein, H. (1958). Laplace, Engels i nasi współcześni [Laplace, Engels and our Contemporaries]. *Studia Filozoficzne* **1**, 143-174 .

Eilstein, H. (1961). Przyczynki do koncepcji materii jako bytu fizycznego [Towards the Concept of Matter as Physical Being]. In: *Jedność materialna świata* [Material Unity of the World], ed. H. Eilstein. Warsaw: Książka i Wiedza.

Giedymin, J. (1964). *Problemy–założenia–rozstrzygnięcia* [Problems–Assumptions–Resolutions]. Poznań: PWN.

Greniewski, H. (1955). *Elementy logiki indukcji* [Elements of the Logic of Induction]. Warszawa: PWN.

Kmita. J. (1971). *Z metodologicznych problemów interpretacji humanistycznej* [On Methodological Problems of Humanistic Interpretation]. Warszawa: PWN.

Kmita. J., Nowak L., and Topolski, J. (1975). Against the False Alternatives. *Poznań Studies in the Philosophy of the Sciences and the Humanities* **l**, 2, 1-10.

Koj, L. (1971). *Semantyka a pragmatyka* [Semantics and Pragmatics]. Warszawa: PWN.

Kotarbiński. T. (1966). *Gnosiology: The Scientific Approach to the Theory of Knowledge*, transl. by O. Wojtasiewicz. Oxford and New York: Pergamon Press.

Kotarbiński. T. (1965). *Praxiology: An Introduction to the Sciences of Efficient Action*, transl. by O. Wojtasiewicz. Oxford and New York: Pergamon Press.

Krajewski. W. (1977). *Correspondence Principle and the Growth of Science*. Boston and Dordrecht: Reidel

Kubiński, T. (1958). An Essay in the Logic of Questions. In: *Atti del XII Congresso Internazionale di Filosofia*. V. Florence: Sansoni, pp. 315-322.

Kubiński, T. (1968). The Logic of Questions. In: R. Klibansky (Ed.). *Contemporary Philosophy. A Survey*. Florence: La nuova Italia, pp. 185-189.

Lange. O. (1970). *Papers in Economics and Sociology 1930-1960*. Transl. and ed. by P.F. Knightsfield. Oxford and New York: Pergamon Press.

Lange. O. (1965). *Wholes and Parts. A General Theory of System Behaviour*. Oxford and New York: Pergamon Press.

Nowak. L. (1974). *Zasady marksistowskiej filozofii nauki* [Principles of Marxist Philosophy of Science]. Warszawa: Państwowe Wydawnictwo Naukowe.

Nowak. S. (1976). *Understanding und Prediction. Essays in the Methodology of Social and Behavioral Theories*. Boston and Dordrecht: Reidel.

Nowiński. C. (1967). Biologie, théories du développement et dialectique. In: J. Piaget (Ed.) *Logique et connaissance scientifique*. Encyclopédie de la Pleiade. Paris: Gallimard, pp. 862-892.

Ossowski, S. (1962). *O osobliwościach nauk społecznych* [On the Pecularities of the Social Sciences]. Warszawa: PWN.

Pelc. J. (1971). *Studies in Functional Logical Semiotics of Natural Language*. The Hague: Mouton.

Pelc. J. (Ed.) (1971). *Semiotyka polska 1894-1969* [Polish Semiotics 1894-1969]. Warszawa: PWN.

Przełęcki, M. (1969). *The Logic of Empirical Theories*. London: Routledge and Kegan Paul, & New York: Humanities Press.

Przełęcki, M., Szaniawski, K., and Wójcicki. R. (Eds.) (1974). *Formal Methods in the Methodology of Empirical Sciences. Proceedings of the Conference for Formal Methods in the Methodology of the Empirical Sciences*. Warsaw, June 17-21 (*Synthese Library* 103). Warszawa: PWN & Dordrecht: Reidel.

Przełęcki, M., and R. Wójcicki (1969). The Problem of Analyticity. *Synthese* 19, 3-4, 372-399.

Przełęcki, M. and R. Wójcicki (Eds.) (1977). *Twenty Five Years of Logical Methodology in Poland* (*Synthese Library* 87). Warszawa: PWN & Dordrecht: Reidel.

Schaff, A. (1970). *Geschichte und Wahrheit*, ed. by E. M. Szarota. Frankfurt and Zurich: Europa.

Schaff, A. (1962). *Introduction to Semantics*, transl. by O. Wojtasiewicz. Warszawa: PWN & New York: Macmillan.

Schaff, A. (1971). *Theorie der Wahrheit. Versuch einer marxistischen Analyse*. Vienna: Europa.

Such, J. (1975). *Problemy weryfikacji wiedzy* [Problems of Verification of Knowledge]. Warsaw: PWN.

Suszko. R. (1968). Formal Logic and the Development of Knowledge. In: I. Lakatos and A. Musgrave (Eds.). *Problems in the Philosophy of Science* Vol. III. Amsterdam: North-Holland, 210-222.

Szaniawski, K. (1977). Types of Information and Their Role in the Methodology of Science. In: M. Przełęcki, K. Szaniawski, and R. Wójcicki (Eds.). *Formal Methods in the Methodology of Empirical Sciences. Proceedings of the Conference for Formal Methods in the Methodology of Empirical Sciences*. Warszawa, June 17-21, 1974 (*Synthese Library* 103), Warszawa: PWN & Dordrecht: Reidel, pp. 297-308.

Szumilewicz. I. (1972). *O kierunku upływu czasu* [On the Direction of the Flow of Time]. Warszawa: PWN.

Wójcicki, R. (1974). *Metodologia formalna nauk empirycznych* [Formal Methodology of Empirical Sciences]. Wrocław: Zakład Narodowy im. Ossolińskich.

Wolniewicz, B. (1969). A Parallelism Between Wittgensteinian and Aristotelian Ontologies. In: R.S. Cohen and M.W. Wartofsky (Eds.). *Boston Studies in the Philosophy of Science* Vol. IV. Dordrecht: Reidel, pp. 208-217.

Wolniewicz, B. (1972). The Notion of Fact as a Modal Operator. *Teorema*. Num. monográfico, 59-66.

Poznań Studies in the Philosophy
of the Sciences and the Humanities
2001, vol. 74, pp. 267-285

Izabella Nowakowa

MAIN ORIENTATIONS IN THE CONTEMPORARY POLISH
PHILOSOPHY OF SCIENCE

Introduction

I. It is worth stressing how numerous the orientations within the Polish philosophy of science are, which to an extent results from the heterogeneity of the Polish philosophy in general. Although this heterogeneity received a natural boost after the transition to democracy in 1989, it did exist under real socialism as well. Alongside the "official philosophy" of the system, i.e. the Leninist version of Marxism, other interpretations of Marxism existed within its bounds, and outside Marxism there existed the positivist, the Thomist and the phenomenological orientations.

However, not every philosophical orientation leads to a theory of science. In the contemporary Polish philosophy of science, which – as a matter of convention – will be taken here to cover the last four decades, many valuable monographs were written dealing with the problems of correspondence, reduction, simplicity, the status of theoretical terms, the conditions for valid measurement, and other important aspects of scientific methodology (an overview of which is contained in the *Introduction* to this volume, authored by Władysław Krajewski and Jacek J. Jadacki). My aim, however, is not to present the state of the art in research on science, but rather certain theories of science which have emerged in Poland in the past decades. Thus, I am concerned with theories which present some comprehensive vision of science, and which – on the other hand – have managed to attract a number of followers who are both advancing those theories and applying them to the research practice of empirical sciences, in particular natural sciences. The leading conceptions of this kind are the three theories stemming from the tradition of the Lvov-Warsaw School:

- Marian Przełęcki's conception of the indeterminacy of empirical theories;
- Ryszard Wójcicki's conception of approximation;
- Elżbieta Kałuszyńska's conception of methodological relativism;

as well as two theories based on otherwise different interpretations of the
Hegelian-Marxian tradition:
- Jerzy Kmita's historical epistemology;
- Leszek Nowak's idealizational conception of science.

Each of those conceptions has been enriched over time with more or less
plentiful contributions which extended them to fields that were not taken
into account in their original basic versions, and which applied the
assumptions of those conceptions to numerous problems of special nature.

II. Every serious conception starts from a surprising observation which
invalidates what had been deemed obvious and which at the same time
allows to cast a new light on the whole field. But the condition is that we
approach the datum obtained as primary, i.e. that, in a way, we suspend all
our beliefs within the field in question and start the painstaking,
systematic reconstruction thereof, based on the "primary datum" in
question. Thus, the stages of a theoretical discovery are the following:

- firstly, one suspends currently operative stereotypes in order to make
  the basic observation; this is the "primary datum" stage;
- next, based on the clearly delineated theoretical assumptions, one
  submits a conceptualization of the "primary datum"; this basic
  thesis is always a profound idealization, which allows one to
  disregard additional considerations disturbing the clarity of the
  initial intuition expressed in the thesis; this is the construction
  stage;
- then, one demonstrates the explanatory power of the basic thesis by
  applying it to various special aspects wherever the disturbances are
  weak enough to be negligible (this is what applications are about),
  and – furthermore – by rejecting the initial idealizing assumptions
  and modifying the thesis so that it applies to fields to which the
  basic, idealized version does not apply even roughly (this is what
  the concretization of the initial thesis is about); this constitutes the
  explanatory stage;
- the conception itself is a chain of theses, starting from the initial
  and simplest one, and leading through more concrete ones, which
  allow one to explain more and more and in an ever greater detail;
  however, no conception allows one to explain everything in a given
  field exhaustively; therefore, serious scholars welcomes and
  appreciates a critique of their conceptions, because demonstrating
  the divergencies between the conception and the world allows them
  to discern simplifications which had previously escaped them and
  hence to work on further concretizations which bring the new
  version of the conception even closer to reality; this is the stage of
  expanding the conception.

This is, roughly, an ideal of theoretical work, applying to the philosophy of science as well (Nowak 1980, pp. 106*ff*, 194*ff*). We will more or less follow this model in presenting various orientations below.

1. Conceptions drawing on the traditions of the Lvov-Warsaw School

1.1. The conception of the indeterminacy of empirical theories

(1) As the philosophy of science is concerned with the structure and functions of a theory, it assumes a single, specific understanding of theories; thus, against the grain of the basic ideas of the semantic theory of models, it assumes naive semantic naturalism. This critical constatation raised against the standard philosophy of science laid the foundations for a certain basic observation that Marian Przełęcki made in the early 1960s.

This observation was as follows. There is no doubt that the authors of an empirical theory have a certain interpretation in mind; thus, they assign it a specified model (an intended model). But how can one identify such a model, given that a large class of model is always, and as a matter of principle, assigned to the theory? Since Kazimierz Ajdukiewicz's works (e.g. 1931) the understanding of a language entails accepting certain highlighted sentences (postulates) of that language. Only persons who tend to recognize the postulates of a language as true can prove that they understand that language. For example, whoever refuses to acknowledge the sentence "A square is a square" as true simply does not understand the meaning of the word "is". This requirement is by far too weak. A set of models of a language which satisfy its postulates always comprises multiple members, and hence does not single out any one intended model. And based on one of the fundamental theses of logical semantics (the isomorphism theorem), whichever object x we take into account, for every predicate P of that language the complete set of models (postulates) of that language will comprise a model whereby x fall under P as well as a model whereby x does not fall under P. Thus, it turns out that if we define a language solely by supplying meaning postulates, all its (non-empty) predicates will be (completely) fuzzy. Such a language cannot be recognized to be empirical, and no empirical theory can be expressed in that language. In order to be elevated to the status of an empirical language, a language must satisfy the following condition: there are objects which under every interpretation of its certain (non-empty) predicates fall under those predicates, and those which do not fall under them under any of the models of the language. This requirement is somehow satisfied in empirical sciences. But how?

Here is an idea on how to answer that question, an idea which laid the foundations for the theory of science constructed by Przełęcki. As long as we attempt to provide interpretation for terms of a language by verbal means, we merely identify a set of meaning postulates of that language and we are become helpless in the face of the irremovable multiplicity of its interpretations. Therefore, we must resort to non-verbal means. The so-called ostensive definitions are of key importance in this context (Kotarbińska 1959). We specify certain predicates by pointing out which objects fall under them and which do not, and this is the only way. One can do this only with reference to certain exemplary cases; thus, such predicates will irretrievably remain fuzzy. Specifying the meaning of expressions by appealing to other expressions has its limits; in empirical science those limits are constituted by non-verbal means, which necessarily carry the stigma of fuzziness. That is the "primary datum" of the theory of science in question.

(3) The next element of the logical structure of the author's reasoning is a highly technical, and to a large extent formal conceptualization of the initial intuition sketched above. The predicates constructed in a way presented above correspond with (purely) observational concepts. But an empirical language contains both (purely) observational and theoretical predicates. Thus, a question arises about the logical links between the observational and the theoretical constituent of that language. A systematic answer to this question is provided by Przełęcki's works, including his basic monograph (1969; extended Polish edition 1988), which make use of the conceptual apparatus and the theorems of the theory of models. The basic desideratum is that the meaning postulates of an empirical language encompass both theoretical and observational postulates, and that theoretical terms are interpreted in a way that renders true the meaning postulates in which observational terms retain their ordinary interpretation (i.e. one supplied by non-verbal means). The richness of the forms of relationships between predicates of both kinds is discussed systematically and in detail. The analysis is comprehensive and provides positive solutions, but it does not shy away from pointing out undesirable consequences of those solutions, from exposing difficulties and raising new problems.

(4) So much about the theory itself, the core of which was published in Przełęcki (1969). This extended construction was applied by Przełęcki to rather numerous special issues. The issues of interpreting fuzzy terms, substantiating observational sentences, the existence of theoretical objects, the continuity of concepts across the changing theories which encompass those concepts, and others – these are all classic concerns of the philosophy of science, which Przełęcki addresses anew in terms of his own theory, unfailingly making new and important observations. Also, he extends his conceptual machinery beyond the philosophy of science, into

other fields of philosophy. And it turns out that we can gain new understanding in those other fields based on, for example, the peculiarity of the axiological discourse or the mechanism of the functioning of metaphors. And in his most recent writings (1976, 1986, 1990, 1992 and others), Przełęcki uses his conceptual apparatus to address the classic range of metaphysical problems.

Przełęcki's theory of science itself is valid only subject to certain far-reaching simplifications. These include restricting the language of the theory to qualitative terms, which eradicates the mathematical apparatus from the theories under consideration. In his article (1974) the author attempts to extend his conception towards a more realistic formulation of the theory which would comprise quantitative magnitudes as well. This called for modifying the concept of the intended model of a theory – in the extended sense, only those models can be regarded as intended ones in which not only logical concepts are interpreted in a predefined standard way, but mathematical concepts as well. In his article (1978) Przełęcki takes one further step by analyzing the inevitably approximate nature of empirical magnitudes. In doing so, he demonstrates that the fundamental reason for that is not so much the imprecision of the measuring acts, as is commonly believed, but the fuzziness of relevant concepts, i.e. the phenomenon which he made the centrepiece of his understanding of empirical sciences.

Another simplification that Przełęcki points out in the conclusion of his book (1969) is the fact that it ignores the issue of changes in a theory, and only deals with a "distilled profile" of a theory. In his article of 1980, he conceptualizes the phenomenon of the changeability of theories, and demonstrates that the different varieties of the correspondence relationship between subsequent theories can be interpreted in terms of his philosophy of science. Significant contributions to Przełęcki's theory come from Williams (1973, 1974).

1.2. Wójcicki's Theory of Approximation

An alternative theory of science has been worked out by Ryszard Wójcicki. While Przełęcki's theory can be said to extend the ideas espoused by the neopositivist philosophy of science into the areas that were not previously explored by this orientation, Wójcicki's theory is antipositivist on principle, and its refutation of the traditional ideas of neopositivism gets constantly sharper as the theory develops. In criticising neopositivism, Wójcicki draws on the conventionalist tradition in the philosophy of science, which has some presence in the Lvov-Warsaw School as well, mainly owing to Ajdukiewicz's conceptions from before 1939 and to pragmatist tendencies that were also present in this tradition (cf. e.g. Poznański and Wundheiler 1933).

(1) Here is Wójcicki's crucial intuition:

[T]he empirical theories are formulated, as a rule, in languages which operate [...] magnitudes [...] of the "exact" type: the exact location, the exact mass, the exact GDP, the exact population of a town, etc. However, the results of empirical research are always of approximate nature. The outcome of the measurement of the location of an object is an area in which a given object occurs, the result of a measurement of mass is an interval approximating the "real" value, and even when we speak of the population of a given town, the number may be supplied only as a certain approximation [Wójcicki 1974a, p. 162].

This intuition is expressed in purely objective terms and explicated by the author as follows.

Let P be a numerical parameter of the form:

$$P: Q^1 x \dots x \, Q^m \dashrightarrow R$$

defined on the m-element Cartesian product and taking values from the set of numbers R. For each x, the set $<x>$ is defined entirely arbitrarily provided that x itself belongs to $<x>$; $<x>$ is termed the neighbourhood of x. Then, a parameter $<P>$ is defined as

$$<P>: Q^1 x \dots x \, Q^m \dashrightarrow g(R),$$

where $<P>$ assigns m-tuple of objects not a single number but an interval of numbers embracing at any rate the number attached to this m-tuple by the parameter P. In other words, $<P>$ differs from P in that it takes on values from the set of intervals of numbers from R, whereas P takes on values from the set R itself provided that

$$<P>(a^1 \dots a^m) = <x> \text{ iff } P(a^1 \dots a^m) = x.$$

Consider the pair of parameters $(P, <P>)$. The parameter P is termed an idealization of $<P>$ and $<P>$ is named an approximation of P.

Speaking of approximate parameters we shall mean parameters that are approximations of some other parameters. And, similarly, speaking of exact, or point, parameters... we shall mean parameters that are idealizations of some other parameters. Needless to say, both a given exact parameter may have many different approximations and a given approximate parameter may be equipped in many different idealizations [Wójcicki 1974, p. 164].

The goal of the author is

to construct the notion of approximate truth which follows the idea that statements being of the form of mathematical equations are normally approximate statements, that is they are true within certain intervals of accuracy [Wójcicki 1969, p. 495].

The notion of approximate truth is constructed, roughly, in the following way. Let F_i be a physical magnitude. According to the author's explication, this means that F_i is a family of classes of abstraction determined by a certain equivalence relation defined on the set of physical objects W; s is a scale of this magnitude. Function p is a measurement of $\mathbf{F}^i$

iff p projects the set W in the class of intervals $g(R)$ on the condition that the number attached by the scale s to the class of abstraction $[a]$ generated by an member a of W belongs to the interval $p^s(a)$, that is $s([a]) \in p^s(a)$. A pair $f = (F, p^s)$ is labelled an operative magnitude.

(2) The main application of the ideas outlined above concerns a generalization of the semantic concept of truth. Let it X be the universe of physical objects. Then the system $X = (X, f^1 \dots f^n)$ where $f^i = (F^i, p^{si})$ is termed an operative system. Simplifying a little, an idealization of the system X is $X' = (X', F'^1 \dots F'^m)$ which satisfies the following condition: for every F'^i and every $a^1 \dots a^r \in X$: $F'^i(a^1 \dots a^r) \in f^i(a^1 \dots a^r)$.

A statement Z is true in the structure (operative system) X iff it is true (in the sense of the semantic definition of truth) in each idealization of X. On the other hand, Z is false in X iff it is false (in the sense of the semantic definition of truth) in every idealization of the structure X. Also, Z may be an undetermined statement in X iff Z is neither true nor false in X. And finally, a theory T is approximatively true in the structure X iff there exists such an idealization of X that all the statements from T are true in X (Wójcicki 1972, pp.41-42).

(3) This conception initiated a range of applications and expansions. Based on this conception, numerous key methodological issues were considered, such as the issues of the structure of magnitude (Kałuszyńska 1983), of inter-theory relations (Żytkow 1976), and especially of the correspondence of theories (Garstka 1974, Nadel-Turoński 1974) and of metaphors (Nadel-Turoński 1976). In all these areas new and interesting results were accomplished. But it must be stressed that the conception was evolving to match the increasingly antipositivist attitude of the author (Wójcicki 1982, 1991). E. Kałuszyńska's conception, which was inspired by R. Wójcicki's work, bears testimony to where this evolution is heading.

1.3. Elżbieta Kałuszyńska's methodological relativism

(1) Elżbieta Kałuszyńska's writings (with her 1994 monograph as her central contribution) fit within the most recent orientation in the philosophy of science which took upon itself the task of finally coming to grips with the neopositivist tradition of the founding fathers of our discipline. "Finally" means going back not only to the methodological theses of the Vienna Circle – which have been a typical object of criticism, initiated by the members of the Circle themselves – but their philosophical assumptions as well.

Those assumptions primarily concern the way in which the philosophy of science was practised by neopositivists, and in which it is often being practised to this day. This way of philosophising, the trademark of which is that the methodologists construct their own language based on the language of logic used for science, has been termed by Kałuszyńska logicism in the philosophy of science. And it is this logicist programme in methodology that Kałuszyńska opposes in her recent writings. She chooses Sneed-Balzer-Moulines's structuralist philosophy of science as the main object of her criticism, and reveals the non-formal components of its purportedly purely formal constructions. She strives to demonstrate that the most recent incarnation of logicism in methodology assumes much more than logic with the set theory, and that what it assumes in addition in fact means assuming illegally (i.e. contrary to verbal declarations) that those non-formal components are in principle irreducible to formal assumptions. Thus, as Kałuszyńska seems to be asserting, the logicist programme in the philosophy of science initiated by neopositivism sustains final defeat: the methodology of science cannot be made a superstructure imposed over logic.

The other assumption of the traditional philosophy of science that she subjects to criticism is representationalism, i.e. the belief that when investigating science one is allowed to assume that the cognizing subject is not limited by the restrictions resulting form the biological, social and ultimately historical constitution of humans, and that as a result their constructions bear no signs whatsoever of being human constructions. "What is assumed here", Kałuszyńska writes, "is the accessibility and transparency of reality. A 'mirror theory' reflects a 'given' reality" (1994, p. 59). Thus, the author opposes the practice of doing the philosophy of science as a theory of any cognizing subject, and calls for transforming it into a theory of human cognition which takes into account what is already known, thanks to the various scientific disciplines, about the factors specific to "subjects equipped with human cognitive capacities" (1994, p. 24).

The third basic assumption of the philosophy of science, as practised up till now, which the author subjects to criticism touches the realm of metaphysics and concerns the understanding of reality. The philosophy of science assumes that what we cognize is the actual world. But the cognizing subject is faced not directly with reality, but with a certain human articulation thereof. Kałuszyńska terms this articulation the world of science, which she understands as "a complex, multi-plane system of abstract objects that are interrelated by a network of dependencies". The world of science undergoes changes every time one creates a theory which "locates [...] the objects that it investigates within this system [and] [...] enriches it by adding new qualities and new relationships" (1994, p. 224).

In sum, the tradition of our discipline – as Kałuszyńska asserts – is linked with the study of the subject/object distinction, whereby the state of the subject neither disturbs the epistemic operation of cognition nor affects the other part of the distinction, leaving the metaphysical status of reality untouched. One studies this configuration believing that the relationship of the object's reflection in a subject is essentially of logical nature and that, therefore, appropriate means for its description are supplied by a logic which models both the structure of reality and cognitive constructions. And all that must be jettisoned through a thorough revision of the philosophy of science. Kałuszyńska's most recent work, and especially her monograph *Modele teorii empirycznych* [Models of Empirical Theories] (1994) which was quoted above, is a contribution to establishing the "critical philosophy of science" in the future.

(2) I used the term deliberately, and this is because what we are dealing with in the philosophy of science nowadays is, in a way, a local imitation of the development process that metaphysics went through. According to Nowak's (1998) formulation, metaphysical doctrines share a certain common structure: they identify beings which exist under any circumstances and then postulate a relation that something must exhibit with respect to something else that had already been identified as a being, in order for it to be included under beings itself. The metaphysical doctrine of existence identifies basic beings and defines relationships between beings. Thus, it follows the following formula: (i) there are such and such beings, (ii) if y is in relation R to the being x, then y is a being as well. Philosophy has long been dominated by the realcentrist paradigm, under which the basic beings were constituted by objects that were external to humans, be it material or ideal objects. The remaining lot, including the human soul, had to be legitimated in the light of the thesis of the existence of external reality. Plato's system serves as an example. In his system, the idea of harmony ("goodness") constitutes the basic being, and something exists if it participates in the existing being. The new-era breakthrough in philosophy was brought about by the anthropocentric paradigm, whereby man replaces the external reality. Descartes elaborated his system on the basis of the initial assumption that it is *cogitationes* that exist *prima facie*, and the existence of everything else, not excluding God and matter, must be derivable from this basic fact. Thus, Descartes was the true author of the Copernican revolution in philosophy. New versions of the anthropocentrist paradigm were subsequently created by Locke with his sense data, Kant with his categories of reason, and Husserl with his phenomena.

The trend that we are witnessing in the most recent developments in the philosophy of science can be described as a repetition of that pattern, leading from the realcentrist treatment of the subject/object distinction, whereby the subject essentially disappears as it is merely a mirror reflection of the object, towards the anthropocentric approach to that distinction, whereby the objective reality disappears and becomes merely something constructed by the subject. Accordingly, in much of today's philosophy of science we see the disappearance of the category of truth in the classic sense ("Truths are not discovered, truths are constructed"), which is being superseded by *consensus*. Furthermore, there is dwindling belief in the possibility of demonstrating the validity of scientific cognition ("anything goes") or even merely of cognition through reason ("bye-bye reason"). Thus, the philosophy of science is progressing from naive realism towards Feyerabend's or Hacking's anthropocentric relativism.

(3) The positive proposal for a new philosophy of science submitted by Kałuszyńska does not yield to the lure of easy iconoclasm spawning vague catch-all observations that are more useless than helpful in the area of the methodology of science, which is, after all, a fairly concrete field. Rather than turn out provocative aphorisms such as "Reality is anthropomorphic creation" (Hacking), she strives to find some route that would lead from the "world of science" to reality. Reality is no more than "potentiality", but it is beyond us, and hence is not a subject's free creation. On the other hand, reality is not "given" to subjects either – they give it articulation when constructing an abstract "world of science", and only then is it possible to "disclose properties that are potentially present in reality" (Kałuszyńska 1994, p. 9). This articulation is not arbitrary; conversely, it stems from our biological and social endowment. As the author vividly puts it, "the Baconian idols (particularly the tribal idol) do not so much 'falsify' cognition as make it impossible" (1994, p. 20).

The use of such non-standard philosophical assumptions on both sides, so to speak, has afforded solutions that are relevant to strictly methodological issues. More specifically, Kałuszyńska proposes a certain approach to the relationship between a theory and reality under which "empirical theories, while describing abstract objects, most certainly refer to real objects" (1994, p. 224). Furthermore, a theory sets out a class of abstract models of phenomena. They are to be obtained from theory upon acceptance of certain additional assumptions. A theory is true if it permits the construction of the models of all the phenomena which "it should be describing in the opinion of experts in a given field" (1994, p. 259). Also, she reinterprets in her own terms the principle of correspondence, which "imposes on a new theory the obligation to encompass the phenomena described by the outgoing theory, and hence the necessity to treat it as its own model" (1994, p. 258). And so on, and so forth. All this is guided by the belief that the inevitable dispelling of the myth of reality as the actuality that is described directly by science, as well as the inclusion of "the human factor" in science, does not necessarily have to lead to nihilism in Feyerabend's style. In a science like this – a human creation with a specific history, dealing with a world that it itself constructed – not everything is permitted. What is not permitted, and why, must be determined by the "critical philosophy of science", to which Kałuszyńska's recent work has contributed.

2. Conceptions drawing on the Hegelian-Marxian tradition

As is well known to those minds that are not dimmed by current politics, Marxist philosophy is rich with profound ideas, as can be expected from the inheritor of Hegelianism; at the same time – when it comes to the style of practising philosophy – it is also susceptible to analytical inspirations as can be expected from an orientation advocating a scientific approach. This is evidenced by the cognitive success of "analytical Marxism" in English-speaking countries in the past three decades. Contrary to popular belief, this also applies to the Marxist philosophy practised under "real socialism", even though there is no doubt that the links with the ideology of "real socialism" have seriously weakened it. For historical reasons, the Polish real socialism was relatively liberal in nature (Nowak 1983), and the Polish philosophy continued to comprise different orientations. It is undoubtedly thanks to the influence of the Lvov-Warsaw School that analytical Marxism was initiated in Poland – by Suszko (1957), Rogowski (1964) or Lange (1962), among others – and had already been quite well developed when the seminal works of the Anglo-Saxon version of "analytical Marxism" were written. And the Polish philosophy of science, among others, bears testimony to this.

2.1. Jerzy Kmita's historical epistemology

(1) Jerzy Kmita's historical epistemology (1976, 1979, 1985) is mainly oriented at explaining the peculiarities of the humanities, and has been applied to natural sciences only occasionally and in rather general terms. Since the present volume is devoted to the philosophy of exact sciences, Kmita's conception will only be given a cursory glance. The reason why it is discussed at all is because I aim to present a complete picture of the contemporary Polish theories of science.

Kmita's starting point is – on the micro level – the Weberian concept of rational action, and – on the macro level – the Marxian concept of social practice. The former is explicated in terms of the model of rational action originating from the theory of decision, and it serves to introduce the concepts of cultural action, and communicative action in particular (Kmita & Nowak 1968), as well to reconstruct Dilthey's and Weber's intuitions regarding the peculiarities of humanist understanding, as contrasted with the type of explanation employed by natural sciences (Kmita 1971). The latter is understood as a multi-plane social structure, whereby the reproduction of the dominant relations of production constitutes the ultimate functional logic. Social consciousness adapts to thus defined practice by regulating relevant functional sub-structures. Social practice evolves to match changes in the division of production labour, and new areas of social consciousness emerge accordingly.

(2) When a new chapter in social practice, i.e. research practice, came into being in the modern era, it was accompanied by the emergence of the corresponding theoretical-methodological consciousness. Various stages of research practice are matched by various corresponding types of consciousness that regulate them. Those types range from the consciousness that is indispensable for a collector of empirical facts, through the consciousness needed for conducting isolated explications of those facts, to the consciousness that is necessary for connecting those explications into broader theoretical systems. The task of the philosophy of science is to reconstruct those historically variable forms of scholars' real consciousness and to reveal their link with the current stage of the research practice in which they are involved. Such an understanding of the philosophy of science turns out to be a "theoretical history" of research consciousness, and the changes that it undergoes are not so much a result of approaching "the truth in science" as a reflection of the changes in the consciousness of the scholars themselves. Their consciousness, in turn, adjusts to the changes in their daily research practice. Inductivism, instrumentalism, hypotheticism, etc., are not so much products of the philosophers of science as an indication of the tendency to conform to the research practice of science. The fact that those orientations occur one after another within the consciousness that is specific to the community of philosophers of science does not result from chance, but from historical transformations in research practice which are matched by changes in the consciousness that regulates that practice. And all that, in turn, starts from changes in the economic structure of the society.

(3) These ideas were expanded and applied in numerous works, the most important of which include the monographs by Zamiara (1974), Zgółka (1976, 1980) and Kubicki (1991), as well as collections of articles by Kmita i Zamiara (1989) and Zeidler-Janiszewska (1996).

2.2. *Leszek Nowak's idealizational conception of sciences*

(1) Nowak (1968, pp. 80*ff*) formulates the initial intuition concerning idealization in the following way. Take any scientific law, for example the "law of ideal gases", and try to apply it to the so-called reality. It obtains immediately that the law does not hold good in the world whose

description it is supposed to be. The worlds that science is about are quite different from the world we live in. The world of physics is one of bodies devoid of geometrical dimensions, submerged in inertial systems, moving in a frictionless way, etc. The world of economics is one of balanced economies where the supply of commodity equals exactly the social demand for it, where there is a uniform growth of technology, etc. The world of the humanities is one in which there are only those who consistently follow their values. And so on, and so forth. Indeed, science is about the ideal worlds and the physical world in which we live is more or less distant from its scientific idealizations.

The method of idealization is basically a method of deformation, that is, as it were, it transforms our world into ideal worlds. That is why idealization does not resemble what is sometimes considered to be a paradigmatic example of science, namely a chronicle. It is not so that the theoretician observes facts, notices their appearances, looks for recurrences and similarities in order to find some general rules. As the author pointed out later (Nowak 1977, pp. 344*ff*), it is rather the caricature which seems to be the paradigmatic example of science. Cartoonist leave out some details of the object presented, thus stressing what they claim to be characteristic of it. They do, roughly, a similar thing that theoreticians do: all of them deform the objects of their study. Science consists of the same procedure which we find in caricature. Both deform the world which we inhabit.

(2) The main ideas of the idealizational approach to science are the following (Nowak 1971, 1972, 1980).

I. Idealization of a given phenomenon presupposes that a theorist distinguishes some factors suspected to influence a given phenomenon and divides them into two groups: those believed to be secondary and principal for the phenomenon under investigation. Idealizing conditions are assumptions (e.g. "the resistance of air does not occur", "demand equals supply", "man does not accept contradictions") which are undertaken in order to neglect the secondary factors and thus to simplify the empirical subject matter. By introducing idealizing assumptions the pure idea of a given type of phenomena is formed. It is deprived some of the empirical features, i.e. those which are considered to be secondary. What remains in the idea still contains merely what is considered the principal properties of the empirical original. A hypothesis about the idea which connects various aspects of its equipment is termed an idealizational statement. Such a hypothesis is a conditional stating that under such and such idealizing conditions a certain formula holds (e.g. if anything is a portion of ideal gas, then $pv = kt$ holds for it). If an idealizational statement is valid at all, it is for the pure idea alone.

II. The idealizational statement is concretized by gradually admitting the previously neglected secondary properties and modifying its formula. The laws become more and more complicated and therefore ever closer to the empirical reality. For example, the first concretization of Clapeyron's law reads: if anything is a portion of half-ideal gas, then $p(v - a) = kt$. The concretization procedure continues until the most realistic idealizational statement (the final concretization of the idealizational law) becomes a sufficient approximation of the empirical phenomena. This means that the theoretician takes a risk to admit that if all the remaining idealizing conditions of the final concretization of the initial law are removed at one and the same time, the theoretical outcomes derived from its formula will not deviate from empirical data by more than the level of admissible discrepancies generally accepted in the given domain of research. The approximation of an idealizational statement is a factual statement containing merely realistic assumptions in its antecedent. For instance, if anything is a portion of real gas, then $p(v - a) \approx kt$. The idealizational structure t is a sequence of universal conditionals $T^k, T^{k-1}, ..., T^i, T^{i-1}, AT^{i-1}$ (the indexes $k, k-1, ..., i$ show the amount of the initial idealizing conditions still in force) where T^{k-1} is a concretization of T^k, ..., T^{i-1} is a concretization of T^i and, finally, the factual statement AT^{i-1} is an approximation of T^{i-1}.

III. In a case when the approximation of the final concretization is rejected, the idealizational structure itself is not rejected but improved by continuing the process of concretization. It is rejected only at the point where no concretization proves to be able to cover the discrepancies between the theoretical, predicted outcomes and the actual data. This outcome testifies against the researchers' decision to count as principal such and such factors and neglect the remaining influences as secondary or even against their initial enlisting the adopted group of influences. Thus, the initial idealizational structure is replaced with a new proposal based on another division of the same factors upon principal and secondary or on another list of influences. This trial and error procedure lasts until some idealizational structure gives an admissible discrepancy with the known empirical facts.

IV. Deductively ordered bodies of the already accepted idealizational structures are idealizational theories. Groups of statements made at the same level of idealization, i.e. equipped with the same idealizing conditions, are termed models. The structure of a scientific theory t is given by a sequence of models $M^k, M^{k-1}, ..., M^i, AM^i$, where M^k is the most abstract model equipped with k idealizing conditions, $M^{k-1} ...M^i$ contain their subsequent concretizations and, finally, AM^i contains approximations of the statements of the least abstract model M^i to the empirical reality.

V. Even if the given idealizational theory approximately satisfies the empirical facts that are already known, it might happen that it becomes falsified by the newly discovered facts, that is by allowing too great discrepancies with the new data or experiments. If new secondary factors responsible for the deviations of t from the data are discovered, then a new corrected version t' of the theory t is formed. t' is composed of models M^n, M^{n-1}, ..., M^{rk}, ..., M^{ri}, AM^{ri}, where $n > k$ (the new secondary factors that have been discovered) but the formulas of the model M^k continue to appear in M^n, and the further models from M^{rk} onwards differ from those of t due to corrections introduced which account for the working of the newly discovered secondary factors. This connection between t' and t is termed the dialectical correspondence (Nowakowa 1972, 1975a-b, 1994)[1]. It consists in correcting the initial theory as the previous view on the principal factors (the "essence") of the explained phenomena holds unchanged; what changes and becomes expressed in the new theory t' is the view of how the same essence is manifested in the phenomena.

VI. Dialectical refutation consists in rejecting the previous view of what is principal. What changes in the new structure s is the formula of the initial law. The new structure s must have, however, something in common with the old one, t. The relative success of t in coming closer and closer to the actual world testifies to the fact that its law possesses some explanatory power, although it is far from being sufficient. That is why the old formula is to appear as a component of one of the concretizations of the new structure s (Paprzycka 1990). Then, s is replaced with s', s''... dialectically corresponding to it, etc. What changes in the new theory t'' are formulae from the initial model; the old formulae are to appear as components of the derivative models of the new theory t''.

(3) The above claims constitute what may be called the core of the idealizational approach to science. There are also numerous special variants based on some derivative constructions. For example, a conception of *ceteris paribus* statements (Patryas 1976, 1982), elaboration of the procedures of proto-idealization (Brzeziński 1978, 1985a), of stabilization (Chwalisz 1979, Zielińska 1981), of aggregation (Łastowski 1987), of normalization (Nowak 1991a), of isolation (Nowak 1997a) etc.[2] And all that profited in reconstructions of numerous pieces of the research practice from physics (Such 1974, 1978, 1990, Nowak 1973, 2000a; Nowakowa 1975a, 1994; Patryas 1976; Krajewski 1974a-b, 1977, 1982a; Zielińska 1981, 1986; Kuipers 1985; Boscarino 1990; Kupracz 1992; and others) and biology (Łastowski 1977, 1982, 1987, 1991, 1994; Łastowski and Nowak

---

[1] An alternative approach to correspondence has been elaborated upon in the idealizational terms by Krajewski (1974a, 1977).

[2] More detailed review of various expansions of the idealizational approach to science may be found in Nowak (1992).

1982; Piontek 1985; Kośmicki 1986, 1988; Nowak 2000b, and others), not to mention social sciences.

Let us finally add that recently a certain metaphysical conception doing justice to the cognitive significance of the method of idealization has been elaborated upon (Nowak 1995, 1997b, 1998).

Conclusion

This overview demonstrates that the Polish philosophy of science has developed some comprehensive approaches to science, some of them competing with one another, which is only natural. The special nature of empirical science consists in the fact that – to quote the various approaches – *it is based on a language that is inevitably a fuzzy one* (Przełęcki), *uses approximation with facts* (Wójcicki), *deals with the human articulation of the so-called empirical world* (Kałuszyńska), *regulates research practice subjected to production practice* (Kmita), and *consists in idealization referring to ideal worlds and a return to empirical reality through concretizations* (Nowak). It must be pointed out at least in some cases those approaches are peculiar to the Polish philosophy of science, and that some of them have even anticipated some trends in the foreign philosophy of science.

REFERENCES

Boscarino, G. (1990). Absolute space and idealization in Newton. In: Brzeziński, Coniglione *et al.* (1990), 131-149.

Brzeziński, J. (1978). *Metodologiczne i psychologiczne wyznaczniki procesu badawczego w psychologii* [Methodological and Psychological Determinants of the Research Process in Psychology]. Poznań: Wydawnictwo Uniwersytetu Poznańskiego.

Brzeziński, J. (1985a). The Protoidealizational Model of the Investigative Process in Psychology. In: Brzeziński (Ed.) (1985b), 36-57.

Brzeziński, J. (Ed.) (1985b), *Consciousness: Methodological and Psychological Approaches (Poznań Studies in the Philosophy of the Sciences and the Humanities* 8). Amsterdam–Atlanta (GA): Rodopi.

Brzeziński, J., Fr. Coniglione, T. A. F. Kuipers and L. Nowak (Eds.) (1990a), *Idealization-I: General Problems (Poznań Studies in the Philosophy of the Sciences and the Humanities* 16). Amsterdam–Atlanta (GA): Rodopi.

Brzeziński, J., Fr. Coniglione, T. A. F. Kuipers and L. Nowak (Eds.) (1990b), *Idealization-II: Forms and Applications (Poznań Studies in the Philosophy of the Sciences and the Humanities* 17). Amsterdam–Atlanta (GA): Rodopi.

Chwalisz, P. (1979). Stałe w idealizacyjnej koncepcji nauki [Constants in the Idealizational Conception of Science]. In: A. Klawiter and L. Nowak (1979) *Odkrycie, abstrakcja, prawda, empiria, historia a idealizacja* [Discovery, Abstraction, Truth, Empiricism, History *versus* Idealisation]. Warszawa–Poznań: PWN, 99-104.

Eells, E. and T. Maruszewski (Eds.) (1991), *Probability and Rationality. Studies on L. Jonathan Cohen's Philosophy of Science (Poznań Studies in the Philosophy of the Sciences and the Humanities* **21**). Amsterdam–Atlanta (GA): Rodopi.

Kmita, J. (1971). *Z metodologicznych zagadnień interpretacji humanistycznej* [From Methodological Problems of the Humanistic Interpretation]. Warszawa: PWN.

Kmita, J. (Ed.) (1973). *Elementy marksistowskiej metodologii humanistyki* [Elements of the Marxist Methodology of the Humanities]. Poznań: Wydawnictwo Poznańskie.

Kmita, J. (1976). *Szkice z teorii poznania naukowego* (English translation: Essays on the Theory of Scientific Cognition). Warszawa: PWN & Dordrecht: Reidel 1988.

Kmita, J. (1979). *Epistemologia historyczna* (Historical Epistemology). Warszawa: PWN.

Kmita, J. and L. Nowak (1968). *Studia nad teoretycznymi podstawami humanistyki* [Studies in the Theoretical Foundations of the Humanities]. Poznań: Wydawnictwo Uniwersytetu Poznańskiego.

Kośmicki, E. (1986). *Biologiczne koncepcje zachowania* [Biological Conceptions of Behaviour]. Warszawa–Poznań: PWN.

Kośmicki, E. (1988). *Etologiczne i socjobiologiczne rozwinięcia teorii ewolucji. Studium metodologiczne* [Etological and Sociobiological Extensions of the Theory of Evolution]. *Roczniki Akademii Rolniczej* **177**. Poznań: Wydawnictwo Akademii Rolniczej.

Krajewski, W. (1974a). Kopernik i Galileusz versus Arystoteles. Nowa metoda naukowa przeciw dogmatyzmowi i waskiemu empiryzmowi [Copernicus and Galileo against Aristotle. New Scientific Method against Dogmatism and Narrow Empiricism]. *Studia Metodologiczne*, **12**, 3-22.

Krajewski, W. (1974b). Redukcja, idealizacja, korespondencja (Reduction, Idealization, Correspondence). In: Krajewski (1974c), 115-45.

Krajewski, W. (Ed.) (1974c). *Zasada korespondencji w fizyce a rozwój nauki* [The Principle of Correspondence in Physics and the Growth of Science]. Warszawa: PWN.

Krajewski, W. (1976). Correspondence Principle and Idealization. In: Przełęcki, *et al.* (Eds.) (1976), 380-86.

Krajewski, W. (1977). *The Principle of Correspondence and the Growth of Science*. Dordrecht: Reidel.

Krajewski, W. (1982a). *Prawa nauki. Przegląd zagadnień metodologicznych* [Laws of Science. A Survey of Methodological Problems]. Warszawa: KiW.

Krajewski, W. (Ed.) (1982b). *Polish Essays in the Philosophy of the Natural Sciences (Boston Studies in the Philosophy of Science* **68**), Dordrecht: Reidel.

Kuipers, T.A.F. (Ed.) (1985). The Paradigm of Concretization: The Law of van der Waals. In: Brzeziński (1985b), 185-99.

Kupracz, A. (1992). *O dwóch ujęciach idealizacji w naukach empirycznych. Próba analizy porównawczej* [On Two Approaches to Idealization: A Comparative Analysis]. Poznań: Wydawnictwo PAN.

Łastowski, K. (1977). The Method of Idealization in the Populational Genetics, *Poznań Studies in the Philosophy of the Sciences and the Humanities* **3**, 199-212.

Łastowski, K. (1982). The Theory of Development of Species and the Theory of Motion of Socio-Economic Formation. In: Nowak (Ed.) (1982), 122-157.

Łastowski, K. (1987). *Rozwój teorii ewolucji. Studium metodologiczne* [The Development of the Theory of Evolution. A Methodological Study]. Poznań: Wydawnictwo Uniwersytetu Poznańskiego.

Łastowski, K. (1990). On Multi-Level Scientific Theories. In: Brzeziński *et al.* (Eds.) (1990b), 33-59.

Łastowski, K. (1991). Two Models of Evolution in Darwin's Theory. *Variability and Evolution*, **1**, 25-37.

Łastowski, K. (1994). The Idealizational Status of the Contemporary Theory of Evolution In: A. Klawiter (Ed.) *Understanding Idealization (Theoria* **20**), 29-52.

Łastowski, K. and L. Nowak (1982). Galileusz nauk biologicznych [The Galileo of Biological Sciences]. *Kosmos*, **31**, 3-4, 195-210.

Nadel-Turoński, T. (1976). Metafory matematyczne w teorii fizycznej [Mathematical Metaphors in a Physical Theory]. In: Nowak *et al.* (Eds.), *Poznańskie studia z filozofii nauki*. Z.1. *Teoria a rzeczywistość* [Poznań Studies in the Philosophy of Science. F.1. Theory and Reality]. Warszawa–Poznań: PWN, 33-51.

Nowak, L. (1968). *Próba metodologicznej charakterystyki prawoznawstwa* [Essay in the Methodological Characteristics of Jurisprudence]. Poznań: Wydawnictwo Uniwersytetu Poznańskiego.

Nowak, L. (1971). *U podstaw Marksowskiej metodologii nauk* [Foundations of the Marxian Methodology of Science]. Warszawa: PWN.

Nowak, L. (1972a). Theories, Idealization and Measurement. *Philosophy of Science* **39**, 533-547.

Nowak, L. (1973). Popperowska koncepcja praw i sprawdzania [Popperian Conception of Laws and Testing]. In: Kmita (Ed.) (1973), 303-324.

Nowak, L. (1977). *Wstęp do idealizacyjnej teorii nauki* [An Introduction to the Idealizational Theory of Science]. Warszawa: PWN.

Nowak, L. (1980). *The Structure of Idealization. Towards a Systematic Interpretation of the Marxian Idea of Science (Synthese Library* **131**). Dordrecht: Reidel.

Nowak, L. (Ed.) (1982). *Social Classes, Action and Historical Materialism (Poznań Studies in the Philosophy of the Sciences and the Humanities* **6**). Amsterdam–Atlanta (GA): Rodopi.

Nowak, L. (1983). *Property and Power. Towards a non-Marxian Historical Materialism.* Dordrecht: Reidel.

Nowak, L. (1991a). The Method of Relevant Variables and Idealization. In: Eells and Maruszewski (Eds.) (1991), 41-63.

Nowak, L. (1991b). Thoughts Are Facts of Possible Worlds. Truths Are Facts of a Given World. *Dialectica* **45**, 273-287.

Nowak, L, (1992). The Idealizational Approach to Science: A Survey. In: J. Brzeziński and L. Nowak (Eds.). *Idealization III: Approximation and Truth (Poznań Studies in the Philosophy of the Sciences and the Humanities* **25**). Amsterdam–Atlanta (GA): Rodopi, 9-63.

Nowak, L. (1995). Anti-Realism, (Supra-)Realism and Idealization. In: Herfel *et al.* (1995), 225-241.

Nowak, L. (1997a). Uwagi o tak zwanej metodzie izolacji [Remarks on the So-called Method of Isolation]. In: J. Mrozek (Ed.). *Między filozofią nauki a filozofią historii* [Between Philosophy of Science and Philosophy of History]. Gdańsk: Wydawnictwo Uniwersytetu Gdańskiego, 29-38.

Nowak, L. (1997b). On the Concept of Nothingness. *Axiomathes* **8**, 1-3, 381-394.

Nowak, L. (1998). *Byt i myśl. U podstaw negatywistycznej metafizyki unitarnej* [Being and Mind. Foundations of the Negativist Unitarian Metaphysics]. Vol. I: *Nicość i istnienie* [Existence and Nothingness]. Poznań: Zysk.

Nowakowa, I. (1972). *Idealizacja i problem korespondencji praw fizyki* [Idealization and the Problem of Correspondence of Laws of Physics]. Ph.D. thesis, Dept. of Philosophy, Poznań University.

Nowakowa, I. (1974). The Concept of Dialectical Correspondence. *Dialectics and Humanism*, **1**, 3, 51-55.

Nowakowa, I. (1975a). *Dialektyczna korespondencja a rozwój nauki* [The Dialectical Correspondence and the Development of Science], Warszawa-Poznań: PWN.

Nowakowa, I. (1975b), Idealization and the Problem of Correspondence, *Poznań Studies in the Philosophy of the Sciences and the Humanities* **1**, 1, 65-70.

Nowakowa, I. (1982). Dialectical Correspondence and Essential Truth. In: Krajewski (Ed.) (1982b), 135-146.

Nowakowa, I. (1994). *Correspondence and Truth. The Dynamics of Science in the Idealizational Methodology.* Amsterdam–Atlanta (GA): Rodopi.

Paprzycka, K. (1990). Reduction and Correspondence in the Idealizational Approach to Science. In: Brzeziński *et al.* (Eds.) (1990a), 277-86.

Patryas, W. (1976). *Eksperyment a idealizacja* (Experiment and Idealization). Warszawa-Poznań: PWN.

Patryas, W. (1982). The Pluralistic Approach to Empirical Testing and the Special Forms of Experiment. In: W. Krajewski (Ed.) (1982b), 127-134.

Piontek, J. (1985). Koncepcja antropologii Jana Czekanowskiego [The Conception of Anthropology of Jan Czekanowski]. In: Piontek and Malinowski (Eds.) (1985), 135-144.

Piontek, J. and A. Malinowski (1985) (Eds.). *Teoria i empiria w polskiej szkole antropologicznej* [Theory and Experience in the Polish Anthropological School] (*Seria Antropologia* **11**). Poznań: Wydawnictwo Uniwersytetu Poznańskiego.

Przełęcki, M. (1969). *The Logic of Empirical Theories*. London: Routledge and Kegan Paul. (quoted after the German translation in: W. Balzer and M. Heidelberger (Hrsg.) *Zur Logik der empirischen Theorem*. Berlin: Walter de Gruyter 1983, 43-96).

Przełęcki, M. (1988). *Logika teorii empirycznych* [The Logic of Empirical Theories, 2nd ed.], Warszawa: PWN.

Przełęcki, M. (1993). Fikcje dowolnie bliskie realności [Fictions Arbitrarily Close to Reality]. An Interview by J.J. Jadacki. *Filozofia Nauki* **1**, 3-4, 9-19.

Przełęcki, M., K. Szaniawski and R. Wójcicki (Eds.) (1976). *Formal Methods in the Methodology of Empirical Sciences*. Wrocław: Ossolineum & Dordrecht: Reidel.

Rogowski, S. (1963) *Logika kierunkowa a Heglowska teza o sprzeczności zmiany* [The Directional Logic and the Hegelian Thesis of Contradictoriness of Change]. Toruń: Wydawnictwo Uniwersytetu Toruńskiego.

Such, J. (1974). Relacja korespondencji a wynikanie [The Relation of Correspondence and Entailment]. In: Krajewski (Ed.) (1974c), 65-114.

Such, J. (1978). Idealization and Concretization in the Natural Sciences, *Poznań Studies in the Philosophy of the Sciences and the Humanities* **4**, 1-4, 49-74.

Such, J. (1990). The Idealizational Conception of Science and the Law of Universal Gravitation. In: J. Brzeziński *et al.* (Eds.) (1990b), 125-130.

Wójcicki, R. (1974). *Metodologia formalna: metody, pojęcia i zagadnienia* [The Formal Methodology: Methods, Concepts and Problems]. Wrocław: Ossolineum.

Wójcicki, (1979). *Topics in the Formal Methodology of Empirical Sciences*. Dordrecht: Reidel.

Zamiara, K. (1974). *Metodologiczne znaczenie sporu o status poznawczy teorii* [The Methodological Significance of the Realism-Instrumentalism Controversy]. Warszawa: PWN.

Zgółka, T. (1976). *O strukturalnym wyjaśnianiu faktów językowych* [On the Structural Explanation of Linguistic Facts]. Warszawa–Poznań: PWN.

Zielińska, R. (1981). *Abstrakcja, Idealizacja, Uogólnienie* [Abstraction, Idealization, Generalization]. Poznań: Wydawnictwo Uniwersytetu Poznańskiego.

Zielińska, R. (1986). Problem sprzężeń zwrotnych w kategorialnej interpretacji dialektyki [The Problem of Feed back in the Categorial Interpretation of Dialectics] In: Maruszewski (1986), 241-255.

POZNAŃ STUDIES IN THE PHILOSOPHY OF THE SCIENCES AND THE HUMANITIES

Contents of back issues

VOLUME 1 (1975)

Main topics:
The Method of Humanistic Interpretation; The Method of Idealization; The Reconstruction of Some Marxist Theories.
(sold out)

VOLUME 2 (1976)

Main topics:
Idealizational Concept of Science; Categorial Interpretation of Dialectics.
(sold out)

VOLUME 3 (1977)

Main topic:
Aspects of the Production of Scientific Knowledge.
(sold out)

VOLUME 4 (1978)

Main topic:
Aspects of the Growth of Science.
(sold out)

VOLUME 5 (1979)

Main topic:
Methodological Problems of Historical Research.
(sold out)

VOLUME 6 (1982)

SOCIAL CLASSES ACTION & HISTORICAL MATERIALISM

Main topics:
On Classes; On Action; The Adaptive Interpretation of Historical Materialism; Contributions to Historical Materialism.
(sold out)

VOLUME 7 (1982)

DIALECTICAL LOGICS FOR THE POLITICAL SCIENCE
(Edited by Hayward R. Alker, Jr.)

VOLUME 8 (1985)

CONSCIOUSNESS: METHODOLOGICAL AND PSYCHOLOGICAL APPROACHES
(Edited by Jerzy Brzeziński)

VOLUME 9 (1986)

THEORIES OF IDEOLOGY AND IDEOLOGY OF THEORIES
(Edited by Piotr Buczkowski and Andrzej Klawiter)

VOLUME 10 (1987)

WHAT IS CLOSER-TO-THE-TRUTH?
A PARADE OF APPROACHES TO TRUTHLIKENESS
(Edited by Theo A.F. Kuipers)

VOLUME 11 (1988)

NORMATIVE STRUCTURES OF THE SOCIAL WORLD
(Edited by Giuliano di Bernardo)

VOLUME 12 (1987)

POLISH CONTRIBUTIONS TO THE THEORY AND PHILOSOPHY OF LAW
(Edited by Zygmunt Ziembiński)

VOLUME 20 (1990)

Jürgen Ritsert
MODELS AND CONCEPTS OF IDEOLOGY

VOLUME 21 (1991)

PROBABILITY AND RATIONALITY
STUDIES ON L. JONATHAN COHEN'S PHILOSOPHY OF SCIENCE
(Edited by Ellery Eells and Tomasz Maruszewski)

VOLUME 22 (1991)

THE SOCIAL HORIZON OF KNOWLEDGE
(Edited by Piotr Buczkowski)

VOLUME 23 (1991)

ETHICAL DIMENSIONS OF LEGAL THEORY
(Edited by Wojciech Sadurski)

VOLUME 24 (1991)

ADVANCES IN SCIENTIFIC PHILOSOPHY
ESSAYS IN HONOUR OF PAUL WEINGARTNER ON THE OCCASION OF
THE 60TH ANNIVERSARY OF HIS BIRTHDAY
(Edited by Gerhard Schurz and Georg J.W. Dorn)

VOLUME 25 (1992)

IDEALIZATION III: APPROXIMATION AND TRUTH
(Edited by Jerzy Brzeziński and Leszek Nowak)

VOLUME 26 (1992)

IDEALIZATION IV: INTELLIGIBILITY IN SCIENCE
(Edited by Craig Dilworth)

VOLUME 27 (1992)

Ryszard Stachowski

THE MATHEMATICAL SOUL.
AN ANTIQUE PROTOTYPE OF THE MODERN MATEMATISATION OF
PSYCHOLOGY

VOLUME 28 (1993)

POLISH SCIENTIFIC PHILOSOPHY:
THE LVOV-WARSAW SCHOOL
(Edited by Francesco Coniglione, Roberto Poli and Jan Woleński)

VOLUME 29 (1993)

Zdzisław Augustynek and Jacek J. Jadacki

POSSIBLE ONTOLOGIES

VOLUME 30 (1993)

GOVERNMENT: SERVANT OR MASTER?
(Edited by Gerard Radnitzky and Hardy Bouillon)

VOLUME 31 (1993)

CREATIVITY AND CONSCIOUSNESS.
PHILOSOPHICAL AND PSYCHOLOGICAL DIMENSIONS
(Edited by Jerzy Brzeziński, Santo Di Nuovo, Tadeusz Marek and
Tomasz Maruszewski)

VOLUME 32 (1993)

FROM ONE-PARTY-SYSTEM TO DEMOCRACY
(Edited by Janina Frentzel-Zagórska)

J. Frentzel-Zagórska, *Introduction*. **Part I: Theoretical Approaches** – Z. Bauman, *A Postmodern Revolution*; L. Holmes, *On Communism, Post-communism, Modernity and Postmodernity*; L. Nowak, *The Totalitarian Approach and the History of Socialism*; J. Pakulski, *East European Revolutions and 'Legitimacy Crisis'*. **Part II: The Transitional Period** – A. Czarnota, M. Krygier, *From State to Legal Traditions? Prospects for the Rule of Law after Communism*; M. Szabó, *Social Protest in a Post-communist Democracy: Taxi Drivers' Demonstration in Hungary*; Z. Bauman, *Dismantling Patronage State*; E. Mokrzycki, *Between Reform and Revolution: Eastern Europe Two Years after the Fall of Communism*; J. Frentzel-Zagórska, *The Road to Democratic Political System in Post-communist Eastern Europe*. **Part III: The Case of Yugoslavia** – R.F. Miller, *Yugoslavia: The End of the Experiment*.

VOLUME 33 (1993)

SOCIAL SYSTEM, RATIONALITY, AND REVOLUTION
(Edited by Marcin Paprzycki and Leszek Nowak)

Introduction. **On the Nature of Social System** – U. Preuss, *Political Order and Democracy. Carl Schmitt and his Influence*; K. Paprzycka, *A Paradox in Hobbes' Philosophy of Law*; S. Esquith, *Democratic Political Dialogue*; E. Jeliński, *Democracy in Polish Reformist Socialist Thought*; K. Paprzycka, *The Master and Slave Configuration in Hegel's System*; M. Godelier, *Lévi-Strauss, Marx and After. A Reappraisal of Structuralist and Marxist Tools for Social Logics*; K. Niedźwiadek, *On the Structure of Social System*; W. Czajkowski, *Social Being and Its Reproduction.* **On Rationality and Captivity** – M. Ziółkowski, *Power and Knowledge*, L. Nowak, *Two Inter-Human Limits to the Rationality of Man*; M. Paprzycki, *The Non-Christian Model of Man: An Attempt at Psychological Explanation*; R. Egiert, *Toward the Sophisticated Rationalistic Model of Man.* **On Social Revolution** – L. Nowak, *Revolution is an Opaque Progress but a Progress Nonetheless*; K. Paprzycka, M. Paprzycki, *How do Enslaved People Make Revolutions?*; G. Tomczak, *Is it Worth Winning a Revolution?*; K. Brzechczyn, *Civil Loops and the Absorption of Elites*; R. McCleary, *What Makes Marxist Historical Materialism Objective?*; G. Kotlarski, *Classes and Masses in Social Philosophy of Rosa Luxemburg.* **On Real Socialism** – E. Gellner, *The Civil and the Sacred*; W. Marciszewski, *Economics and the Idea of Information. Why Socialism must have Collapsed?*; L. Nowak, K. Paprzycka, M. Paprzycki, *On Multilinearity of Socialism*; A. Siegel, *The Overrepression Cycle in the Soviet Union. An Operationalization of a Theoretical Model*; K. Brzechczyn, *The State of the Teutonic Order as a Socialist Society.* **Discussions** – R. McCleary, *Socioanalysis and Philosophy*; W. Heller, *Methodological Remarks on the Public and the Private in Hannah Arendt's Political Philosophy*; K. Brzechczyn, *On Unsuccessful Conquest and Successful Subordination.*

VOLUME 34 (1994)

Izabella Nowakowa

IDEALIZATION V: THE DYNAMICS OF IDEALIZATIONS

Introduction; Chapter I: *Idealization and Theories of Correspondence*; Chapter II: *Dialectical Correspondence of Scientific Laws*; Chapter III: *Dialectical Correspondence in Science: Some Examples*; Chapter IV: *Dialectical Correspondence of Scientific Theories*; Chapter V: *Generalizations of the Rule of Correspondence*; Chapter VI: *Extensions of the Rule of Correspondence*; Chapter VII: *Correspondence and the Empirical Environment of a Theory*; Chapter VIII: *Some Methodological Problems of Dialectical Correspondence.*

VOLUME 35 (1993)

EMPIRICAL LOGIC AND PUBLIC DEBATE.
ESSAYS IN HONOUR OF ELSE M. BARTH
(Edited by E.C.W. Krabbe, R.J. Dalitz, and P.A. Smit)

Part I: Interpersonal Reasoning: Conflicts and Fallacies – T. Govier, *Needing Each Other for Knowledge: Reflections on Trust and Testimony*; J. Woods, *'Secundum quid' as a Research Programme*; G. Nuchelmans, *On the Fourfold Root of the 'argumentum ad hominem'*; F.H. van Eemeren, R. Grootendorst, *The History of the 'argumentum ad hominem' Since the Seventeenth Century*; L.S. van Epenhuysen, *Debate in a Bermuda Triangle of Medical Ethics*; E.C.W. Krabbe, *Reasonable Argument and Fallacies in the Kok-*

Stekelenburg Debate; **Part II: Linguistic and Conceptual Tools** – J.D. North, *Some Weak Links in the Great Chain of Being*; A. Næss, *'You assert this?' An Empirical Study of Weight Expressions*; R. Wiche, *Gerrit Mannoury on the Communicative Functions of Negation in Ordinary Language*; J. Hoepelman, T. van Hoof, *Default and Dogma*; Ch. Goossens, *On the Logic of Nonmoral Commitment*; W. Marciszewski, *Arguments Founded on Creative Definitions*; R. Jorna, *Cognitive Science and Connectivism: Friend and Enemy or Move and Counter-Move, an Application of Empirical Logic*; **Part III: Dialectical Climates and Tempests** – R.H. Johnson, *Dialectical Fields and Manifest Rationality*; P. du Preez, *Reason Which Cannot Be Reasoned With: What Is Public Debate and How Does it Change?*; M.A. Finocchiaro, *Logic, Democracy, and Mosca*; P.A. Smit, *The Logic of Virtue and Terror*; G. van Benthem van den Bergh, *On Obstacles to Public Debate*; J.P. van Bendegem, *Real-Life Mathematics versus Ideal Mathematics: The Ugly Truth*; **Part IV: The Disempowerment of Woman: Strategies and Counter-Moves** – V. Songe-Møller, *The Road of Being and the Exclusion of the Feminine. An Analysis of the Poem of Parmenides*; R.J. Dalitz, *The Subjection of Women in the Contractual Society. An Analysis of Thomas Hobbes' Theory of Agreement*; H. Schröder, *Anti-Semitism and anti-Feminism Again: The Dissemination of Otto Weininger's* Sex and Character *in the Seventies and Eighties*; J.R. Richards, *Traditional Spheres and Traditional Logic*.

VOLUME 36 (1994)

MARXISM AND COMMUNISM: POSTHUMOUS REFLECTIONS ON POLITICS, SOCIETY, AND LAW
(Edited by Martin Krygier)

M. Krygier, *Introduction*; A. Flis, *From Marx to Real Socialism: The History of a Utopia*; P. Marciniak, *The Collapse of Communism: Defeat or Opportunity for Marxism in Eastern Europe*; J. Clark, A. Wildavsky, *Chronicle of a Collapse Foretold: How Marx Predicted the Demise of Communism (Although He Called It "Capitalism")*; L. Nowak, *Political Theory and Socialism. On the Main Paradigms of Political Power and Their Methodological and Historical Legitimation*; E. Mokrzycki, *Marxism, Sociology, and "Real Socialism"*; R. Bäcker, *The Collapse of Communism and Theoretical Models*; A. Zybertowicz, *Three Deaths an Ideology: The Withering Away of Marxism and the Collapse of Communism. The Case of Poland*; M. Krygier, *Marxism, Communism, and the Rule of Law*; A. Czarnota, *Marxism, Ideology, and Law*; G. Skąpska, *The Legacy of Anti-Legalism*; A. Sajo, *Law and the Legal Scholarship in the Happiest Barrack and Among the Hungry Liberated: Personal Recollections*.

VOLUME 37 (1994)

THE SOCIAL PHILOSOPHY OF AGNES HELLER
(Edited by John Burnheim)

J. Burnheim, *Introduction*; M. Vajda, *A Lover of Philosophy – A Lover of Europe*; P. Despoix, *On the Possibility of a Philosophy of Values. A Dialogue within the Budapest School*; M. Jay, *Women in Dark Times: Agnes Heller and Hannah Arendt*; J.P. Arnason, *The Human Condition and the Modern Predicament*; R.J. Bernstein, *Agnes Heller: Philosophy, Rational Utopia and Praxis*; Z. Bauman, *Narrating Modernity*; P. Beilharz, *Theories of History – Agnes Heller and R.G. Collingwood*; P. Wolin, *Heller's Theory of Everyday Life*; P. Harrison, *Radical Philosophy and the Theory of Modernity*; A.J. Jacobson, *The Limits of Formal Justice*; P. Murphy, *Civility and Radicalism*; P. Murphy, *Pluralism and Politics*; V. Camps, *The Good Life: A Moral Gesture*; L. Boella, *Philosophy Beyond the Baseless and Tragic Character of Action*; G. Márkus, *The Politics of Morals*; A. Heller, *A Reply to My Critics*; *The Bibliography of Agnes Heller*.

VOLUME 38 (1994)

IDEALIZATION VI: IDEALIZATION IN ECONOMICS
(Edited by Bert Hamminga and Neil B. De Marchi)

Introduction – B. Hamminga, N. De Marchi, *Preface*; B. Hamminga, N. De Marchi, *Idealization and the Defence of Economics: Notes Toward a History*. **Part I: General Observations on Idealization in Economics** – K.D. Hoover, *Six Queries about Idealization in an Empirical Context*; B. Walliser, *Three Generalization Processes for Economic Models*; S. Cook, D. Hendry, *The Theory of Reduction in Econometrics*; M.C.W. Janssen, *Economic Models and Their Applications*; A.G. de la Sienra, *Idealization and Empirical Adequacy in Economic Theory*; I. Nowakowa, L. Nowak, *On Correspondence between Economic Theories*; U. Mäki, *Isolation, Idealization and Truth in Economics*. **Part II: Case Studies of Idealization in Economics** – N. Cartwright, *Mill and Menger: Ideal Elements and Stable Tendencies*; W. Balzer, *Exchange Versus Influence: A Case of Idealization*; K. Cools, B. Hamminga, T.A.F. Kuipers, *Truth Approximation by Concretization in Capital Structure Theory*; D.M. Hausman, *Paul Samuelson as Dr. Frankenstein: When an Idealization Runs Amuck*; H.A. Keuzenkamp, *What if an Idealization is Problematic? The Case of the Homogeneity Condition in Consumer Demand*; W. Diederich, *Nowak on Explanation and Idealization in Marx's Capital*; G. Jorland, *Idealization and Transformation*; J. Birner, *Idealizations and Theory Development in Economics. Some History and Logic of the Logic Discovery*. **Discussions** – L. Nowak, *The Idealizational Methodology and Economics. Replies to Diederich, Hoover, Janssen, Jorland and Mäki*.

VOLUME 39 (1994)

PROBABILITY IN THEORY-BUILDING.
EXPERIMENTAL AND NON-EXPERIMENTAL APPROACHES
TO SCIENTIFIC RESEARCH IN PSYCHOLOGY
(Edited by Jerzy Brzeziński)

Part I: Probability and the Idealizational Theory of Science – M. Gaul, *Statistical Dependencies, Statements and the Idealizational Theory of Science*. **Part II: Probability – Theoretical Concepts in Psychology – Measurement** – D. Wahlstein, *Probability and the Understanding of Individual Differences*; B. Krause, *Modeling Cognitive Learning Steps*; D. Heyer, R. Mausfeld, *A Theoretical and Experimental Inquiry into the Relation of Theoretical Concepts and Probabilistic Measurement Scales in Experimental Psychology*. **Part III: Methods of Data Analysis** – T.B. Iwiński, *Rough Set Methods in Psychology*; W. Koutstaal, R. Rosenthal, *Contrast Analysis in Behavioral Research*. **Part IV: Artifacts in Psychological Research and Diagnostic Assessment** – D.B. Strohmetz, R.L. Rosnow, *A Mediational Model of Research Artifacts*; J. Brzeziński, *Dimensions of Diagnostic Space*.

VOLUME 40 (1995)

THE HERITAGE OF KAZIMIERZ AJDUKIEWICZ
(Edited by Vito Sinisi and Jan Woleński)

Preface; J. Giedymin, *Ajdukiewicz's Life and Personality*; K. Ajdukiewicz, *My Philosophical Ideas*; L. Albertazzi, *Some Elements of Transcendentalism in Ajdukiewicz*; T. Batóg, *Ajdukiewicz and the Development of Formal Logic*; A. Church, *A Theory of the Meaning of Names*; M. Czarniawska, *The Way from Concept to Thought. Does it Exist in Ajdukiewicz's Semantical Theory?*; E. Dölling, *Real Objects and Existence*; J. Giedymin, *Radical Conventionalism, Its Background and Evolution: Poincaré, Le Roy, Ajdukiewicz*; J.J.

Jadacki, *Definition, Explication and Paraphrase in the Ajdukiewiczian Tradition*; G. Küng, *Ajdukiewicz's Contribution to the Realism/Idealism Debate*; J. Maciaszek, *Problems of Proper Names: Ajdukiewicz and Some Contemporary Results*; W. Marciszewski, *Real Definitions and Creativity*; K. Misiuna, *Categorial Grammar and Ontological Commitment*; L. Nowak, *Ajdukiewicz and the Status of the Logical Theory of Natural Language*; A. Olech, *Several Remarks on Ajdukiewicz's and Husserl's Approaches to Meaning*; A. Orenstein, *Existence Sentences*; J. Pasek, *Ajdukiewicz on Indicative Conditionals*; R. Poli, *The Problem of Position: Ajdukiewicz and Leibniz on Intensional Expressions*; M. Przełęcki, *The Law of Excluded Middle and the Problem of Idealism*; H. Skolimowski, *Ajdukiewicz, Rationality and Language*; K. Szaniawski, *Ajdukiewicz on Non-deductive Inference*; A. Varzi, *Variable-Binders as Functors*; V. Vasyukov, *Categorial Semantics for Ajdukiewicz-Lambek Calculus*; H. Wassing, *On the Expressiveness of Categorial Grammar*; J. Woleński, *On Ajdukiewicz's Refutation of Scepticism*; *Bibliography of Ajdukiewicz*; *Bibliography on Ajdukiewicz (writings in western languages)*.

VOLUME 41 (1994)

HISTORIOGRAPHY BETWEEN MODERNISM AND POSTMODERNISM.
CONTRIBUTIONS TO THE METHODOLOGY OF
THE HISTORICAL RESEARCH
(Edited by Jerzy Topolski)

Editor's Introduction; J. Topolski, *A Non-postmodernist Analysis of Historical Narratives*; F.R. Ankersmit, *Historism, Postmodernism and Historiography*; D. Carr, *Getting the Story Straight: Narrative and Historical Knowledge*; W. Wrzosek, *The Problem of Cultural Imputation in History. Culture versus History*; J. Tacq, *Causality and Virtual Finality*; G. Zalejko, *Soviet Historiography as a "Normal Science"*; H. Mamzer and J. Ostoja-Zagórski, *Deconstruction of Evolutionist Paradigm in Archaeology*; N. Lautier, *At the Crossroad of Epistemology and Psychology: Prospects of a Didactic of History*; Teresa Kostyrko, *Remarks on "Aesthetization" in Science on the Basis of History*.

VOLUME 42 (1995)

IDEALIZATION VII: IDEALIZATION, STRUCTURALISM,
AND APPROXIMATION
(Edited by Martti Kuokkanen)

Idealization, Approximation and Counterfactuals in the Structuralist Framework – T.A.F. Kuipers, *The Refined Structure of Theories*; C.U. Moulines and R. Straub, *Approximation and Idealization from the Structuralist Point of View*; I.A. Kieseppä, *A Note on the Structuralist Account of Approximation*; C.U. Moulines and R. Straub, *A Reply to Kieseppä*; W. Balzer and G. Zoubek, *Structuralist Aspects of Idealization*; A. Ibarra and T. Mormann, *Counterfactual Deformation and Idealization in a Structuralist Framework*; I.A. Kieseppä, *Assessing the Structuralist Theory of Verisimilitude*. **Idealization, Approxima-tion and Theory Formation** – L. Nowak, *Remarks on the Nature of Galileo's Methodological Revolution*; I. Niiniluoto, *Approximation in Applied Science*; E. Heise, P. Gerjets and R. Westermann, *Idealized Action Phases. A Concise Rubicon Theory*; K.G. Troitzsch, *Modelling, Simulation, and Structuralism*; V. Rantala and T. Vadén, *Idealization in Cognitive Science. A Study in Counterfactual Correspondence*; M. Sintonen and M. Kiikeri, *Idealization in Evolutionary Biology*; T. Tuomivaara, *On Idealizations in Ecology*; M. Kuokkanen and M. Häyry, *Early Utilitarianism and Its Idealizations from a Systematic Point of View*. **Idealization, Approximation and Measurement** – R. Westermann, *Measurement-Theoretical Idealizations and Empirical Research Practice*; U. Konerding,

Probability as an Idealization of Relative Frequency. A Case Study by Means of the BTL-Model; R. Suck and J. Wienöbst, *The Empirical Claim of Probality Statements, Idealized Bernoulli Experiments and their Approximate Version*; P.J. Lahti, *Idealizations in Quantum Theory of Measurement*.

VOLUME 43 (1995)

Witold Marciszewski and Roman Murawski
MECHANIZATION OF REASONING IN A HISTORICAL PERSPECTIVE

Chapter 1: *From the Mechanization of Reasoning to a Study of Human Intelligence*; Chapter 2: *The Formalization of Arguments in the Middle Ages*; Chapter 3: *Leibniz's Idea of Mechanical Reasoning at the Historical Background*; Chapter 4: *Between Leibniz and Boole: Towards the Algebraization of Logic*; Chapter 5: *The English Algebra of Logic in the 19th Century*; Chapter 6: *The 20th Century Way to Formalization and Mechanization*; Chapter 7: *Mechanized Deduction Systems*.

VOLUME 44 (1995)

THEORIES AND MODELS IN SCIENTIFIC PROCESSES
(Edited by William Herfel, Władysław Krajewski,
Ilkka Niiniluoto and Ryszard Wójcicki)

Introduction; **Part 1. Models in Scientific Processes** – J. Agassi, *Why there is no Theory of Models?*; M. Czarnocka, *Models and Symbolic Nature of Knowledge*; A. Grobler, *The Representational and the Non-Representational in Models of Scientific Theories*; S. Hartmann, *Models as a Tool for the Theory Construction: Some Strategies of Preliminary Physics*; W. Herfel, *Nonlinear Dynamical Models as Concrete Construction*; E. Kałuszyńska, *Styles of Thinking*; S. Psillos, *The Cognitive Interplay Between Theories and Models: the Case of 19th Century Optics*. **Part 2. Tools of Science** – N.D. Cartwright, T. Shomar, M. Suarez, *The Tool-Box of Science*; J. Echeverria, *The Four Contexts of Scienctific Activity*; K. Havas, *Continuity and Change; Kinds of Negation in Scientific Progress*; M. Kaiser, *The Independence of Scientific Phenomena*; W. Krajewski, *Scientific Meta-Philosophy*; I. Niiniluoto, *The Emergence of Scientific Specialties: Six Models*; L. Nowak, *Antirealism, (Supra-) Realism and Idealization*; R.M. Nugayev, *Classic, Modern and Postmodern Scientific Unification*; V. Rantala, *Translation and Scientific Change*; G. Schurz, *Theories and Their Applications – a Case of Nonmonotonic Reasoning*; W. Strawiński, *The Unity of Science Today*; V. Torosian, *Are the Ethics and Logic of Science Compatible?* **Part 3. Unsharp Approaches in Science** – E.W. Adams, *Problems and Prospects in a Theory of Inexact First-Order Theories*; W. Balzer, G. Zoubek, *On the Comparision of Approximative Empirical Claims*; G. Cattaneo, M. Luisa Dalla Chiara, R. Giuntini, *The Unsharp Approaches to Quantum Theory*; T.A.F. Kuipers, *Falsification Versus Effcient Truth Approximation*; B. Lauth, *Limiting Decidability and Probability*; J. Pykacz, *Many-Valued Logics in Foundations of Quantum Mechanics*; R.R. Zapatrin, *Logico-Algebraic Approach to Spacetime Quantization*.

VOLUME 45 (1995)

COGNITIVE PATTERNS IN SCIENCE AND COMMON SENSE.
GRONINGEN STUDIES IN PHILOSOPHY OF SCIENCE,
LOGIC, AND EPISTEMOLOGY
(Edited by Theo A.F. Kuipers and Anne Ruth Mackor)

L. Nowak, *Foreword*; **General Introduction** – T.A.F. Kuipers and A.R. Mackor, *Cognitive Studies of Science and Common Sense*; **Part I: Conceptual Analysis in Service of Various Research Programmes** – H. Zandvoort, *Concepts of Interdisciplinarity and Environmental Science*; R. Vos, *The Logic and Epistemology of the Concept of Drug and Disease Profile*; R.C. Looijen, *On the Distinction Between Habitat and Niché, and Some Implications for Species' Differentiation*; G.J. Stavenga, *Cognition, Irreversibility and the Direction of Time*; R. Dalitz, *Knowledge, Gender and Social Bias*; **Part II: The Logic of the Evaluation of Arguments, Hypotheses, Rules, and Interesting Theorems** – E.G.W. Krabbe, *Can We Ever Pin One Down to a Formal Fallacy?*; T.A.F. Kuipers, *Explicating the Falsificationist and Instrumentalist Methodology by Decomposing the Hypothetico-Deductive Method*; A. Keupink, *Causal Modelling and Misspecification: Theory and Econometric Historical Practice*; M.C.W. Janssen and Y.-H. Tan, *Default Reasoning and Some Applications in Economics*; B. Hamminga, *Interesting Theorems in Economics*; **Part III: Three Challenges to the Truth Approximation Programme** – S.D. Zwart, *A Hidden Variable in the Discussion About 'Language Dependency' of Truthlikeness*; H. Hettema and T.A.F. Kuipers, *Sommerfeld's Atombau: A Case Study in Potential Truth Approximation*; R. Festa, *Verisimilitude, Disorder, and Optimum Prior Probabilities*; **Part IV: Explicating Psychological Intuitions** – A.R. Mackor, *Intentional Psychology is a Biological Discipline*; J. Peijnenburg, *Hempel's Rationality. On the Empty Nature of Being a Rational Agent*; L. Guichard, *The Causal Efficacy of Propositional Attitudes*; M. ter Hark, *Connectionism, Behaviourism and the Language of Thought*.

VOLUME 46 (1996)

POLITICAL DIALOGUE: THEORIES AND PRACTICE
(Edited by Stephen L. Esquith)

Introduction – S.L. Esquith, *Political Dialogue and Political Virtue*. **Part I: The Modern Clasics** – A.J. Damico, *Reason's Reach: Liberal Tolerance and Political Discourse*; T.R. Machan, *Individualism and Political Dialogue*; R. Kukla, *The Coupling of Human Souls: Rousseau and the Problem of Gender Relations*; D.F. Koch, *Dialogue: An Essay in the Instrumentalist Tradition*. **Part II: Toward a Democratic Synthesis** – E. Simpson, *Forms of Political Thinking and the Persistence of Practical Philosophy*; J.B. Sauer, *Discoursee, Consensus, and Value: Conversations about the Intelligible Relation Between the Private and Public Spheres*; M. Kingwell, *Phronesis and Political Dialogue*, R.T. Peterson, *Democracy and Intellectual Mediation – After Liberalism and Socialism*. **Part III: Dialogue in Practice** – S. Rohr Scaff, L.A. Scaff, *Political Dialogue in the New Germany: The Burdens of Culture and an Asymmetrical Past*; J.H. Read, *Participation, Power, and Democracy*; S.E. Bennett, B. Fisher, D. Resnick, *Speaking of Politics in the United States: Who Talks to Whom, Why, and Why Not*; J. Forester, *Beyond Dialogue to Transformative Learning: How Deliberative Rituals Encourage Political Judgement in Community Planning Processes*; A. Fatić, *Retribution in Democracy*.

VOLUME 47 (1996)

EPISTEMOLOGY AND HISTORY. HUMANITIES AS A PHILOSOPHICAL PROBLEM AND JERZY KMITA'S APPROACH TO IT
(Edited by Anna Zeidler-Janiszewska)

A. Zeidler-Janiszewska, *Preface*. **Humanistic Knowledge** – K.O. Apel, *The Hermeneutic Dimension of Social Science and its Normative Foundation*; M. Czerwiński, *Jerzy Kmita's Epistemology*; L. Witkowski, *The Frankfurt School and Structuralism in Jerzy Kmita's Analysis*; A. Szahaj, *Between Modernism and Postmodernism: Jerzy Kmita's Epistemology*; A. Grzegorczyk, *Non-Cartesian Coordinates in the Contemporary Humanities*; A. Pałubicka, *Pragmatist Holism as an Expression of Another Disenchantment of the World*; J. Sójka, *Who is Afraid of Scientism?*; P. Ozdowski, *The Broken Bonds with the World*; J. Such, *Types of Determination vs. the Development of Science in Historical Epistemology*; P. Zeidler, *Some Issues of Historical Epistemology in the Light of the Structuralist Philosophy of Science*; M. Buchowski, *Via Media: On the Consequences of Historical Epistemology for the Problem of Rationality*; B. Kotowa, *Humanistic Valuation and Some Social Functions of the Humanities*. **On Explanation and Humanistic Interpretation** – T.A.F. Kuipers, *Explanation by Intentional, Functional, and Causal Specification*; E. Świderski, *The Interpretational Paradigm in the Philosophy of the Human Sciences*; L. Nowak, *On the Limits of the Rationalistic Paradigm*; F. Coniglione, *Humanistic Interpretation between Hempel and Popper*; Z. Ziembiński, *Historical Interpretation vs. the Adaptive Interpretation of a Legal Text*; W. Mejbaum, *Explaining Social Phenomena*; M. Ziółkowski, *The Functional Theory of Culture and Sociology*; K. Zamiara, *Jerzy Kmita's Social-Regulational Theory of Culture and the Category of Subject*; J. Brzeziński, *Theory and Social Practice. One or Two Psychologies?*; Z. Kwieciński, *Decahedron of Education (Components and aspects). The Need for a Comprehensive Approach*; J. Paśniczek, *The Relational vs. Directional Concept of Intentionality*. **The Historical Dimension of Culture and its Studies** – J. Margolis, *The Declension of Progressivism*; J. Topolski, *Historians Look at Historical Truth*; T. Jerzak-Gierszewska, *Three Types of the Theories of Religion and Magic*; H. Paetzold, *Mythos und Moderne in der Philosophie der symbolishen Formen Ernst Cassirers*; M. Siemek, *Sozialphilosophische Aspekte der Übersetzbarkeit*. **Problems of Artistic Practice and Its Interpretation** – S. Morawski, *Theses on the 20th Century Crisis of Art and Culture*; A. Erjavec, *The Perception of Science in Modernist and Postmodernist Artistic Practice*, G. Dziamski, *The Avant-garde and Contemporary Artistic Consciousness*; H. Orłowski, *Generationszugehörigkeit und Selbsterfahrung von (deutschen) Schriftstellern*; T. Kostyrko, *The "Transhistoricity" of the Structure of Work of Art and the Process of Value Transmission in Culture*; G. Banaszak, *Musical Culture as a Configuration of Subcultures*; A. Zeidler-Janiszewska, *The Problem of the Applicability of Humanistic Interpretation in the Light of Contemporary Artistic Practice*; J. Kmita, *Towards Cultural Relativism "with a Small 'r'"*; The Bibliography of Jerzy Kmita.

VOLUME 48 (1996)

THE SOCIAL PHILOSOPHY OF ERNEST GELLNER
(Edited by John A. Hall and Ian Jarvie)

J.A. Hall, I. Jarvie, *Preface*; J.A. Hall, I. Jarvie, *The Life and Times of Ernest Gellner*. **Part 1: Intelectual Background** – J. Musil, *The Prague Roots of Ernest Gellner's Thinking*; Ch. Hahn, *Gellner and Malinowski: Words and Things in Central Europe*; T. Dragadze, *Ernest Gellner in Soviet East*. **Part 2: Nations and Nationalism** – B. O'Leary, *On the Nature of Nationalism: An Appraisal of Ernest Gellner's Writings on Nationalism*; K. Minogue, *Ernest Gellner and the Dangers of Theorising Nationalism*; A.D. Smith, *History and*

Modernity: Reflection on the Theory of Nationalism; M. Mann, *The Emergence of Modern European Nationalism*; N. Stagardt, *Gellner's Nationalism: The Spirit of Modernisation?* **Part 3: Patterns of Development** – P. Burke, *Reflections on the History of Encyclopaedias*; A. MacFarlane, *Ernest Gellner and the Escape to Modernity*; R. Dore, *Soverein Individuals*; S. Eisenstadt, *Japan: Non Axial Modernity*; M. Ferro, *L'Indépendance Telescopée: De la Décolonisation a L'Impérialisme Multinational*. **Part 4: Islam** – A. Hammoudi, *Segmentarity, Social stratification, Political Power and Sainthood: Reflections on Gellner's Theses*; H. Munson, Jr., *Rethinking Gellner's Segmentary Analysis of Morocco's Aiť Atta*; J. Baechler, *Sur le charisme*; Ch. Lindholm, *Despotism and Democracy: State and Society in PreModern Middle East*; H. Munson, *Jr. Muslism and Jew in Morocco: Reflections on the Distinction between Belief and Behaviour*; T. Asad, *The Idea of an Anthropology of Islam*. **Part 5: Science and Disenchantment** – P. Anderson, *Science, Politics, Enchantment*; R. Schroeder, *From the Big Divide to the Rubber Cage: Gellner's Conception of Science and Technology*; J. Davis, *Irrationality in Social Life*. **Part 6: Relativism and Universals** – J. Skorupski, *The Post-Modern Hume: Ernest Gellner's 'Enlightenment Fundamentalism'*; J. Wettersten, *Ernest Gellner: A Wittgensteinian Rationalist*; I. Jarvie, *Gellner's Positivism*; R. Boudon, *Relativising Relativism: When Sociology Refutes the Sociology of Science*; R. Aya, *The Empiricist Exorcist*. **Part 7: Philosophy of History** – W. McNeill, *A Swang Song for British Liberalism?*; A. Park, *Gellner and the Long Trends of History*; E. Leone, *Marx, Gellner, Power*; R. Langlois, *Coercion, Cognition and Production: Gellner's Challenge to Historical Materialism and Post-Modernism*; E. Gellner, *Reply to my Critics*; I. Jarvie, *Complete Bibliography of Gellner's Work*.

VOLUME 49 (1996)

THE SIGNIFICANCE OF POPPER'S THOUGHT
(Edited by Stefan Amsterdamski)

Karl Popper's Three Worlds – J. Watkins, *World 1, World 2 and the Theory of Evolution*; A. Grobler, *World 3 and the Cunning of Reason*. **The Scientific Method as Ethics** – J. Agassi, *Towards Honest Public Relations of Science*; S. Amsterdamski, *Between Relativism and Absolutism: the Popperian Ideal of Knowledge*. **The Open Society and its Prospects** – E. Gellner, *Karl Popper – The Thinker and the Man*; J. Woleński, *Popper on Prophecies and Predictions*.

VOLUME 50 (1996)

THE IDEA OF THE UNIVERSITY
(Edited by Jerzy Brzeziński and Leszek Nowak)

Introduction – K. Twardowski, *The Majesty of the University*. **I** – Z. Ziembiński, *What Can be Saved of the Idea of the University?*; L. Kołakowski, *What Are Universities for?*; L. Gumański, *The Ideal University and Reality*; Z. Bauman, *The Present Crisis of the Universities*. **II** – K. Ajdukiewicz, *On Freedom of Science*; H. Samsonowicz, *Universities and Democracy*; J. Topolski, *The Commonwealth of Scholars and New Conceptions of Truth*; K. Szaniawski, *Plus ratio quam vis*. **III** – L. Koj, *Science, Teaching and Values*; K. Szaniawski, *The Ethics of Scientific Criticism*; J. Brzeziński, *Ethical Problems of Research Work of Psychologists*. **IV** – J. Goćkowski, *Tradition in Science*; J. Kmita, *Is a "Creative Man of Knowledge" Needed in University Teaching?*; L. Nowak, *The Personality of Researchers and the Necessity of Schools in Science*. **Recapitulation** – J. Brzeziński, *Reflections on the University*.

VOLUME 51 (1997)

KNOWLEDGE AND INQUIRY: ESSAYS ON JAAKKO HINTIKKA'S
EPISTEMOLOGY AND PHILOSOPHY OF SCIENCE
(Edited by Matti Sintonen)

M. Sintonen, *From the Science of Logic to the Logic of Science*. **I: Historical Perspectives** –
Z. Bechler, *Hintikka on Plentitude in Aristotle*; M.-L. Kakkuri-Knuuttila, *What Can the
Sciences of Man Learn from Aristotle?*; M. Kusch, *Theories of Questions in German-
Speaking Philosophy Around the Turn of the Century*; N.-E. Sahlin, *'He is no Good for My
Work': On the Philosophical Relations between Ramsey and Wittgenstein*. **II: Formal Tools:
Induction, Observation and Identifiability** – T.A.F. Kuipers, *The Carnap-Hintikka
Programme in Inductive Logic*; I. Levi, *Caution and Nonmonotic Inference*; I. Niiniluoto,
Inductive Logic, Atomism, and Observational Error; A. Mutanen, *Theory of Identifiability*.
III: Questions in Inquiry: The Interrogative Model – S. Bromberger, *Natural Kinds and
Questions*; S.A. Kleiner, *The Structure of Inquiry in Developmental Biology*; A. Wiśniewski,
Some Foundational Concepts of Erotetic Semantics; J. Woleński, *Science and Games*. **IV:
Growth of Knowledge: Explanation and Discovery** – M. Sintonen, *Explanation: The Fifth
Decade*; E. Weber, *Scientific Explanation and the Interrogative Model of Inquiry*; G.
Gebhard, *Scientific Discovery, Induction, and the Multi-Level Character of Scientific
Inquiry*; M. Kiikeri, *On the Logical Structure of Learning Models*. **V: Jaakko Hintikka: Replies.**

VOLUME 52 (1997)

Helena Eilstein
LIFE CONTEMPLATIVE, LIFE PRACTICAL.
AN ESSAY ON FATALISM

Preface. **Chapter One: Oldcomb and Newcomb** – 1. *In the King Comb's Chamber of Game*;
2. *The Newcombian Predicaments*. **Chapter Two: Ananke** – 1. *Fatalism: What It Is Not?*; 2.
Fatalism: What Is It?; 3. *Fatalism and a priori Arguments*; 4. *Fatalism and 'Internal'
Experience*; 5. *Determinism, Indeterminism and Fatalism*; 6. *Transientism, Eternism and
Fatalism*; 7. *Fatalism: What It Does Not Imply?*. **Chapter Three: Fated Freedom** – 1. *More
on Libertarianism*; 2. *On the Deterministic Concept of Freedom*; 3. *Moral Self and
Responsibility in the Light of Probabilism*; 4. *Fated Freedom*. **Chapter Four: The Virus of
Fatalism** – 1. *Fatalism and Problems of Cognition*; 2. *The Virus of Fatalism: Why Mostly
Harmless?*

VOLUME 53 (1997)

Dimitri Ginev
A PASSAGE TO THE HERMENEUTIC PHILOSOPHY OF SCIENCE

Preface; **Introduction**: *Topics in the Hermeneutic Philosophy of Science*; **Chapter 0**: *On the
Limits of the Rational Reconstruction of Scientific Knowledge*; **Chapter 1**: *On the
Hermeneutic Nature of Group Rationality*; **Chapter 2**: *Towards a Hermeneutic Theory of
Progressive Change in Scientific Development*; **Chapter 3**: *Beyond Naturalism and
Traditionalism*; **Chapter 4**: *A Critical Note on Normative Naturalism*; **Chapter 5**: *Micro and
Macrohermeneutics of Science*; *Concluding Remarks*.

VOLUME 54 (1997)

IN ITINERE. EUROPEAN CITIES AND THE BIRTH OF MODERN SCIENTIFIC PHILOSOPHY
(Edited by Roberto Poli)

Introduction – R. Poli, *In itinere. Pictures from Central-European Philosophy.* **Stages of the Tour** – K. Schuhmann, *Philosophy and Art in Munich around the Turn of the Century*; M. Libardi, *In itinere: Vienna 1870-1918*; W. Baumgartner, *Nineteenth-Century Würzburg: The Development of the Scientific Approach to Philosophy*; L. Dappiano, *Cambridge and the Austrian Connection*; J. Sebestik, *Prague Mosaic. Encounters with the Prague Philosophers*; J.J. Jadacki, *Warsaw: The Rise and Decline of Modern Scientific Philosophy in the Capital City of Poland*; J. Woleński, *Lvov*; L. Albertazzi, *Science and the Avant-Garde in Early Nineteenth-Century Florence*; F. Minazzi, *The Presence of Phenomenology in Milan between the Two World Wars. The Contribution of Antonio Banfi and Giulio Preti.*

VOLUME 55 (1997)

REALISM AND QUANTUM PHYSICS
(Edited by Evandro Agazzi)

Introduction – E. Agazzi. **Part One: Philosophical Considerations** – P. Horwich, *Realism and Truth*, E. Agazzi, *On the Criteria for Establishing the Ontological Status of Different Entities*; A. Baltas, *Constraints and Resistance: Stating a Case for Negative Realism*; M. Paty, *Predicate of Existence and Predicability for a Theoretical Object in Physics.* **Part Two: Observability and Hidden Entities** – F. Bonsak, *Atoms: Lessons of a History*; A. Cordero, *Arguing for Hidden Realities*; B. d'Espagnat, *On the Difficulties that Attributing Existence to "Hidden" Quantities May Raise*; M. Pauri, *The Quantum, Space-Time and Observation.* **Part Three: Applications to Quantum Physics** – D. Albert, *On the Phenomenology of Quantum-Mechanical Superpositions*; G.C. Ghiraldi, *Realism and Quantum Mechanics*; M. Crozon, *Experimental Evidence of Quark Structure Inside Hadrons.*

VOLUME 56 (1997)

IDEALIZATION VIII: MODELLING IN PSYCHOLOGY
(Edited by Jerzy Brzeziński, Bodo Krause and Tomasz Maruszewski)

Part I: Philosophical and Methodological Problems of Cognition Process – J. Wane, *Idealizing the Cartesian-Newtonian Paradigm as Reality: The Impact of New-Paradigm Physics on Psychological Theory*; E. Hornowska, *Operationalization of Psychological Magnitudes. Assumptions-Structure-Consequences*; T. Bachmann, *Creating Analogies – On Aspects of the Mapping Process between Knowledge Domains*; H. Schaub, *Modelling Action Regulation.* **Part II: The Structure of Ideal Learning Process** – S. Ohlson, J.J. Jewett, *Ideal Adaptive Agents and the Learning Curve*; B. Krause, *Towards a Theory of Cognitive Learning*; B. Krause, U. Gauger, *Learning and Use of Invariances: Experiments and Network Simulation*; M. Friedrich, *"Reaction Time" in the Neural Network Module ART 1.* **Part III: Control Processes in Memory** – J. Tzelgov, V. Yehene, M. Naveh-Benjamin, *From Memory to Automaticity and Vice Versa: On the Relation between Memory and Automaticity*; H. Hagendorf, S. Fisher, B. Sá, *The Function of Working Memory in Coordination of Mental Transformations*; L. Nowak, *On Common-Sense and (Para-) Idealization*; I. Nowakowa, *On the Problem of Induction. Toward an Idealizational Paraphrase.*

VOLUME 57 (1997)

EUPHONY AND LOGOS
ESSAYS IN HONOUR OF MARIA STEFFEN-BATÓG AND TADEUSZ BATÓG
(Edited by Roman Murawski, Jerzy Pogonowski)

Preface. **Scientific Works of Maria Steffen-Batóg and Tadeusz Batóg** – *List of Publications of Maria Steffen-Batóg; List of Publications of Tadeusz Batóg;* J. Pogonowski, *On the Scientific Works of Maria Steffen-Batóg;* Jerzy Pogonowski, *On the Scientific Works of Tadeusz Batóg;* W. Lapis, *How Should Sounds Be Phonemicized?;* P. Nowakowski, *On Applications of Algorithms for Phonetic Transcription in Linguistic Research;* J. Pogonowski, *Tadeusz Batóg's Phonological Systems.* **Mathematical Logic** – W. Buszkowski, *Incomplete Information Systems and Kleene 3-valued Logic;* M. Kandulski, *Categorial Grammars with Structural Rules;* M. Kołowska-Gawiejnowicz, *Labelled Deductive Systems for the Lambek Calculus;* R. Murawski, *Satisfaction Classes – a Survey;* K. Świrydowicz, *A New Approach to Dyadic Deontic Logic and the Normative Consequence Relation;* W. Zielonka, *More about the Axiomatics of the Lambek Calculus.* **Theoretical Linguistics** – J.J. Jadacki, *Troubles with Categorial Interpretation of Natural Language;* M. Karpiński, *Conversational Devices in Human-Computer Communication Using WIMP UI;* W. Maciejewski, *Qualitative Orientation and Gramatical Categories;* Z. Vetulani, *A System of Computer Understanding of Text;* A. Wójcik, *The Formal Development of van Sandt's Presupposition Theory;* W. Zabrocki, *Psychologism in Noam Chomsky's Theory (Tentative Critical Remarks);* R. Zuber, *Defining Presupposition without Negation.* **Philosophy of Language and Methodology of Sciences** – J. Kmita, *Philosophical Antifundamentalism;* A. Luchowska, *Peirce and Quine: Two Views on Meaning;* S. Wiertlewski, *Method According to Feyerabend;* J. Woleński, *Wittgenstein and Ordinary Language;* K. Zamiara, *Context of Discovery – Context of Justification and the Problem of Psychologism.*

VOLUME 58 (1997)

THE POSTMODERNIST CRITIQUE OF THE PROJECT OF ENLIGHTENMENT
(Edited by Sven-Eric Liedman)

S.-E. Liedman, *Introduction;* R. Wokler, *The Enlightenment Project and its Critics;* M. Benedikt, *Die Gegenwartsbedeutung von Kants aufklärender Akzeptanz und Zurückweisung des Modells der Naturwissenschaft für zwischenmenschliche Verhältnisse: Verfehlte Beziehungen der Geisterwelt Swedenborgs;* S.-E. Liedman, *The Crucial Role of Ethics in Different Types of Enlightenment (Condorcet and Kant);* S. Dahlstedt, *Forms of the Ineffable: From Kant to Lyotard;* P. Magnus Johansson, *On the Enlightenment in Psycho-Analysis;* E. Lundgren-Gothlin, *Ethics, Feminism and Postmodernism: Seyla Benhabib and Simone de Beauvoir;* E. Kiss, *Gibt es ein Projekt der Aufklärung und wenn ja, wie viele (Aufklärung vor dem Horizont der Postmoderne);* E. Kennedy, *Enlightenment Anticipations of Postmodernist Epistemology;* L. Nowak, *On Postmodernist Philosophy: An Attempt to Identify its Historical Sense;* M. Castillo, *The Dilemmas of Postmodern Individualism.*

VOLUME 59 (1997)

BEYOND ORIENTALISM. THE WORK OF WILHELM HALBFASS AND ITS IMPACT ON INDIAN AND CROSS-CULTURAL STUDIES
(Edited by Eli Franco and Karin Preisendanz)

E. Franco, K. Preisendanz, *Introduction and Editorial Essay on Wilhelm Halbfass;* Publications by Wilhelm Halbfass; W. Halbfass, *Research and Reflections: Responses to my*

Respondents. Beyond Orientalism? Reflections on a Current Theme. **Part I: Cross-Cultural Encounter and Dialogue** – F. X. Clooney, SJ, *Wilhelm Halbfass and the Openness of the Comparative Project*; F. Dallmayr, *Exit from Orientalism: Comments on Wilhelm Halbfass*; S. D. Serebriany, *Some Marginal Notes on India and Europe*; R. Sen (née Mookerjee), *Some Reflections on India and Europe: An Essay in Understanding*; K. Karttunen, *Greeks and Indian Wisdom*; D. Killingley, *Mlecchas, Yavanas and Heathens: Interacting Xenologies in Early Nineteenth-Century Calcutta*; W. Halbfass, *Research and Reflections: Responses to my Respondents. Cross-Cultural Encounter and Dialogue.* **Part II: Issues of Comparative Philosophy** – J. Nath Mohanty, *Between Indology and Indian Philosophy;* J. S. O'Leary, *Heidegger and Indian Philosophy*; S. Rao, *"Subordinate" or "Supreme"? The Nature of Reason in India and the West*; R. Iveković, *The Politics of Comparative Philosophy*; B.-A. Scharfstein, *The Three Philosophical Traditions*; W. Halbfass, *Research and Reflections: Responses to my Respondents. Issues of Comparative Philosophy.* **Part III: Topics in Classical Indian Philosophy** – J. E. M. Houben, *Bhartṛhari's Perspectivism: The Vṛtti and Bhartṛhari's Perspectivism in the First kāṇḍa of the Vākyapadīya*; J. Bronkhorst, *Philosophy and the Vedic Exegesis in the Mīmāṃsā*; J. Taber, *The Significance of Kumārila's Philosophy*; K. Harikai, *Kumārila's Acceptance and Modification of Categories of the Vaiśeṣika School*; V. Lysenko, *The Vaiśeṣika Notions of ākāsa and diś from the Perspective of Indian Ideas of Space*; B. M. Perry, *Early Nyāya and Hindu Orthodoxy: ānvīkṣikī and adhikāra*; W. Halbfass, *Research and Reflections: Responses to my Respondents: Topics in Classical Indian Philosophy.* **Part IV: Developments and Attitudes in Neo-Hinduism** – A. O. Fort, *Jīvanmukti and Social Service in Advaita and Neo-Vedānta*; S. Elkman, *Religious Plurality and Swami Vivekananda.* **Part V: Indian Religion, Past and Present** – M. Hara, *A Note on dharmasya sūkṣmā gatiḥ*; A. Wezler, *The Story of Aṇī-Māṇḍavya as told in the Mahābhārata: Its significance for Indian Legal and Religious History*; Y. Grinshpon, *Experience and Observation in Traditional and Modern Pātañjala Yoga*; F. J. Korom, *Language Belief and Experience in Bengali Folk Religion*; W. Halbfass, *Research and Reflection: Responses to my Respondents. Developments and Attitudes in Neo-Hinduism; Indian religion, Past and Present.*

VOLUME 60 (1998)

MARX'S THEORIES TODAY
(Edited by Ryszard Panasiuk and Leszek Nowak)

R. Panasiuk, *Introduction*; **Part I: On Dialectics and Ontology** – S.-E. Liedman, *Engels and the Laws of Dialectics*; R. Panasiuk, *On Dialectics in Marxism Again*; R. Albritton, *The Unique Ontology of Capital*; R. Washner, *It is not Singularity that Governs the Nature of Things. The Principle of Isolated Individual and Its Negation in Marx's Doctoral Thesis*; **Part II: On Historical Materialism and Social Theories** – Z. Cackowski, *The Continuing Validity of the Marxian Thought*; P. Casal, *From Unilineal to Universal Historical Materialism*; I. Hunt, *A Dialectical Interpretation and Resurrection of Historical Materialism*; W. Krajewski, *The Triumph of Historical Materialism*; L. Nowak, *The Adaptive Interpretation of Historical Materialism: A Survey. On a Contribution to Polish Analytical Marxism*; M. Kozłowski, *A New Look at Capitalism. Between the Decommunisation of Marx's and the Defeudalisation of Hegel's Visions of Capitalism*; F. Moseley, *An Empirical Appraisal of Marx's Economic Theory*; Ch. Bertram and A. Carling, *Stumbling into Revolution. Analytical Marxism, Rationality and Collective Action*; K. Graham, *Collectives, Classes and Revolutionary Potential in Marx*; U. Himmelstrand, *How to Become and Remain a Marxicising Sociologist. An Egocentric Report*; **Part III: On Axiology and the Socialist Project** – P. Kamolnick, *Visions of Social Justice in Marx: An Assessment of Recent Debates in Normative Philosophy*; W. Schmied-Kowarzik, *Karl Marx as a Philosopher of Human Emancipation*; H. J. Sandkühler, *Marx – Welche Rationalität? Epistemische Kontexte und Widersprüche der Transformation von Philosophie in Wissenschaft*; J. Kmita, *The*

Production of "Rational Reality" and the "Systemic Coercion"; J. Bidet, *Metastructure and Socialism*; T. Andreani, *Vers une Issue Socialiste à la Crise du Capitalisme*; W. Becker, *The Bankruptsy of Marxism. About the Historical End of a World Philosophy*; D. Aleksandrowicz, *Myth, Eschatology and Social Reality in the Light of Marxist Philosophy.*

VOLUME 61 (1997)

REPRESENTATIONS OF SCIENTIFIC REALITY
CONTEMPORARY FORMAL PHILOSOPY OF SCIENCE IN SPAIN
(Edited by Andoni Ibarra and Thomas Mormann)

Introduction – A. Ibarra, T. Mormann, *The Long and Winding Road to Philosophy of Science in Spain.* **Part 1: Representation and Measurement** – A. Ibarra, T. Mormann, *Theories as Representations*; J. Garrido Garrido, *The Justification of Measurement*; O. Fernandez Prat, D. Quesada, *Spatial Representations and Their Physical Content*; J.A. Díez Calzada, *The Theory-Net of Interval Measurement Theory.* **Part 2: Truth, Rationality, and Method** – J.C. García-Bermejo Ochoa, *Realism and Truth Approximation in Economic Theory*; W.J.Gonzáles, *Rationality in Economics and Scientific Predictions*; J.P. Zamora Bonilla, *An Invitation to Methodonomics.* **Part 3: Logics, Semantics and Theoretical Structures** – J.L.Falguera, *A Basis for a Formal Semantics of Linguistic Formulations of Science*; A. Sobrino, E. Trillas, *Can Fuzzy Logic Help to Pose Some Problems in the Philosophy of Science?*; J. de Lorenzo, *Demonstrative Ways in Mathematical Doing*; M. Casanueva, *Genetics and Fertilization: A Good Marriage*; C.U. Moulines, *The Concept of Universe from a Metatheoretical Point of View.*

VOLUME 62 (1998)

IN THE WORLD OF SIGNS
(Edited by Jacek Juliusz Jadacki and Witold Strawiński)

Introduction. *How to Move in the World of Signs.* **Part I: Theoretical Semiotics** – A. Bogusławski, *Conditionals and Egocentric Mental Predicates*; W. Buszkowski, *On Families of Languages Generated by Categorial Grammar*; K.G. Havas, *Changing the World – Changing the Meaning. On the Meanings of the "Principle of Non-Contradiction"*; H. Hiż, *On Translation*; S. Marcus, *Imprecision, Between Variety and Uniformity: The Conjugate Pairs*; J. Peregrin, P. Sgall, *Meaning and "Propositional Attitudes"*; O.A. Wojtasiewicz, *Some Applications of Metric Space in Theoretical Linguistic.* **Part II: Methodology** – E. Agazzi, *Rationality and Certitude*; I. Bellert, *Human Reasoning and Artifical Intelligence. When Are Computers Dumb in Simulating Human Reasoning?*; T. Bigaj, *Analyticity and Existence in Mathematics*; G.B. Keene, *Taking up the Logical Slack in Natural Languages*; A. Kertész, *Interdisciplinarity and the Myth of Exactness*; J. Srzednicki, *Norm as the Basis of Form*; J.S. Stepanov, *"Cause" in the Light of Semiotics*; J.A. Wojciechowski, *The Development of Knowledge as a Moral Problem.* **Part III: History of Semiotics** – E. Albrecht, *Philosophy of Language, Logic and Semiotics*; G. Deledalle, *A Philosopher's Reply to Questions Concerning Peirce's Theory of Signs*; J. Deledalle-Rhodes, *The Transposition of Linguistic Sign in Peirce's Contributions to "The Nation"*; R. E. Innis, *From Feeling to Mind: A Note on Langer's Notion of Symbolic Projection*; R. Kevelson, *Peirce's Semiotics as Complex Inquiry: Conflicting Methods*; J. Kopania, *The Cartesian Alternative of Philosophical Thinking*; Xiankun Li, *Why Gonsung Long (Kungsun Lung) Said "White Horse Is Not Horse"*; L. Melazzo, *A Report on an Ancient Discussion*; Ding-fu Ni, *Semantic Thoughts of J. Stuart Mill and Chinese Characters*; I. Portis-Winner, *Lotman's Semiosphere: Some Comments;* J. Réthoré, *Another Close Look at the Interpretant*; E. Stankiewicz, *The Semiotic Turn of Breal's "Semantique"*; **Part IV: Linguistic** – K. Heger, *Passive and Other Voices Seen from an*

Onomasiological Point of View; L.I. Komlószi, *The Semiotic System of Events, Intrinsic Temporal and Deictic Tense Relations in Natural Language. On the Conceptualization of Temporal Schemata*; W. M. Osadnik, E. Horodecka, *Polysystem Theory, Translation Theory and Semiotics*; A. Wierzbicka, *THINK – a Universal Human Concept and a Conceptual Primitive*. **Part V: Cultural Semiotics** – G. Bettetini, *Communication as a Videogame*; W. Krysiński, *Joyce, Models, and Semiotics of Passions*; H. Książek-Konicka, *"Visual Thinking" in the Poetry of Julian Przyboś and Miron Białoszewski*; U. Niklas, *The Space of Metaphor*; M. C. Ruta, *Captivity as Event and Metaphor in Some of Cervantes' Writings*; E. Tarasti, *From Aestetics to Ethics: Semiotics Observations on the Moral Aspects of Art, Especially Music*; L. Tondl, *Is It Justified to Consider the Semiotics of Technological Artefacts?*; V. Voigt, *Poland, Finland and Hungary (A Tuatara's View)*; T.G. Winner, *Czech Poetism: A New View of Poetics Language*; J. Wrede, *Metaphorical Imagery – Ambiguity, Explicitness and Life*. **Part VI: Psycho-Socio-Semiotics** – E.M. Barth, *A Case Study in Empirical Logic and Semiotics. Fundamental Modes of Thought of Nazi Politician Vidkun Quisling, Based on Unpublished Drafts and Notebooks*; P. Bouissac, *Why Do Memmes Die?* W. Kalaga, *Threshold of Signification*, A. Podgórecki, *Do Social Sciences Evaporate?*

VOLUME 63 (1998)

IDEALIZATION IX: IDEALIZATION IN CONTEMPORARY PHYSICS
(Edited by Niall Shanks)

N. Shanks, *Introduction*; M. Bishop, *An Epistemological Role for Thought Experiments*; I. Nowak & L. Nowak, *"Models" and "Experiments" as Homogeneous Families of Notions*; S. French & J. Ladyman, *A Semantic Perspective on Idealization in Quantum Mechanics*; Ch. Liu, *Decoherence and Idealization in Quantum Measurement*; S. Hartmann, *Idealization in Quantum Field Theory*; R. F. Hendry, *Models and Approximations in Quantum Chemistry*; D. Howard, *Astride the Divided Line: Platonism, Empiricism, and Einstein's Epistemological Opportunism*; G. Gale, *Idealization in Cosmology: A Case Study*; A. Maidens, *Idealization, Heuristics and the Principle of Equivalence*; A. Rueger & D. Sharp, *Idealization and Stability:A Perspective from Nonlinear Dynamics*; D. L. Holt & R. G. Holt, *Towards a Very Old Account of Rationality in Experiment: Occult Practices in Chaotic Sonoluminescence.*

VOLUME 64 (1998)

PRAGMATIC IDEALISM. CRITICAL ESSAYS ON NICHOLAS RESCHER'S SYSTEM
OF PRAGMATIC IDEALISM
(Edited by Axel Wüstehube and Michael Quante)

Introduction: A. Wüstehube, *Is Systematic Philosophy still Possible?*; T. Airaksinen, *Moral Facts and Objective Values*; L. Rodríguez Duplá, *Values and Reasons*; G. Gale, *Rescher on Evolution and the Intelligibility of Nature*; J Kekes, *The Nature of Philosophy*; P. Machamer, *Individual and Other-Person Morality: A Plea for an Emotional Response to Ethical Problems*; D. Marconi, *Opus Incertum*; M. Marsonet, *Scientific Realism and Pragmatic Idealism*; R. Martin, *Was Spinoza a Person?*; H. Pape, *Brute Facts, Real Minds and the Postulation of Reality: Resher on Idealism and the Ontological Neutrality of Experience*; J. C. Pitt, *Doing Philosophy: Rescher's Normative Methodology*; L. B. Puntel, *Is Truth "Ideal Coherence"?*; M. Quante, *Understanding Conceptual Schemes: Rescher's Quarrel with Davidson*; A. Siitonen, *The Ontology of Facts and Values*; M. Willaschek, *Skeptical Challenge and the Burden of Proof: On Rescher's Critique of Skepticism*; N. Rescher, *Responses.*

VOLUME 65 (1999)

THE TOTALITARIAN PARADIGM AFTER THE END OF COMMUNISM.
TOWARDS A THEORETICAL REASSESSMENT.
(Edited by Achim Siegel)

A. Siegel, **Introduction**: *The Changing Fortunes of the Totalitarian Paradigm in Communist Studies*. **On Recent Controversies Over The Concept Of Totalitarianism** – K. von Beyme, *The Concept of Totalitarianism – A Reassessment after the End of Communist Rule*; K. Mueller, *East European Studies, Neo-Totalitarianism and Social Science Theory*; L. Nowak, *A Conception that is Supposed to Correspond to the Totalitarian Approach to Realsocialism*; E. Nolte, *The Three Versions of the Theory of Totalitarianism and the Significance of the Historical-Genetic Version*; E. Jesse, *The Two Major Instances of Totalitarianism: Observations on the Interconnection between Soviet Communism and National Socialism*. **Classic Concept Of Totalitarianism: Reassessment And Reinterpretation** – J.P. Arnason, *Totalitarianism and Modernity: Franz Borkenau's "Totalitarian Enemy" as a Source of Sociological Theorizing on Totalitarianism*; A. Sölner, *Sigmund Neumann's "Permanent Revolution": A Forgotten Classic of Comparative Research into Modern Dictatorship*; F. Pohlmann, *The "Seeds of Destruction" in Totalitarian Systems. An Interpretation of the Unity in Hannah Arendt's Political Philosophy*; W.J. Patzelt, *Reality Construction under Totalitarianism: An Ethnomethodological Elaboration of Martin Draht's Concept of Totalitarianism*; A. Siegel, *Carl Joachim Friedrich's Concept of Totalitarianism: A Reinterpretation*; M.R. Thompson, *Neither Totalitarian nor Authoritarian: Post-Totalitarianism in Eastern Europe*.

VOLUME 66 (1999)

Leon Gumański

TO BE OR NOT TO BE? IS THAT THE QUESTION?
AND OTHER STUDIES IN ONTOLOGY, EPISTEMOLOGY AND LOGIC

Preface; *The Elements of a Judgment and Existence*; *Traditional Logic and Existential Presuppositions*; *To Be Or Not To Be? Is That The Question?*; *Some Remarks On Definitions; Logische und semantische Antinomien*; *A New Approach to Realistic Epistemology*; *Ausgewählte Probleme der deontischen Logik*; *An Attempt at the Definition of the Biological Concept of Homology*; *Similarity*.

VOLUME 67 (1999)

Kazimierz Twardowski

ON ACTIONS, PRODUCTS AND OTHER TOPICS IN PHILOSOPHY
(Edited by Johannes Brandl and Jan Woleński)

Introduction; Translator's Note; Self-Portrait (1926/91); Biographical Notes. **I. On Mind, Psychology, and Language:** *Psychology vs. Physiology and Philosophy (1897)*; *On the Classification of Mental Phenomena (1898)*; *The Essence of Concepts (1903/24)*; *On Idio- and Allogenetic Theories of Judgment (1907)*; *Actions and Products (1912)*; *The Humanities and Psychology (1912/76)*; *On the Logic of Adjectives (1923/27)*. **II. On Truth and Knowledge:** *On So-Called Relative Truths (1900)*; *A priori, or Rational (Deductive) Sciences and a posteriori, or Empirical (Inductive) Sciences (1923)*; *Theory of Knowledge. A Lecture Course (1925/75)*. **III. On Philosophy:** *Franz Brentano and the History of Philosophy (1895)*; *The Historical Conception of Philosophy (1912)*; *On Clear and Unclear Philosophical Style (1920)*; *Symbolomania and*

Pragmatophobia (1921); Address at the 25th Anniversary Session of the Polish Philosophical Society (1929/31); On the Dignity of the University (1933). **Bibliography.**

VOLUME 68 (2000)

Tadeusz Czeżowski

KNOWLEDGE, SCIENCE AND VALUES. A PROGRAM FOR SCIENTIFIC PHILOSOPHY
(Edited by Leon Gumański)

L. Guma□ski, *Introduction.* **Part 1: Logic, Methodology and Theory of Science** – *Some Ancient Problems in Modern Form; On the Humanities; On the Method of Analytical Description; On the Problem of Induction; On Discussion and Discussing; On Logical Culture; On Hypotheses; On the Classification of Sentences and Propositional Functions; Proof; On Traditional Distinctions between Definitions; Deictic Definitions; Induction and Reasoning by Analogy; The Classification of Reasonings and its Consequences in the Theory of Science; On the so-called Direct Justification and Self-evidence; On the Unity of Science; Scientific Description.* **Part 2: The World of Human Values and Norms** – *On Happiness; How to Understand "the Meaning of Life" ?; How to Construct the Logic of Goods?; The Meaning and the Value of Life; Conflicts in Ethics; What are Values?; Ethics, Psychology and Logic.* **Part 3: Reality–Knowledge–World** – *Three Attitudes towards the World; On Two Views of the World; A Few Remarks on Rationalism and Empiricism; Identity and the Individual in Its Persistence; Sensory Cognition and Reality; Philosophy at the Crossroads; On Individuals and Existence.* J.J. Jadacki, *Trouble with Ontic Categories or Some Remarks on Tadeusz Cze□owski's Philosophical Views;* W. Mincer, *The Bibliography of Tadeusz Cze□owski.*

VOLUME 69 (2000)

Izabella Nowakowa, Leszek Nowak

THE RICHNESS OF IDEALIZATION

Preface; **Introduction** – *Science as a Caricature of Reality.* **Part I: THREE METHODO-LOGICAL REVOLUTIONS** – *1. The First Idealizational Revolution.Galileo's-Newton's Model of Free Fall; 2. The Second Idealizational Revolution. Darwin's Theory of Natural Selection; 3. The Third Idealizational Revolution. Marx's Theory of Reproduction.* **Part II: THE METHOD OF IDEALIZATION** – *4. The Idealizational Approach to Science: A New Survey; 5. On the Concept of Dialectical Correspondence; 6. On Inner Concretization. A Certain Generalization of the Notions of Concretization and Dialectical Correspondence; 7. Concretization in Qualitative Contexts; 8. Law and Theory: Some Expansions; 9. On Multiplicity of Idealization.* **Part III: EXPLANATIONS AND APPLICATIONS** – *10. The Ontology of the Idealizational Theory; 11. Creativity in Theory-building; 12. Discovery and Correspondence; 13. The Problem of Induction. Toward an Idealizational Paraphrase; 14. "Model(s) and "Experiment(s). An Analysis of Two Homogeneous Families of Notions; 15. On Theories, Half-Theories, One-fourth-Theories, etc.; 16. On Explanation and Its Fallacies; 17. Testability and Fuzziness; 18. Constructing the Notion; 19. On Economic Modeling; 20. Ajdukiewicz, Chomsky and the Status of the Theory of Natural Language; 21. Historical Narration; 22. The Rational Legislator.* **Part IV: TRUTH AND IDEALIZATION** – *23. A Notion of Truth for Idealization; 24. "Truth is a System": An Explication; 25. On the Concept of Adequacy of Laws; 26. Approximation and the Two Ideas of Truth; 27. On the Historicity of Knowledge.* **Part V: A GENERALIZATION OF IDEALIZATION** – *28. Abstracts Are Not Our Constructs. The Mental Constructs Are Abstracts; 29. Metaphors and Deformation; 30. Realism, Supra-Realism and Idealization.* **REFERENCES** – *I. Writings on Idealization; II. Other Writings.*

VOLUME 70 (2000)

QUINE. NATURALIZED EPISTEMOLOGY, PERCEPTUAL KNOWLEDGE AND ONTOLOGY
(Edited by Lieven Decock and Leon Horsten)

Introduction. **Naturalized Epistemology** – T. Derksen, *Naturalistic Epistemology, Murder and Suicide? But what about the Promises!*; Ch. Hookway, *Naturalism and Rationality*; M. Gosselin, *Quine's Hypothetical Theory of Language Learning. A Comparison of Different Conceptual Schemes and Their Logic.* **The Nature of Perceptual Knowledge** – J. van Brakel, *Quine and Innate Similarity Spaces*; D. Koppelberg, *Quine and Davidson on the Structure of Empirical Knowledge*; E. Picardi, *Empathy and Charity*. **Ontology** – S. Laugier, *Quine: Indeterminacy, 'Robust Realism', and Truth*; R. Vergauwen, *Quine and Putnam on Conceptual Relativity and Reference: Theft or Honest Toil?*; I. Douven, *Empiricist Semantics and Indeterminism of Reference*; L. Decock, *Domestic Ontology and Ideology*; P. Gochet, *Canonical Notation, Predication and Ontology*.

VOLUME 71 (2000)

LOGIC, PROBABILITY AND SCIENCE
(Edited by Niall Shanks)

N. Shanks & R.R. Gardner, *Introduction*; C. Morgan, *Canonical Models and Probabilistic Semantics (Commentary by François Lepage; Reply by Morgan)*; F. Lepage, *A Many-Valued Probabilistic Logic (Commentary by Charles Morgan; Reply by Lepage)*; P. Rawling, *The Exchange Paradox, Finite Additivity, and the Principle of Dominance (Commentary by Robert R. Gardner; Reply by Rawling)*; S. Vineberg, *The Logical Status of Conditional and its Role in Confirmation (Commentary by Piers Rawling; Reply by Vineberg)*; D. Mayo, *Science, Error Statistics, and Arguing from Error (Commentary by Susan Vineberg; Reply by Mayo)*; M.N. Lance, *The Best is the Enemy of the Good: Bayesian Epistemology as a Case Study in Unhelpful Idealization (Commentary by Leszek Nowak; Reply by Lance)*; R.B. Gardner & M.C. Wooten, *An Application of Bayes' Theorem to Population Genetics (Commentary by Lynne Seymour; Reply by Gardner and Wooten)*; P.D. Johnson, Jr., *Another Look at Group Selection (Commentary by Niall Shanks; Reply by Johnson)*; C.F. Juhl, *Teleosemantics, Kripkenstein and Paradox (Commentary by Daniel Bonevac; Reply by Juhl)*; D. Bonevac, *Constitutive and Epistemic Principles (Commentary by Mark Lance; Reply by Bonevac)*; O. Bueno, *Empiricism, Mathematical Truth and Mathematical Knowledge (Commentary by Chuang Liu; Reply by Bueno)*; Ch. Liu, *Coins and Electrons: A Unified Understanding of Probabilistic Objects (Commentary by Steven French; Reply by Liu)*; A. Maidens, *Are Electrons Vague Objects? (Commentary by David Over; Reply by Maidens)*.

VOLUME 72 (2000)

ON COMPARING AND EVALUATING SCIENTIFIC THEORIES
(Edited by Adam Jonkisz and Leon Koj)

L. Koj, *Preface*; L. Koj, *Methodology and Values*; L. Koj, *Science as System*; A. Grobler, *Explanation and Epistemic Virtue*; P. Giza, *"Intelligent" Computer System and Theory Comparison*; H. Ogryzko-Wiewiórowski, *Methods of Social Choice of Scientific Theories*; K. Jodkowski, *Is the Causal Theory of Reference a Remedy for Ontological Incommensurability?*; W. Balzer, *On Approximative Reduction*; C. Ulises Moulines, *Is There Genuinely Scientific Progress?*; A. Jonkisz, *On Relative Progress in Science*.

VOLUME 73 (2000)

THE RATIONALITY OF THEISM
(Edited by Adolfo García de la Sienra)

THE MONIST

AN INTERNATIONAL JOURNAL OF GENERAL PHILOSOPHICAL INQUIRY

APRIL 2001 ◆ VOL. 84, NO. 2 ◆ SINGLE ISSUE, US $12

PROBABILITY AS A GUIDE TO LIFE
ADVISORY EDITORS: HENRY E. KYBURG, JR. & MARIAM THALOS

Editor: Barry Smith, University at Buffalo, State University of New York
Single issues, $12 postpaid. Annual subscription: institutions, US $50; individuals, US $30; make checks payable to THE MONIST, 315 Fifth Street, Peru, IL U.S.A. 61354.

Grazer Philosophische Studien
Vol 60 - 2000

Hrsg. von Johannes L. Brandl, Marian David und Leopold Stubenberg.

Amsterdam/Atlanta, GA 2000. 231 pp.   EUR 46,-/US-$ 43.-

Editions Rodopi B.V.
USA/Canada: One Rockefeller Plaza, Ste. 1420, New York, NY 10020,
Tel. (212) 265-6360,
Call toll-free (U.S. only) 1-800-225-3998, Fax (212) 265-6402
All other countries: Tijnmuiden 7, 1046 AK Amsterdam, The Netherlands.
Tel. ++ 31 (0)20 611 48 21, Fax ++ 31 (0)20 447 29 79
Orders-queries@rodopi.nl   www.rodopi.nl

GEORG MISCH

LOGIK UND EINFÜHRUNG
IN DIE GRUNDLAGEN
DES WISSENS

Die Macht der antiken Tradition
in der Logik und die gegenwärtige Lage

STUDIA   CULTUROLOGICA

Sonderheft 1999

Inquiries & Orders : CRITIQUE & HUMANISM Publishing House, 11, Slaveykov Sq.,
Sofia 1000, Bulgaria, Fax /359 2/ 98 00 243, E-mail: kx_ch@cserv.mgu.bg
(Back issues are also available)

What is the Meaning of Human Life?

Raymond Angelo Belliotti

Amsterdam/Atlanta, GA 2001. VIII,176 pp.
(Value Inquiry Book Series 109)
ISBN: 90-420-1296-X   EUR 34,-/US-$ 32.-

This book examines core concerns of human life. What is the relationship between a meaningful life and theism? Why are some human beings radically adrift, without radical foundations, and struggling with hopelessness? Is the cosmos meaningless? Is human life akin to the ancient Myth of Sisyphus? What is the role of struggle and suffering in creating meaning? How do we discover or create value? Is happiness overrated as a goal of life? How, if at all, can we learn to die meaningfully?

Contents:

Editions Rodopi B.V.
USA/Canada: One Rockefeller Plaza, Ste. 1420, New York, NY 10020,
Tel. (212) 265-6360,
Call toll-free (U.S. only) 1-800-225-3998, Fax (212) 265-6402
All other countries: Tijnmuiden 7, 1046 AK Amsterdam, The Netherlands.
Tel. ++ 31 (0)20 611 48 21, Fax ++ 31 (0)20 447 29 79
Orders-queries@rodopi.nl   www.rodopi.nl

THE MONIST

AN INTERNATIONAL JOURNAL OF GENERAL PHILOSOPHICAL INQUIRY

JULY 2001 ◆ VOL. 84, NO. 3 ◆ SINGLE ISSUE, US $12

THE EPIDEMIOLOGY OF IDEAS
ADVISORY EDITOR: DAN SPERBER

Editor: Barry Smith, University at Buffalo, State University of New York
Single issues, $12 postpaid. Annual subscription: institutions, US $50; individuals, US $30; make checks payable to THE MONIST, 315 Fifth Street, Peru, IL U.S.A. 61354.
Website: http://wings.buffalo.edu/philosophy/monist   E-mail: philomon1@netscape.net

Theoretical Interpretations of the Holocaust

Ed. by Dan Stone

Amsterdam/Atlanta, GA 2001. IX,239 pp.
(Value Inquiry Book Services 108)
ISBN: 90-420-1505-5   EUR 45.-/US-$ 42.50

This book aims to show the many resources at our disposal for grappling with the Holocaust as the darkest occurrence of the twentieth century. These wide-ranging studies on philosophy, history, and literature address the way the Holocaust had led to the reconceptualization of the humanities. The scholarly approaches of Pierre Klossowki, Georges Bataille, and Maurice Blanchot are examined critically, and the volume explores such poignant topics as violence, evil, and monuments.

Editions Rodopi B.V.
USA/Canada: One Rockefeller Plaza, Ste. 1420, New York, NY 10020,
Tel. (212) 265-6360, *Call toll-free* (U.S.only) 1-800-225-3998,
Fax (212) 265-6402
All Other Countries: Tijnmuiden 7, 1046 AK Amsterdam, The Netherlands.
Tel. ++ 31 (0)20 6114821, Fax ++ 31 (0)20 4472979
orders-queries@rodopi.nl   **www.rodopi.nl**